Financial Mathematics and Fintech

This series addresses the emerging advances in mathematical theory related to finance and application research from all the fintech perspectives. It is a series of monographs and contributed volumes focusing on the in-depth exploration of financial mathematics such as applied mathematics, statistics, optimization, and scientific computation, and fintech applications such as artificial intelligence, block chain, cloud computing, and big data. This series is featured by the comprehensive understanding and practical application of financial mathematics and fintech. This book series involves cutting-edge applications of financial mathematics and fintech in practical programs and companies.

The Financial Mathematics and Fintech book series promotes the exchange of emerging theory and technology of financial mathematics and fintech between academia and financial practitioner. It aims to provide a timely reflection of the state of art in mathematics and computer science facing to the application of finance. As a collection, this book series provides valuable resources to a wide audience in academia, the finance community, government employees related to finance and anyone else looking to expand their knowledge in financial mathematics and fintech. The key words in this series include but are not limited to:

a) Financial mathematics
b) Fintech
c) Computer science
d) Artificial intelligence
e) Big data

Mohammad Irfan • Khan Muhammad
Nader Naifar • Muhammad Attique Khan
Editors

Applications of Block Chain technology and Artificial Intelligence

Lead-ins in Banking, Finance, and Capital Market

Springer

Editors
Mohammad Irfan
NSB Academy Business School
Bangalore, India

Nader Naifar
Dept of Finance and Investment
Imam Muhammad ibn Saud Islamic University
Riyadh, Saudi Arabia

Khan Muhammad
Dept of Applied Artificial Intelligence
Sungkyunkwan University
Seoul, Korea (Republic of)

Muhammad Attique Khan
Dept of Computer Science
HITEC University
Taxila, Pakistan

ISSN 2662-7167 ISSN 2662-7175 (electronic)
Financial Mathematics and Fintech
ISBN 978-3-031-47323-4 ISBN 978-3-031-47324-1 (eBook)
https://doi.org/10.1007/978-3-031-47324-1

This Springer imprint is published by the registered company Springer Nature Switzerland AG
The registered company address is: Gewerbestrasse 11, 6330 Cham, Switzerland

Contents

Chapter 1
The Impact of Artificial Intelligence on the Future of Computing: A Comparative Study

Tarun Kumar Vashishth, Bhupendra Kumar, Md Shabbir Alam, Vikas Sharma, and Sachin Chaudhary

1.1 Introduction

One sector that stands to undergo significant transformation is the traditional banking system. As the backbone of global financial transactions, traditional banking systems have long played a central role in facilitating economic activities. However, they are not without their limitations, including inefficiencies, high costs, security vulnerabilities, and complex intermediation processes. Blockchain technology offers a decentralized and transparent approach to recording and verifying transactions, utilizing a distributed ledger that is shared among participants in a network. By leveraging cryptography and consensus mechanisms, blockchain introduces trust and transparency into financial interactions, reducing the reliance on intermediaries and potentially streamlining processes.

By examining various dimensions, including trust and transparency, security, cost efficiency, decentralization, and the emergence of cryptocurrencies, this study aims to shed light on the potential benefits and challenges associated with integrating blockchain into traditional banking systems.

One significant aspect to explore is the potential improvement in trust and transparency brought about by blockchain technology. The inherent immutability of blockchain records and Real-time transaction monitoring has the potential to improve the overall integrity of financial transactions, reducing fraud and enhancing customer confidence. Security is another critical consideration. Traditional banking systems face cybersecurity threats, and the storage of sensitive customer data poses

T. K. Vashishth (✉) · B. Kumar · V. Sharma · S. Chaudhary
School of Computer Science and Applications, IIMT University, Meerut, UP, India

M. S. Alam
College of Business Administration, University of Bahrain, Zallaq, Bahrain

M. Irfan et al. (eds.), *Applications of Block Chain technology and Artificial Intelligence*, Financial Mathematics and Fintech,
https://doi.org/10.1007/978-3-031-47324-1_1

risks. Blockchain technology's decentralized nature and cryptographic algorithms offer robust security measures that could potentially mitigate these vulnerabilities.

Cost efficiency is an important factor that can influence the adoption of blockchain by traditional banks. Blockchain technology can eliminate middlemen and streamline the process through smart contracts, thereby reducing transaction costs, increasing efficiency, and increasing profitability. Interest in the emergence of digital currencies such as Bitcoin and Ethereum has also increased in recent years. This digital asset is based on blockchain technology and has the potential to disrupt traditional banking by enabling peer-to-peer transactions and bypassing middlemen. Understanding the impact of cryptocurrencies on traditional banking is key to assessing the full impact of blockchain technology.

However, the integration of blockchain into traditional banking systems is not without challenges. Regulatory frameworks, scalability, interoperability, and data privacy are critical areas that need to be addressed. Blockchain technology's decentralized nature challenges existing regulatory structures, requiring policymakers to adapt and establish frameworks that try to balance customer protection and innovation. Additionally, scalability &interoperability concerns need to be resolved to accommodate the vast volume of transactions processed by traditional banking systems.

1.1.1 Banking System

The banking system is a fundamental component of the global financial infrastructure, serving as the backbone of economic activities and financial transactions. It encompasses a network of financial institutions, including banks, credit unions, and other entities, that offers a variety of financial assistance for people, companies, and governments. Accepting consumer deposits is one of the banking system's main duties. Individuals and businesses deposit their money into bank accounts, which are then used by banks to extend loans and provide credit to borrowers. This deposit-taking function helps individuals and businesses to safeguard their funds and earn interest on their savings, while simultaneously enabling banks to channel these deposits into productive economic activities.

Moreover, the banking system acts as a financial intermediary, connecting savers and borrowers. Banks play a vital role in evaluating the creditworthiness of borrowers, managing risks, and facilitating the flow of funds between surplus and deficit units in the economy. By providing financial intermediation services, banks facilitate lending and borrowing activities, enabling individuals and businesses to invest, expand, and finance various projects.

In addition to deposit-taking and lending, the banking system offers a wide range of financial services. These services include payment processing, such as facilitating electronic funds transfers, issuing checks, and providing debit and credit cards. Banks also offer services such as foreign exchange, trade finance, wealth management, investment banking, and insurance, catering to the diverse financial needs of their customers.

Furthermore, the banking system serves as a custodian of financial information. Banks collect and store vast amounts of customer data, including personal information, transaction histories, and credit profiles. This information is crucial for banks to assess creditworthiness, manage risks, and provide tailored financial solutions. One of the most important duties of the banking system is to guarantee the confidentiality and safety of consumer data. The banking system is highly regulated to maintain financial stability, protect consumers, and prevent money laundering and fraudulent activities. Banks are required to comply with capital adequacy requirements, maintain liquidity buffers, and adhere to anti-money laundering and consumer protection regulations.

1.1.1.1 Traditional Banking System

In businesses to carry out everyday transactions, making commerce more convenient and efficient. Traditional banks provide investment banking, wealth management, foreign exchange, financial services, and other services in addition to these fundamental business operations. These services cater to the diverse needs of individuals, corporations, and institutional clients, offering specialized financial expertise and solutions.

Furthermore, traditional banks serve as custodians of financial information, collecting and maintaining vast amounts of customer data. This information includes personal details, transaction histories, and credit profiles. Banks are responsible for safeguarding this sensitive data and ensuring the privacy and security of their customers.

However, traditional banking systems are not without their challenges. They often face issues related to inefficiency, high operational costs, complex regulatory frameworks, and limited accessibility. The reliance on intermediaries, manual processes, and legacy systems can result in delays, errors, and higher transaction costs (Fig. 1.1).

Moreover, traditional banking systems are susceptible to fraud and security breaches. Cyber attacks, identity theft, and unauthorized access to customer accounts pose significant risks. Maintaining robust security measures and protecting customer data are paramount for traditional banks.

In recent years, the rise of digital technologies and the advent of blockchain technology have presented both possibilities and difficulties for conventional banking systems. The adoption of digital banking solutions has allowed banks to enhance customer experiences, streamline processes, and offer innovative services. Mobile banking applications, online platforms, and digital wallets have transformed the way customers interact with their banks and carry out transactions.

Furthermore, blockchain technology, with its decentralized and transparent nature, can transform current banking practices. Distributed ledger for recording and verifying transactions, blockchain technology can enhance trust, transparency,

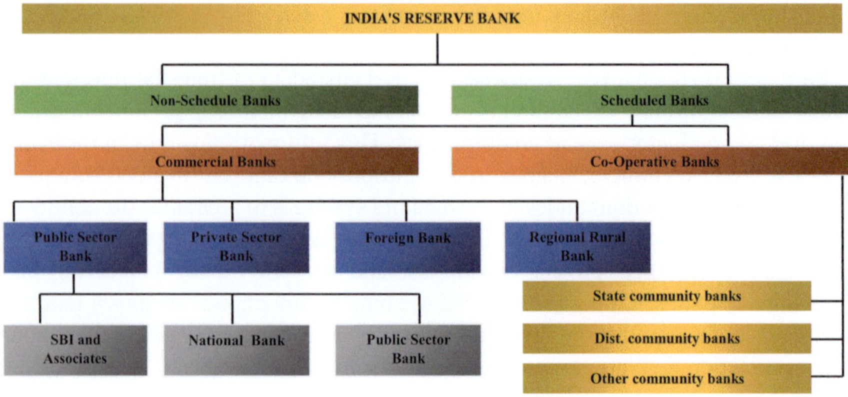

Fig. 1.1 System of banking in India structure

and security. It has the potential to streamline processes, reduce costs, and mitigate fraud risks.

However, integrating blockchain into traditional banking systems also raises regulatory and implementation challenges. Legal structures must change to account for the special features of blockchain technology, such as decentralized governance and cross-border transactions. Additionally, scalability, interoperability, and data privacy issues need to be addressed to ensure seamless integration with existing banking infrastructure.

In conclusion, the traditional banking system plays a crucial role in facilitating economic activities and providing essential financial services. Despite its strengths, it faces challenges related to efficiency, security, and accessibility. The emergence of digital technologies and blockchain technology offers both opportunities and challenges for traditional banks. Embracing digital innovations and leveraging blockchains transformative potential can help traditional banks stay competitive, enhance customer experiences, and adapt to the evolving financial landscape.

Drawback of Traditional Banking

While traditional banking has been a longstanding and reliable system for financial transactions and services, it does come with several drawbacks. Some of the drawbacks of traditional banking include:

(i) **LimitedAccess**: Traditional banking requires physical presence at a bank branch during specific operating hours. This can be inconvenient for individuals who have busy schedules or live in remote areas with limited access to bank branches.
(ii) **Time-consumingProcesses**: Many traditional banking processes involve extensive paperwork, long waiting times, and multiple visits to the bank.

Transactions such as opening an account, applying for loans, or conducting certain financial operations can be time-consuming and cumbersome.

(iii) **Geographical Constraints**: Traditional banks typically have a limited number of branches, which means that individuals living in rural or underserved areas may face difficulties accessing banking services. This can lead to financial exclusion and hinder economic development in those regions.

(iv) **Higher Fees**: Traditional banks often charge various fees for their services, including maintenance fees, ATM withdrawal fees, overdraft fees, and transaction fees. These costs can add up and affect customers, particularly those with lower incomes or who frequently require banking services.

(v) **Limited Innovation**: Traditional banks can be slow to adopt new technologies and innovations compared to fintech companies and online banking platforms. This can result in a lack of convenient digital services and limit access to the latest financial tools and solutions.

(vi) **Strict Eligibility Criteria**: Traditional banks typically have strict eligibility criteria for loans and credit, making it difficult for individuals with limited credit history or lower incomes to access financial products. This can create barriers to obtaining necessary funds for personal or business needs.

(vii) **Lack of Personalization**: Traditional banking often lacks a personalized approach to customer service. Customers may not receive tailored recommendations or personalized financial advice, leading to a less satisfying banking experience.

(viii) **Security Concerns**: While traditional banks have implemented security measures to protect customer accounts, instances of fraud and identity theft still occur. Traditional banking methods such as checks and physical documents can be vulnerable to theft or forgery.

It is important to note that with the rise of technology and digital advancements, many traditional banks are working to address these drawbacks by offering online and mobile banking services, adopting innovative technologies, and improving customer experience.

1.1.1.2 Digital Banking System

The digital banking system represents a transformative shift in the way banking services are delivered and experienced by customers. It leverages technology and digital channels to offer a wide range of financial services through online platforms, mobile applications, and other digital channels. This evolution in banking has revolutionized the industry, providing customers with convenience, accessibility, and personalized experiences.

Online banking is a crucial component of the digital banking system. Through secure websites or mobile apps, customers may access their accounts, examine balances, and carry out various transactions. Customers no longer need to travel to actual bank locations in order to conveniently manage their funds because of online banking.

Mobile banking applications are a significant component of the digital banking system. With smartphones becoming ubiquitous, customers can access their bank accounts, make payments, transfer funds, and perform other financial activities directly from their mobile devices. Mobile banking apps often offer additional features like biometric authentication, budgeting tools, and personalized financial insights.

The digital banking system also includes electronic payment services that simplify and expedite transactions. Customers can make online purchases, pay bills, and transfer funds electronically, reducing the reliance on cash and checks. Peer-to-peer transfers, contactless payments, and mobile wallets are examples of digital payment mechanisms, offer convenience and speed, enhancing the overall customer experience.

Furthermore, digital banking has facilitated the emergence of innovative financial products and services. Fintech companies, in collaboration with traditional banks, are introducing novel solutions to meet evolving customer needs. These include robo-advisory services, automated investment platforms, crowdfunding, and peer-to-peer lending. Open banking, which enables users to securely share their financial data with authorized third-party providers to receive customized financial goods and services, has also been made possible by digital banking.

The digital banking system has empowered customers by providing self-service options and personalized experiences. Customers can customize their banking preferences, set alerts and notifications, and access a wealth of financial information and educational resources. This self-service approach enhances financial literacy and empowers individuals to make informed decisions about their finances.

Moreover, the digital banking system has the potential to promote financial inclusion. It allows individuals who were previously underserved by traditional banking to access basic financial services. Digital banking reduces barriers to entry, such as physical distance and documentation requirements, making banking services more accessible to marginalized populations.

However, the digital banking system also presents challenges and considerations. Cybersecurity and data privacy are critical concerns, as the reliance on digital channels increases the vulnerability to cyber threats and data breaches. Banks must invest in robust security measures, encryption technologies, and fraud detection systems to safeguard customer information and transactions.

The digital banking system requires ongoing investment in technology infrastructure and digital literacy initiatives to ensure broad adoption and usability. Banks need to address digital divides, provide support for older or less tech-savvy customers, and ensure accessibility for individuals with disabilities.

Benefits of Digital Banking

Online banking, usually referred to as digital banking, or electronic banking, has many benefits over conventional banking practices. The following are some benefits of digital banking:

(i) **Facilitation:** If a customer has access to digital banking, it saves time and effort by eliminating the need to go to a physical bank branch.

(ii) **Availability round-the-clock:** Unlike traditional banking, digital banking services are accessible every day of the week, 24 h a day. Customers can check their account balances at any time, even on weekends and holidays, make transfers, pay bills and complete other operations.

(iii) **Cost Savings**: Comparing digital banking to traditional banking, costs are frequently lower or nonexistent. Many online banks offer fee-free accounts, reduced transaction fees, and no or minimal account maintenance fees. This can result in significant cost savings for customers.

(iv) **Enhanced Account Management**: Digital banking provides users with real-time access to their account information. Customers can monitor their transactions, track expenses, view statements, and analyze their financial activity using intuitive interfaces and tools. This enables better financial management and budgeting.

(v) **Quick and Efficient Transactions**: Digital banking allows for fast and seamless transactions. Funds can be transferred between accounts instantly, and bill payments can be made with a few clicks. This eliminates the need for writing and mailing checks or visiting physical branches to conduct transactions.

(vi) **Mobile Banking**: Many digital banking platforms offer dedicated mobile apps that provide even greater convenience. Customers can use their smartphones or tablets to instantly access their accounts, complete transactions, and manage their finances. Mobile banking also often supports additional features like mobile check deposit and biometric authentication for added security.

(vii) **Enhanced Security Measures**: Digital banking platforms employ advanced security measures to protect customer information and transactions. These may include multi-factor authentication, encryption, fraud monitoring, and secure data transmission protocols. In some cases, digital banking can be considered more secure than traditional banking methods, such as paper checks or physical documents.

(viii) **Access to Financial Tools and Services**: Digital banking often integrates with other financial tools and services. Customers can easily apply for loans, credit cards, or investment accounts online, and accesses a wider range of financial products and services. Additionally, digital banking platforms may offer financial planning tools, budgeting apps, and personalized recommendations to help customers manage their finances effectively.

Digital banking has transformed the way individuals and businesses interact with their finances, providing greater accessibility, convenience, and flexibility. However, it is crucial for users to remain vigilant and follow recommended security practices to protect their personal and financial information.

1.1.2 Blockchain Technology

A decentralized and distributed ledger system known as blockchain technology keeps track of and verifies transactions across numerous computers, or nodes. It uses cryptographic algorithms to ensure security and immutability of data. Some are key points are following for block-chaining (Fig. 1.2).

(i) **Decentralization**: Unlike traditional centralized systems, blockchain operates in a decentralized manner, where no single entity has full control. Transactions are validated and recorded by multiple participants in the network, creating a transparent and tamper-resistant system.
(ii) **Transparency and Immutability**: Blockchain provides transparency as every transaction is recorded on a public ledger, visible to all participants. Once recorded, transactions are nearly impossible to alter, providing a high level of data integrity.
(iii) **Security:** Blockchain protects data using advanced encryption techniques to prevent unauthorized access and manipulation. A proof-of-work or proof-of-stake consensus protocol ensures the authenticity of transactions and maintains the integrity of the blockchain.

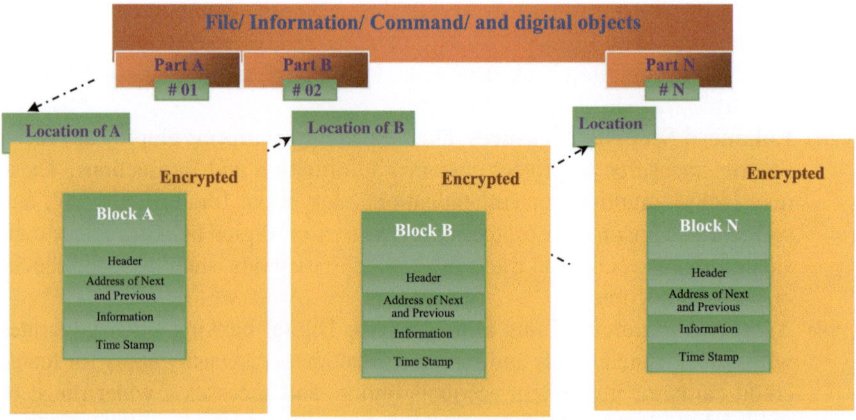

Fig. 1.2 Show how a Digital object is stored at various places, with encrypted blocks of chain, no single point know where the whole object is stored. All sections take part to accomplish a single operation, smoothly and securely. It is a link list that is controlled by numbers of computers/nodes, encrypted links, and encrypted operation for store digital objects over the distributed network

(iv) **Effective Agreements:** When specific criteria are met, effective agreements—which are self-executing contracts written in code—automatically take effect. Peer-to-peer transactions are made possible by these contracts, which do away with the need for middlemen.

(v) **Use Cases**: Blockchain technology has applications beyond cryptocurrencies. It can be used for decentralized finance (DeFi), and more. Blockchain's ability to provide transparency, security, and trust has opened up numerous possibilities across various industries.

(vi) **Challenges**: Blockchain technology faces challenges compatibility, energy use, and legal frameworks, and privacy concerns. Scaling blockchain networks to handle large transaction volumes while maintaining efficiency remains a significant challenge.

(vii) **Public vs. Private Blockchains:** Open to all users and supported by a distributed infrastructure of nodes, public blockchains like Bitcoin and Ethereum are available to all users. Blockchains that are private or permissioned limit access to authorized users and are frequently utilized within organizations in particular purposes.

(viii) **Blockchain compatibility**: It refers is the capacity of several blockchains to interact and communicate with one another. Standards and protocols are being developed to facilitate seamless integration between various blockchain platforms.

1.1.3 Digital Currency

One of the most well-known and widely adopted digital currencies is Bitcoin, which was introduced in 2009. Bitcoin's creation marked the beginning of a new era in financial transactions, offering a decentralized and secure alternative to traditional monetary systems. Since then, numerous other digital currencies have emerged, each with its unique features and use cases.

Digital currencies provide several key advantages over traditional currencies. First and foremost, they offer increased security using cryptography. Transactions conducted using digital currencies are encrypted and verified through complex mathematical algorithms, ensuring the integrity of the transaction data and protecting against fraud and counterfeiting.

The potential for greater financial transaction efficiency is a key benefit of digital currency. Due to intermediaries and complicated settlement procedures, cross-border transactions in traditional banking systems can be time-consuming and expensive. Peer-to-peer transactions using digital currencies are nearly instantaneous, requiring no middlemen and lowering transaction costs.

Additionally, digital currencies provide for increased financial inclusion, especially in places with limited access to traditional banking services. With access to the internet and a digital wallet, individuals can participate in the global economy, conduct transactions, and store value without the need for a traditional bank account.

Moreover, digital currencies provide transparency and traceability in transactions. Blockchain technology, which underlies many digital currencies, creates a distributed ledger where transaction records are immutably stored. This transparency can help combat illicit activities, as transactions can be easily audited and traced back to their source.

Despite the numerous benefits, digital currencies face challenges and considerations. One notable challenge is regulatory oversight. Governments and regulatory bodies are still grappling with how to effectively regulate digital currencies, as they can pose risks related to money laundering, tax evasion, and consumer protection. Striking a balance between innovation and regulation remains a key focus for policymakers.

Additionally, the volatility of digital currencies is another consideration. The value of digital currencies can experience significant fluctuations, making them subject to speculation and potential financial risks. This volatility can be attributed to factors such as market demand, regulatory changes, and technological developments.

Looking to the future, digital currencies hold the potential to reshape the financial landscape further. Central banks are exploring the concept of Central Bank Digital Currencies (CBDCs), which would be digital representations of fiat currencies issued and regulated by central banks. CBDCs aim to combine the benefits of digital currencies, such as efficiency and traceability, with the stability and credibility of traditional fiat currencies (Fig. 1.3).

In conclusion, digital currencies have emerged as a transformative force in the financial world. With their potential for enhanced security, efficiency, financial inclusion, and transparency, digital currencies offer new possibilities for individuals

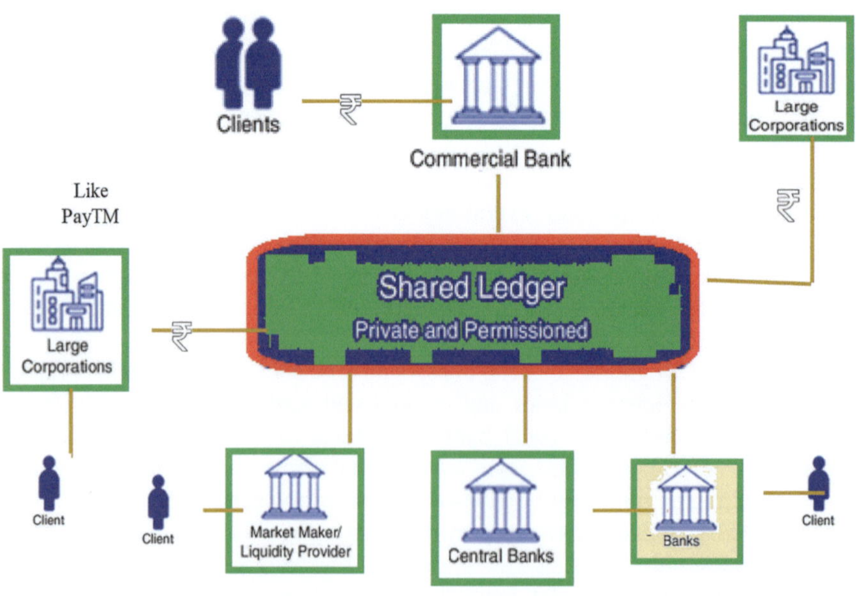

Fig. 1.3 Digital banking system

and businesses worldwide. However, navigating the regulatory landscape and addressing volatility remain important considerations as the future of digital currencies continues to unfold.

1.2 Literature Review

The impact of blockchain technology on traditional banking systems is a topic that has received significant attention in recent years. A number of scholars have conducted research and have provided a comparative analysis of blockchain-based banking systems and traditional banking systems. Some studies are-

Nakamoto (2008) Bitcoin whitepaper provides a detailed technical description of the Bitcoin protocol and outlines the benefits of a decentralized, peer-to-peer payment system. Since its publication, the Bitcoin network has grown significantly and has spawned an entire ecosystem of cryptocurrencies and blockchain-based applications.

Yermack (2017) provides a thoughtful analysis of blockchain technology's possible effects on corporate governance. The author suggests that while there are challenges to the adoption of blockchain technology in this area, the potential benefits are significant and should not be ignored.

Al-Turjman and Almsafir (2019) provides a comprehensive review of the potential use of blockchain technology for improving the security and privacy of IoT systems. The authors suggest that while there are challenges to the adoption of blockchain technology in this area, the potential benefits are significant and should be explored further.

Chiu and Koeppl (2017) provides a comprehensive examination of the economic aspects of cryptocurrencies, shedding light on their potential benefits, challenges, and implications for the broader economy.

Iansiti and Lakhani (2017) It makes the case that blockchain technology has the power to transform numerous industries by offering a safe and open platform for transacting, and it incites businesses and decision-makers to look into its possible uses. Overall, "The Truth about Blockchain" is ahighly informative and accessible overview of blockchain technology, highlighting its fundamental features, advantages, and challenges. The article provides a useful starting point for anyone interested in understanding the potential applications of blockchain technology.

Mazières and Kohli (2018) provides a comprehensive overview of the SCP, highlighting its unique design features and potential applications. The article offers valuable insights into the challenges and opportunities associated with achieving consensus in decentralized networks, and it provides a useful starting point for further research in this area.

Raskin (2017) concludes by emphasizing the transformative potential of blockchain technology in banking and finance, while acknowledging the need for further research, experimentation, and collaboration to overcome the challenges and fully realize its benefits. Overall, "The Application of Blockchain Technology in Banking and Finance" provides an overview of the potential impact of blockchain technology on the banking and finance industry. It highlights the benefits, challenges, and real-world use cases, contributing to the broader understanding of how blockchain can revolutionize financial services.

Swan (2015) concludes by summarizing the key takeaways and demonstrating how blockchain technology has the potential to change the world's economy. Overall, "Blockchain: Blueprint for a New Economy" is a valuable resource for anyone seeking to understand the principles, applications, and implications of blockchain technology. The book provides a comprehensive overview of the technology and its potential, making it an essential read for entrepreneurs, investors, developers, and policymakers alike.

Zheng et al. (2018) concludes with a summary of the key takeaways and a discussion of the future of blockchain technology. The authors emphasize the need for continued research and development to address the challenges associated with blockchain and to fully realize its potential as a transformative technology. Overall, "Blockchain Challenges and Opportunities: A Survey" provides a comprehensive overview of the state of blockchain technology, highlighting both the challenges and opportunities that the technology presents. The article is a valuable resource for anyone seeking to understand the potential impact of blockchain on various industries and domains.

Catalini and Gans (2019) explores the potential challenges and limitations of blockchain, including issues related to scalability, interoperability, and governance. The authors highlight the need for further research and development to address these challenges and fully realize the potential of blockchain. Overall, "Some Simple Economics of Blockchain" provides a valuable contribution to the growing body of research on the economics of blockchain technology. The paper highlights the potential economic implications of blockchain and identifies key areas for future research and development.

Overall, the literature suggests that blockchain technology has the potential to disrupt traditional banking systems and improve the efficiency and security of banking services.

1.3 Discussion (Fig. 1.4)

1.3.1 Advantage of Blockchain in Banking

The impact of blockchain technology on traditional banking systems has already begun to reshape the financial landscape, and its future scope is vast and promising. Here is a comparative analysis of the potential future impacts of blockchain technology on traditional banking systems:

(i) **Enhanced Security and Trust**: Blockchain's decentralized and immutable nature ensures enhanced security and trust in financial transactions. In the

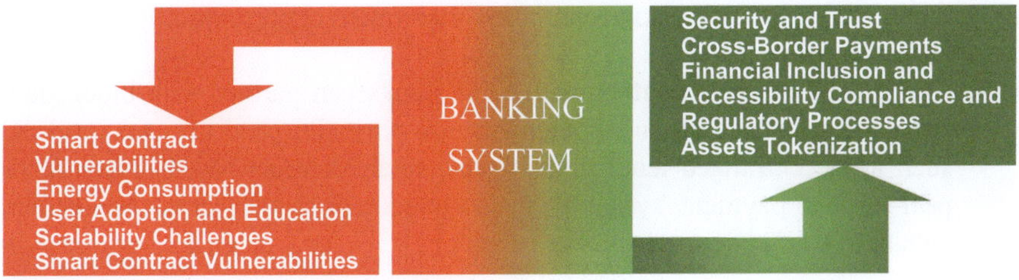

Fig. 1.4 Impact of digitization on banking system

future, traditional banks can leverage blockchain to strengthen their security measures, reducing the risk of fraud, data breaches, and cyberattacks. This technology can enable secure and transparent verification of transactions, identities, and ownership, fostering a higher level of trust between banks and their customers.

(ii) **Streamlined Invoices**: Traditional banking systems often involve multiple intermediaries, resulting in lengthy and expensive processes. With blockchain, financial institutions can leverage smart contracts and distributed ledger technology to streamline cross-border payments, making them faster, more efficient, and cost-effective.

(iii) **Financial Inclusion and Accessibility**: One of the significant benefits of blockchain technology is its potential to provide financial services to the unbanked and underbanked populations. By utilizing blockchain-based solutions, traditional banks can extend their services to individuals and businesses in remote areas or countries with limited access to banking infrastructure. This can lead to greater financial inclusion and economic empowerment worldwide.

(iv) **Improved Compliance and Regulatory Processes**: Blockchain's ability to provide a transparent and immutable record of transactions can simplify compliance and regulatory processes for banks. In the future, blockchain-based solutions can automate regulatory reporting, ensuring accuracy, consistency, and real-time monitoring of transactions. Additionally, smart contracts can be programmed to enforce regulatory compliance automatically, reducing the need for manual intervention.

(v) **Efficient Identity Management**: Blockchain-based identity management systems have the potential to revolutionize customer onboarding and KYC (Know Your Customer) processes. Traditional banks can utilize blockchain to create a secure and decentralized identity verification system. This would allow customers to maintain control over their personal data while granting banks access to necessary information, simplifying the onboarding process and reducing the risk of identity theft.

(vi) **TokenizationofAssets**: In the future, traditional banks can leverage blockchain to create digital tokens representing these assets. This would facilitate fractional ownership, enhance liquidity, and enable efficient transfer and trading of assets. It could also open up new investment opportunities for individuals and institutions.

(vii) **Collaboration with Fintech Startups**: Banks can partner with blockchain companies to develop and implement new financial products and services, such as decentralized lending platforms, smart contract-based insurance or peer-to-peer payment systems. Such collaborations can drive innovation, enhance customer experience, and create new revenue streams.

1.3.2 Disadvantage of Blockchain in Banking

While blockchain technology offers numerous advantages for the banking industry, it is essential to consider the potential disadvantages and challenges it may present. Here are some disadvantages of blockchain in banking:

(i) **The adaptability Issues:** Due to its distributed nature, blockchain technology has difficulties with scaling. The blockchain network may become slower and less effective as more transactions are made. This poses a challenge for banking systems that handle a large volume of transactions daily. Resolving scalability issues while maintaining the security and decentralization of the blockchain remains a significant hurdle.

(ii) **Regulatory Uncertainty**: The regulatory landscape surrounding blockchain technology is still evolving. Banks need to comply with various regulations and standards to ensure customer protection, prevent money laundering, and adhere to know-your-customer (KYC) requirements. However, blockchain's decentralized and pseudonymous nature can create challenges in meeting these regulatory obligations, especially in terms of identity verification and transaction monitoring.

(iii) **Energy Consumption**: Blockchain networks demand a lot of processing power and energy, especially those that use proof-of-work (PoW) consensus methods like Bitcoin. Concerns concerning blockchain mining's environmental impact have been highlighted because to its energy-intensive nature. When adopting blockchain technology, banks must take sustainability into account and figure out how to reduce the carbon footprint of maintaining blockchain networks.

(iv) **Governance and Legal Issues**: Blockchain networks typically operate in a decentralized manner without a central authority controlling the system. This lack of centralized governance can pose challenges when it comes to resolving disputes, enforcing legal contracts, and managing the overall system. In the banking sector, where legal frameworks and contractual obligations are crucial, adapting blockchain technology to align with existing legal and governance structures can be complex.

(v) **Integration with Legacy Systems**: Many traditional banks have complex legacy systems and infrastructure that are not easily compatible with blockchain technology. Integrating blockchain into existing banking systems may require significant time, effort, and cost. Banks need to carefully plan the integration process to ensure smooth transitions, minimize disruptions, and maintain the security and integrity of customer data.

(vi) **User Adoption and Education**: Encouraging user adoption and educating customers about the benefits and proper usage of blockchain-based services can be a challenge. Banks need to invest in customer education initiatives to ensure a seamless transition to blockchain-based banking solutions and address any concerns or misconceptions.

(vii) **Potential for Smart Contract Vulnerabilities**: While smart contracts offer automation and efficiency, they are not immune to vulnerabilities or bugs. Inadequate coding, programming errors, or malicious attacks can exploit smart contracts, leading to financial losses or legal disputes. Banks must carefully audit and test smart contracts to ensure their reliability, security, and compliance with regulatory requirements.

It's important to note that many of these challenges are being actively addressed and mitigated as the technology matures. Blockchain is an evolving field, and ongoing research, development, and industry collaboration are essential to overcome these disadvantages and unlock the full potential of blockchain in the banking sector.

1.4 Conclusion

In conclusion, the impact of blockchain technology on traditional banking systems has been substantial and continues to evolve. Through a comparative analysis, it becomes evident that blockchain technology brings several benefits and challenges to the traditional banking landscape.

Blockchain technology offers increased transparency, security, and efficiency in financial transactions. The distributed ledger system eliminates the need for intermediaries, streamlines processes, and reduces costs. Smart contracts enable automated and secure execution of agreements, reducing the potential for errors and disputes. Moreover, the immutability of blockchain records enhances the security and auditability of transactions.

However, the integration of blockchain technology into traditional banking systems is not without its challenges. Regulatory frameworks and compliance requirements need to be addressed to ensure a smooth transition. Scalability issues must be overcome to handle large-scale transaction volumes effectively. Interoperability between different blockchain platforms and legacy systems is another important aspect that needs to be considered.

Despite these challenges, forward-thinking banks have started exploring and implementing blockchain technology to enhance their services. Collaboration with partnerships between traditional banks and fintech companies have emerged as a strategy to harness the potential of blockchain technology effectively. As the technology matures and regulatory frameworks adapt, we can anticipate further integration and innovative use cases in the banking sector.

It is important for traditional banks to actively monitor and understand the evolving landscape of blockchain technology. By embracing the opportunities it presents and addressing the associated challenges, banks can position themselves at the forefront of digital transformation, delivering enhanced financial services and meeting the evolving needs of customers in the digital age.

1.5 Future Scope

The future scope of blockchain technology in traditional banking systems is incredibly vast and holds immense potential for transforming the financial landscape. As blockchain continues to mature, several exciting opportunities lie ahead. One prominent area of exploration is enhanced interoperability between different blockchain networks and traditional banking systems. Developing standardized protocols and frameworks that enable seamless communication and data exchange will facilitate the integration of blockchain into existing banking infrastructure.

Central Bank Digital Currencies (CBDCs) represent another exciting avenue for future development. Central banks worldwide are exploring the idea of issuing digital currencies using blockchain technology. Implementing CBDCs would revolutionize the monetary system, enabling faster and more efficient transactions while maintaining regulatory oversight. Extensive research and development efforts will be dedicated to designing secure and scalable CBDC platforms that seamlessly integrate with traditional banking systems.

Decentralized Finance (DeFi) is another area that holds tremendous potential. DeFi platforms built on blockchain enable individuals to engage in financial activities, such as lending, borrowing, and trading, without intermediaries. The future scope involves exploring the integration of DeFi applications into traditional banking systems, creating hybrid models that combine the advantages of decentralized finance with the stability and regulatory compliance of traditional banks. Future research will focus on addressing scalability challenges and developing interoperable blockchain networks that enable seamless transactions between different banking systems. Identity verification is another critical aspect where blockchain can make a significant impact. Blockchain-based identity solutions have the potential to simplify and secure Know Your Customer (KYC) processes used by banks. Future developments will focus on building robust, privacy-preserving identity solutions that put individuals in control of their personal data while ensuring compliance with regulatory requirements.

To support the widespread adoption of blockchain technology in traditional banking systems, clear and adaptable regulatory frameworks are necessary. Collaborative partnerships between regulators, banks, and blockchain technology providers will be vital in establishing comprehensive regulatory frameworks that foster innovation while ensuring financial stability and consumer protection.

Addressing scalability and energy efficiency challenges is also crucial for the future of blockchain in banking. Research will focus on developing innovative solutions such as layer-two scaling techniques, sharing, and improved consensus mechanisms to enhance the scalability and energy efficiency of blockchain networks, making them suitable for large-scale banking operations.

In conclusion, the future scope of blockchain technology in traditional banking systems is promising and diverse. Further exploration and development in areas such as interoperability, CBDCs, DeFi integration, cross-border payments, identity solutions, regulatory frameworks, scalability, and energy efficiency will shape the transformation of traditional banking systems, enabling a more efficient, secure, and inclusive financial ecosystem.

References

Al-Turjman F, Almsafir MK (2019) Blockchain technology for enhancing the security and privacy of internet of things (IoT) systems: a review. J Netw Comput Appl 126:46–63

Catalini C, Gans JS (2019) Some simple economics of blockchain. NBER Working Paper No. 22952

Chiu MH, Koeppl T (2017) Cryptocurrencies' financial affairs—besides Btc. Financ Horiz Bull 31(2):61–84

Iansiti M, Lakhani KR (2017) The truth about blockchain. Harv Bus Rev 95(1):118–127

Mazières D, Kohli M (2018) The stellar consensus protocol: a federated model for internet-level consensus. In Proceedings of the 26th USENIX security symposium, pp 489–507

Nakamoto S (2008) Bitcoin: a peer-to-peer electronic cash system. Retrieved from https://bitcoin.org/bitcoin.pdf

Raskin M (2017) The application of blockchain technology in banking and finance. In: Chaudhry S, Gandhimathi P, Elchouemi A (eds) Innovations in financial services: artificial intelligence, blockchain, and the future of fintech. Springer, pp 191–201

Swan M (2015) Blockchain: blueprint for a new economy. O'Reilly Media

Yermack D (2017) Corporate Governance and Blockchains. Rev Financ 21(1):7–31

Zheng Z, Xie S, Dai HN, Chen X, Wang H (2018) Blockchain challenges and opportunities: a survey. Int J Web Grid Serv 14(4):352–375

Chapter 2
Apparent Advantages and Negative Facet of Block Chain in Banking Sector: An Innovative Theoretical Perspective

Diksha Verma, Pooja Kansra, and Shad Ahmad Khan

2.1 Introduction

Banking companies have shifted themselves from traditional mode to digitization mode using different models like mobile banking, omnichannel banking etc. Not only banking sector but other industries also are shifting themselves towards digitization mode. As per latest survey, it has been found that technology is going to grow to $ 67.4 billion in 2026 from $ 4.9 billion in 2021. In spite of numerous challenges, Banks have initiated use of Blockchain technology in a whole hearted manner (Chirag 2023). Blockchain has the capability of transforming the financial market in to a digitized hub by creating a transparent and public ledger for recording the transactions (Gupta 2021). According to the latest scenario, Economies are witnessing a pioneering gesture related to industrial technologies which is popularly known as "Industry 4.0". Earlier the concept of Industry 4.0 was acknowledged with manufacturing sector but nowadays along with the mechanized sector, many service industries lining from telecom to banking are expectant of the remuneration in the same field. This digitalization has resulted in invention of new business models. Service sector nowadays is contributing a majority of share in GDP of the developed and developing economies. The industries related to service sector are either trying or in line of trying digitalization as a new practice of doing business. Various

D. Verma
Mittal School of Business, Lovely Professional University, Bengaluru, Karnataka, India

Chandigarh Business School of Administration, CGC Landran, Ajitgarh, Punjab, India

P. Kansra
Mittal School of Business, Lovely Professional University, Bengaluru, Karnataka, India

S. A. Khan (✉)
College of Business, University of Buraimi, Al Buraimi, Oman

M. Irfan et al. (eds.), *Applications of Block Chain technology and Artificial Intelligence*, Financial Mathematics and Fintech,
https://doi.org/10.1007/978-3-031-47324-1_2

options such as cloud computing, internet of things, artificial intelligence, 3D Printing and block chain are the new normal of doing business and fall under the span of industry 4.0 revolutions (Garg et al. 2019).

Block chain is one such element in industrial revolution that has expected considerable concentration in financial field of technology (FinTech) (Guo and Liang 2016). It is basically a data base that stores the transaction record digitally, connected mutually in a single list named as chain. Block chain is a system that abridges the likelihood of occurrence of hacking, cheating etc. of stored information (Ravi 2021). Currently, International bodies such as IMF and UN and also various developed countries like the United States, the United Kingdom and Japan, have started focusing on growth and enlargement of block chains in various areas. Likewise, India, Russia, China and South Africa have boosted the effort of research on block chain technology (Guo and Liang 2016).

Banking industry is nowadays shifting towards modern banking from traditional business. This area has also put forward their shoes in to the block chain side. Banks majorly replicate transactions of their current assets on the block chain resulting in efficacy for their solutions in relation to block chain. A kind of software developed for holding the transactions related to transfer of digital assets to the customers digitally is typically called block chain in banking (Ravi 2021). Block chain has the prospective to diminish and vanish the existing middle men and also civilizing recording of transactions by implementing transparency and traceability of transactions digitally which ultimately impact advancement of traditional way of banking to modern banking. Globally $ 20 billion amount could be saved before year 2022, if block chain is implemented in banking, according to a statement by Accenture (Garg et al. 2019).

Different uses of Blockchain technology associated with banking sector are discussed below. These uses ultimately are making banking industry decentralized. Transfer of payment from one point to the destination is made easy and less costly with the help of Bitcoin, Ether etc. These are the developed open blockchains which can be used by bank customers to send or receive money without any additional charges. Secondly, it takes around 3–4 days to clear or settle one problematic transaction because of various channels involved in settling the transaction. But blockchain performs as a decentralized ledger that keeps record of transaction in transparent and open manner making it accessible for all. Therefore blockchain has another advantage of settling and clearing the problematic transaction in few seconds of time. If a customer is involved in buying or selling of securities through banking institutions as intermediaries, banks also have to maintain the record of debts, stocks, brokers, exchanges n order to execute the actual transaction. Blockchain has another advantage of creating a decentralized database that keeps the record of digital securities and making the transaction easy for customers. Banks in order to provide loans to their customers has to go through multiple rounds of checking credit score, income of the consumers, possession standing, debts payoff ratio etc. Blockchain has a transparent investment friendly system through which personal loans can be given to the bank customers making it convenient for clients to apply for loan (Gupta 2021; Chirag 2023).

Numerous challenges and difficulties are associated with the invention of a new technology whether in any field. Although the vocation related to block chain is in progress and efforts are being done to present a simplistic image of organization's issues yet more is required. Frequently it is quoted that dependence on middlemen diminishes with the use of block chain technology in various sectors and the trust is build by the block chain on its own still there is a high need for supremacy to formulate and implement such a scenario in which block chain can operate with the organizations (Janssen et al. 2020).

Description of block chain's scientific position has been discussed and explored theoretically from different perspectives still any endeavor is being made in previous review related to positive or negative role of implementation of block chain in banking industry where other variables in banking are also being associated time to time such as omni channels, bancassurance etc. (Garg et al. 2019). Despite the fact that block chain technology has just started to become known nowadays, still there is a need for extensive overview of the concept for practical espousal in the banking industry (Janssen et al. 2020). The current study is an attempt to add to this conversation by providing a conceptual model on block chain with special relevance to banking sector. Firstly discussion of theoretical background of Block chain in relation to the banking sector will be initiated depending on the previous literature and after that a effort will be made to identify the apparent advantages of implementing block chain technology in banking industry giving boost to industry revolution 4.0 (Garg et al. 2019).

Study has been tailored into five unrelated sections. Section I defines the basic information about the notion and conception of block chain technology and its effect general effect on banking sector followed by section II which discusses about the available literature on the same concept. Section III is detailed narration about the apparent advantages of implementing block chain technology in banking industry and Section IV includes conversation about the disadvantages and negative aspect of implementing technology of block chain in the field of Banking (Oberoi and Kansra 2019).

2.2 Review of Literature

2.2.1 Block Chain Technology

According to Hackett (2016), corroboration of various records and data available in digital mode can be consummate by examining the similar set of connections since they are allied to internet connectivity and all peer structure has comprehensive way in to all digitally available records. Therefore, block chain is exclusively a series of blocks linked collectively with the application of computational algorithms (Woodside et al. 2017).

According to Yoo (2017), a scattered practice of preparing ledgers in which all the participants simultaneously share records of transactions among them is typically called Block chain. A data base that is basically dispersed and keeps records of incessantly developing list of data which is difficult to tamper and adjustment is Block chain. The collection of various transactions is often known as Blocks and concept wise the same is called block chain under which various blocks are symmetrically assembled one after another.

According to Cocco et al. (2017), Block chain technology has the power of advancing the whole environment related to financial systems and also improvising the efficacy of present financial structure. Block chain technology helps in controlling wastage of conservative energy and might set at the place of consumptive systems of energy that basically hold up the fiat currency. Because all the records available digitally are assembled cryptographically under a particular heading along with the address of those coupled with the transaction, the chain is concurrently modernized at every single nodule. Hence, it is grueling for hackers to fetch the data available online or records out of any nodule or adapted system (Golosova and Romanovs 2018).

2.2.2 Block Chain Technology in Banking

Guo and Liang (2016) discussed about numerous challenges witnessed by traditional banking sectors. Commercial banks are more required to bank upon the latest technologies introduced nowadays in order to be victorious and conquer the aggressive market. Block chain is one of such technologies that could be worn to meet the present challenges of emerging innovative techniques of serving customers. Block chain ultimately aids in dropping the cost for banks by providing more efficient services. Block chain in collaboration with the banking sector has enhanced the operating system of banks in a different manner. Block chain helped in reducing the need for intermediaries for transactions related to banks and other financial institutions (Cocco et al. 2017).

According to Garg et al. (2019), currently merchant financing, loyalty and trust programs for bank customers and syndicated loans are certain tools under block chain technology being tested by Indian Banking Sector on regular basis. Various banks such as Axis Bank, SBI, Yes Bank and Reserve Bank of India are actually testing block chain technology in practical scenario.

Hence this paper is trying to centre upon the apparent advantages of implementing Block chain technology in banking sector and also the drawbacks and limitations of using bock chain technology in the same field. Despite the fact that, prospect are astonishing, appliance of block chain is yet at embryonic stage and have not been fully employed (Olaf et al. 2017).

2.3 Apparent Advantages of Implementing Block Chain Technology in Banking

Currently the key compassion of every association is to make initiatives for saving environment. Every organization endeavours to develop ecological presentation by launching numerous programs so that surroundings could be protected. For the purpose of achieving same many institutions are conducting experiments and implementing block chain technology so that intention can be achieved. Block chain technology has the capability to associate green technologies with its services which lead to development of economy in a better way. Block chain technology is intended to provide peer to peer arrangement and do not depend on a solitary middle power. Banks related transactions are processed with the aid of transmit system. It transmits the communications with the help of nodes. Every node has imitation of the block chain which is basically connected with other nodes. Every node is secure as its existence is unknown to the other ultimately securing the block chain from exterior assaults. Various crypto currency networks depend on defined algorithms, consent mechanism like PoW, Bitcoin network etc. Block chain technology has prompt a lot curiosity something like its prospect succession and the significant advantages that it might get in the framework of the reassign of material goods and belongings contained by industrial system. Though, present be numerous apprehension about the block chain technology, like its probable and potential capability to accomplish the dispensation velocity desirable for an mechanized defrayal domicile, to lesser overheads in contrast to established compensation structure, and to include the enhance the shattered removal possessions when superior deal capacity are concerned (Cocco et al. 2017).

It could be seen that vicinity where block chain which is more vigorously practical in the fiscal zone is enhancing for defrayal, transfer of funds, contracts related to securities etc. According to the literature it has been observed that in various countries also validation process for holding assets are normally used with the intention of keeping block chain in dominating part. The initiative to commence a clogged or confidential disseminated ledger that does not depart in the course of the central bank is pacing in costs amid banks. Domestic fiscal establishment too require mutual stroke by fiscal association throughout a block chain conglomerate to relate block chain technology with financial segment. End user's desires and industrial expansion are varying. Equivalently, when the prospect to contravene the information detained by folks has stretched, the necessity for block chain technology is sturdily budding for the reason that the efforts of the association could be protected (Yoo 2017).

Block chain technology is one such kind of technology that basically unites with multiplicity of circumstances. It majorly target on fiscal and monetary transactions leading to accomplishment of asset digitization. This stridently augments the procedure effectiveness of the clearance and defrayal of economic possessions subsequent to dealings, at the same time as reducing overheads. Block chain technology ultimately solve numerous of problems related to banking sector in various ways.

Obviously, block chain has come up with an only back up plan for all the banking related issues. Many banking industries have started using block chain technology as their core operation. Banking sector is utilizing its resources on conducting research activities related to block chain technology. This will facilitate them to turn into the lead the way of industrial appliance that can escort and contribute in the creation of novel industry background, therefore, endlessly humanizing the aptitude and content of high-value pecuniary services and nurturing innovative impetus for development in the sector (Guo and Liang 2016).

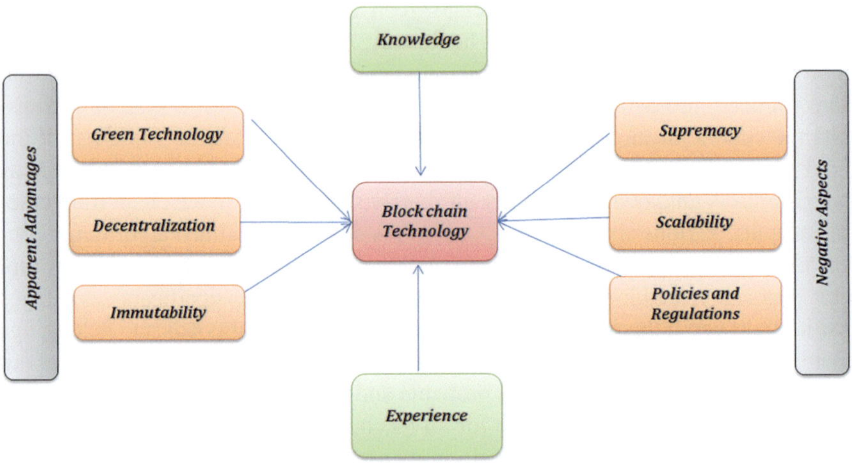

Source: Model based on Review of literature

2.4 Negative Aspect of Implementing Block Chain Technology in Banking Industry

Decentralized composition of Block chain provides unattired defies for financial and regulatory associations and bodies looking forward adopting technology as its controller. Presently there are three main block chain restrictions such as shortage of supremacy, lack of scalability and risks associated with dictatorial fulfillment.

2.4.1 Shortage of Supremacy

However the decentralized character of block chain technology is somewhat its advantageous feature still it poses various restrictions for the financial segment. In the absence of central decision maker, a severe hurdle is members who participate in block chain powered transactions may have uneven objectives. Block chain is somewhat related to DAO (*Decentralized Autonomous Organization)* which aims at removing the need for any authority by inventing an automatic decision making structure with numerous rules and regulations. However the hacking of DAO is quite effortless. Since the susceptibility in the code following the DAO formed by German startup was there, the arrangement was lacerated, and approximately a $50 million value of crypto currencies was theft. Sequentially reimbursement to the DAO sponsor, the mainstream of the public nominated 'hard fork' (*a variation in the code of conduct which formulates formerly legal blocks void*), parting the remaining society unappreciated. This state hoisted various red flags. The whole arrangement was done to assess efficient and reasonable authority with no interference of human being. Nevertheless, it became evident that structure is yet exposed to lobbyists and customary major proportion bases appointment system. This leads to unacceptability for various financial organizations.

2.4.2 Lack of Scalability

Owing to the reality that block chain is a disseminated structure, its snowballing dispensation influence in a straight line, based upon the calculative command of the strategy concerned. In contrast to Visa's 1700 dealings for every second, block chain is competent of procedure about 4.6 dealings for every second normally. These fissures entail a gigantic confront for the embracing of block chain for banking sector at universal level. With the mounting recognition of block chain, the dilemma of scalability turns into perceptible. Various scaling methods have been adopted but every method comes up with certain restraints most popular of which is 'sharding'. For the purpose of Block chain technology, sharding basically segregates load data into parts transversely P2P arrangements related to particular shards. Main reason behind applicability of this method is making every node free from the requisite of block chain, concluding increase in number of the deals. The most important dispute of sharding is allied to protection and announcement. When a block chain is gash into shards, every one perform the same as a personage block chain system. This adjoins an additional cover of complication intended for the person since it demands an out of the ordinary contact criteria. The supplementary defy is precaution. Due to splitting, the muddle influence of the block chain nodes diminishes. This results in easy approach hackers to lacerate whichever solitary slice, which predictably escort to an eternal thrashing of records. Even though the intelligent brains are finding the way out to this dilemma, merely the era spirit enlighten if it is truly achievable to slice block chain in to various divisions.

2.4.3 *Policies and Regulations*

Various Financial Organizations also witness a shortage of lucidity for rules and procedures. Presently there are no such rules for crypto currencies and digital dealings. It will not be practically possible for financial and fiscal associations to implement block chain until a structured roadmap and scaffold is promised. Even though the endeavor of block chain is to save statistics, the act out system may spectacularly bind the espousal of block chain in universal banking. The European assembly unconfined a work on block chain, targeting solving this quarrel. Although it provided profusion of approach to deal with the enduring collide of issues, it is further possible that profound variations for block chain technology will be essential to unravel the entire distress.

2.5 Conclusion

Block chain which is sort of disseminated and record maintenance system, it jointly provide conviction, ambiguity, protection and statistics reliability devoid of enclose to exercise involvement of outsiders various studies conducted so far shows that block chain associates with the technical phase. On the other hand, discussion done in this paper basically reveals about conceptual and deep view of block chain technology adoption particularly in banking. Adoption of Block chain technology in banking sector has become the need of the hour, although it embraces numerous advantages and negative aspects of implementing it. One of the major advantages discussed during the study creation of green technology with using block chain technology in banking area. For protection and security of data also, block chain cooperate a gigantic and immense character. This paper furthermore discussed about the limits and shortcoming of executing block chain technology in banking area. Famine of scalability, suitable acquaintance about regulations and convention and be deficient in superiority were the main inconvenience on which the light was thrown. However various programs are being in use which is specially focused on overcoming the limitations of block chain technology implementation. In the last, it can be said that before actual implementation of Blockchain technology in the banking industry, it has to pass through different criteria. For actual adoption and implementation of blockchain technology there is a need by banking companies to firstly create the infrastructure in which blockchain can run and operate to solve the problems of clients in effective and efficient manner. If it will be done, in the long run Blockchain is going to transform the banking industry up to next level.

Conflict of Interest The authors declare that they have no conflict of interest.

Ethical Approval The present study is a critical review based, so it doesn't encompass any reports based on animals.

Funding None.

References

Chirag (2023) Why are banks adopting blockchain technology?. Appinventv. Retrieved from https://appinventiv.com/blog/blockchain-in-banking/#:~:text=Blockchain%20technology%20in%20banking%20revolutionizes,assets%20%E2%80%9Coff%2Dchain%E2%80%9D

Cocco L, Marchesi M, Pinna A (2017) Banking on blockchain: costs savings thanks to the blockchain technology. Future Internet 9(25):1–20

Garg S, Kaur K, Kaddoum G, Choo KKR (2019) Toward secure and provable authentication for Internet of Things: Realizing industry 4.0. IEEE Internet Things J 7(5):4598–4606

Golosova J, Romanovs A (2018, November) The advantages and disadvantages of the blockchain technology. In 2018 IEEE 6th workshop on advances in information, electronic and electrical engineering (AIEEE). IEEE, pp 1–6

Guo Y, Liang C (2016) Blockchain application and outlook in the banking industry. Financ Innov 2:1–12

Gupta T (2021) Top 5 uses of Blockchain in banking. DQINDIA ONLINE. Retrieved from https://www.dqindia.com/top-5-uses-blockchain-banking/

Hackett R (2016) Wait, what is Blockchain? Accessed on February 2021, http://fortune.com/2016/05/23/block chain-definition/

Janssen M, Weerakkody V, Ismagilova E, Sivarajah U, Irani Z (2020) A framework for analysing blockchain technology adoption: Integrating institutional, market and technical factors. Int J Inf Manag 50:302–309

Oberoi S, Kansra P (2019) Factors influencing medical tourism in India: a critical review. SAMVAD 17:09–16

Olaf A, Hagen H, Hajj J (2017) 2017 Technology trends. PwC. Accessed on February 2021. https://www.strategyand.pwc.com/trend/2017-technology-trends

Ravi H (2021) Innovation in banking: fusion of artificial intelligence and blockchain. Asia Pacific Journal of Innovation and Entrepreneurship 15(1):51–61

Woodside JM, Augustine FK Jr, Giberson W (2017) Blockchain technology adoption status and strategies. J Int Technol Inf Manag 26(2):65–93

Yoo S (2017) Blockchain based financial case analysis and its implications. Asia Pacific J Innov Entrep 11(3):312–321

Chapter 3
Revolutionizing Rural Finance: Exploring the Impact of FinTech on Financial Inclusion in India

Swati Gupta, Rajeev Srivastava, Zakir Hossen Shaikh, and Mohammad Irfan

3.1 Introduction

The advent of technology has proven to be both a challenge and an opportunity for businesses today. Indeed, the digital revolution has profoundly altered the broader corporate environment, including the banking and financial services industries. The use of technology, regardless, how basic it is, is motivated by the desire to simplify operations and enhance user experience. However, many of these service providers, notably banking and financial institutions had been confronted with the transition from manual to automated operations, popularly known as FinTech.

Roots of FinTech has been claimed somewhere in 1800s, particularly with invention of 'panteligraph' by 'Giovanni Caselli' during 1860s, for verifying signatures in banking and use of 'charge plates and charge coins' in late 1800s till early 1960s. This was followed by 'Fedwire' (1918–1970) initially used by US Federal Reserve Bank with telegraphy as 'Real Time Gross Settlement' fund transfer system, 1970 onwards this started using telex and finally computers' network. The earliest documented connection between finance and technology can be traced back to John Maynard Keynes' publication, 'The Economic Consequences of the Peace', which was released in December 1919 (Taskinsoy 2020). The advent of credit cards in

S. Gupta (✉)
UWSB, Karnavati University, Uvarsad, Gujarat, India

R. Srivastava
IMS Unison University, Dehradun, Uttarakhand, India

Z. H. Shaikh
Bahrain Training Institute, Ministry of Education, Isa Town, Kingdom of Bahrain

M. Irfan
NSB Academy, Business School, Bangalore, Karnataka, India

M. Irfan et al. (eds.), *Applications of Block Chain technology and Artificial Intelligence*, Financial Mathematics and Fintech,
https://doi.org/10.1007/978-3-031-47324-1_3

1950 by Diner's Club Inc. and subsequently by American Express in 1958 is widely regarded as a significant milestone in the FinTech revolution. In 1967 establishment of first ATM by Barclays Bank was another step forward. Establishment of NASDAQ in 1971 was an introduction to new era of FinTech, followed by introduction of online trading by TradePlus in 1982 and online banking in US and Bank of Scotland in 1983. The new millennium with the help of AI, Machine Learning, IoT has already introduced the world to unbelievable FinTech innovations which has not only made human life easy and comfortable, but also brought financial services within reach of all. Fintech services include lending platforms that provide quick access to loans, payment systems that enable digital transactions, international money transfer services that make sending money overseas easier and more affordable, personal finance tools for managing money, consumer banking services that provide online and m-banking and ATM access, insurance services offering new and innovative products, and equity financing services that offer new ways for start-ups to raise capital and go public. The advent of smartphones and mobile applications has enabled individuals to access information and perform tasks on-the-go, without the need for a stationary desk or laptop. The utilisation of m-banking has become a prevalent aspect of contemporary daily life.

COVID-19 pandemic has not only changed human lifestyle but also influenced attitude towards technology adoption. FinTech has got a launching boost because of COVID-19 protocol imposed by governments. To discourage physical touch, most of the governments throughout the world, encouraged touch less transaction with the use of technology. Hence, being handy and easy to use, user friendly mobile application based services are gaining popularity. On one hand pandemic have compelled people to adopt technology for their transactions, on other hand financial institution, particularly banks have left with no choice, but to join hands with FinTech start-ups and share information for win-win situation. These days, mobile based banking applications have become most popular FinTech products.

Existing pre pandemic research has explored determinants of m-banking adoption using both qualitative and quantitative information. This work is expected to be a post pandemic watershed moment in the advancement of this field of inquiry. Thus, this study offers a chance to take a step back and assess the cumulative information gained from diverse academic databases. This study is significant because of the contradictory findings from previous studies (Malaquias et al. 2018; Singh and Srivastava 2014; Sulaiman et al. 2007) and need to test these findings in post pandemic environment. As a result, this article aims to perform an in-depth examination of the variables influencing adoption of m-banking particularly in post pandemic era and propose a sustainable model.

Despite the rapid expansion of rural mobile phone networks, there remains a dearth of network service infrastructure in rural areas of India. This inadequacy represents a necessary but insufficient condition for the growth of m-banking in both rural and urban areas. The utilisation of m-banking has the capacity to enhance the accessibility of financial services in rural regions. However, there exists a dearth of information concerning the determinants that impact its adoption and usage among rural residents, as well as the impediments to its proliferation. Although there exists a substantial body of literature that investigates the adoption of m-banking across various contexts and through diverse constructs, the bulk of the present-day, pertinent research in India is predominantly theoretical. Consequently, there is a scarcity of associated empirical research that delves into the acceptance of m-banking by rural residents. Consequently, it is imperative to comprehend the motivating factors that prompt individuals to embrace m-banking. Thus, the objective of this research is to establish the components. Hence, it is imperative for providers of m-banking services to have a comprehensive understanding of the dynamics involved in this nascent trend. The purpose of this research is to learn what factors have the most impact on rural Indian customers' decision to use m-banking.

This paper aims to investigate the factors that contribute to the acceptance and long-term usage of m-banking. The study presents an empirical model specifically focused on the adoption of m-banking in rural areas of India. The primary goal of this research is to enhance the existing body of literature on m-banking and provide guidance to FinTech start-ups, considering the complexities of mobile technology and the wide array of services available. This will be accomplished by examining and evaluating the current understanding of m-banking and its adoption among various segments of the population in emerging economies. The ultimate objective is to consolidate and integrate multiple research sources into a cohesive knowledge base, while also examining and analyzing the methodologies, frameworks, and paradigms employed in this field.

3.2 Literature Review and Theoretical Framework

3.2.1 Background

The establishment of trust is a crucial prerequisite for the acceptance and utilisation of M-banking. This is because trust is a fundamental element of any business collaboration (Kim et al. 2009) and serves as a critical factor in m-commerce by mitigating uncertainty. The utilisation of m-banking technology has the capacity to enhance individuals' standard of living due to its convenient accessibility, secure nature, portability, user-friendly interface, and cost-effectiveness (Mago and Chitokwindo 2014). According to Tiwari et al. (2006), the implementation of m-banking has a positive impact on the productivity of banks. There exist a multitude of factors that impact the level of trust that consumers place in m-banking services.

M-banking services provide individuals with the ability to perform various financial transactions. These transactions encompass tasks such as accessing account information, transferring funds, requesting a chequebook, initiating or renewing fixed deposits, setting up a dematerialized account, opening a loan account, facilitating utility payments, and similar activities (Chawla and Joshi 2017). The

Technology Acceptance Model (TAM), introduced by Gefen et al. (2003), and the Unified Theory of Acceptance and Use of Technology (UTAUT), proposed by Im et al. (2011), are theoretical frameworks that have been established to understand and analyze the factors influencing the acceptance and utilization of m-banking services. These models have gained widespread recognition and are frequently utilised to forecast the level of acceptance of novel technological innovations. The effectiveness of both models can be attributed to the proposed modifications that have undergone thorough scrutiny and improvement through advancements and innovations.

According to Kumbhar (2011), M-banking is a low-cost and simple method of providing financial services to non-banked Indian clients. There are several elements that influence the acceptance of facility by local consumers in India. Due to a lack of technological awareness, the use of cell phones as an online banking tool suffers. However, according to www.statista.com, there were 242.92 million mobile phone internet users in 2015, which is expected to treble by 2022 to 486.70 million. In 2019, mobile connections accounted for 73% of total online traffic in India. As a result, this appears to be an untapped market for India's financial services business. Furthermore, a large proportion of rural Indians should have a basic understanding of m-banking's use and execution (Singh and Malik 2019).

Riquelme and Rios (2010) studied the impact of gender on m-banking uptake further. According to the study, risk, utility, and social norm have the most influence on m-banking uptake. They discovered that women value simplicity of use and social standing, but males value relative benefit as major factors influencing m-banking usage.

According to the (IMRB 2016), Indian Internet penetration is now about 31%. Rural India, which has a population of approximately 906 million people according to the 2011 census, has just 163 million Internet users, accounting for roughly 17% of the total. As a result, there are still over 750 million prospective consumers in rural India who have yet to become internet users; if only they could be adequately identified. Rural India is thought to have jumped ahead of urban India in adopting cell phones as a platform for Internet access due to lower device penetration (Chircu and Mahajan 2009). The Covid-19 shutdown is now responsible for some of this leapfrogging.

According to Gupta et al. (2022), TAM is one of the most contested models for predicting the extent to which a modern technology notion will be embraced. The model provides an overview of customer attitudes about m-banking implementation. Trust in e-commerce, on the other hand, has arisen as an important factor influencing consumer confidence in the right development of security-assurance promises made by facilitators, which are mostly publicised on websites (Kim 2006) and thus confirmed through analytical testing. Similarly, banks must present compelling reasons to clients in order to foster trust and accept technological development (Malaquias 2016).

Nouri and Soltani (2019) posit that the perceived value and perceived ease of use of m-banking are positively influenced by various factors, including personal, societal, organisational, and technical aspects. The attitudes of individuals towards m-banking are significantly influenced by their perceived value and perceived ease of use. The research categorises crucial factors for effective technology implementation models into four primary domains, namely personal, social, organisational, and technical factors, which have been verified to influence the propensity to adopt m-banking. Mirhossein conducted a study on the factors that affect the level of acceptance of m-banking technologies among clients, using a similar approach. The study's results indicate that a multitude of variables identified in the model, whether stated or implied, have an impact on the adoption of m-banking by consumers. The most pivotal aspect in this scenario was the anticipated increase.

According to Giovanis et al. (2019), the most effective model for gaining acceptance in the realm of M-banking (MB) is an expanded version of the Decomposed Theory of Planned Behavior (DTPB) that incorporates projected hazard. The primary predictor of consumer intention to adopt MB services is customer disposition, which is measured through three key factors: utility, simplicity, and compatibility of m-banking. In comparison, two less significant acceptance variables include consumer expectations regarding the quality of information, services, and incentives needed for program usage, as well as the influence of internal and external social contexts on M-banking utilization. Lastly, perceived risk negatively impacts the development of behaviors and the readiness to use m-banking resources.

As Gefen et al. suggest, the study's background and literature warrant for deeper examination into the theoretical development of the model (2003a). Thus, the purpose of this article is to capture the nuances brought about by technology advancements in m-banking in India. Furthermore, Chirku and Mahajan (2009) emphasised the importance of focusing on the digital gap, particularly in BRIC countries, of which India is an essential component.

3.2.2 Covid 19 and M-banking in India

India implemented a strict nationwide lockdown on March 25, even when the number of confirmed COVID-19 cases was relatively low, making it one of the earliest and most stringent measures globally. This sudden halt in activities posed significant challenges for individuals, requiring them to meet essential needs and conduct transactions through mobile applications. According to a recent analysis by the World Bank conducted by Mora, low-income countries have been promoting the adoption of digital payment channels in response to the COVID-19 pandemic. The pandemic has accelerated the pace of digital transformation in society, as evidenced by the profound impact of digitization and increased reliance on the internet for communication and purchases.

In India, the combination of the COVID-19 pandemic and demonetization has led to a notable increase in the adoption of digital banking platforms, such as m-banking and mobile wallets, shifting away from traditional physical banking methods. Agarwal et al. reported a significant surge in the usage of various applications like Google Pay, ICICI iMobile, SBI Yono, and mobile wallets like Paytm.

Financial institutions have sought to enhance and expand their operations by employing instructional materials as part of their promotional strategies (Gopal and Schnabl 2022).

Nair et al. (2021) conducted a study in India that suggests the COVID-19 pandemic is likely to increase the utilization of m-banking services. M-banking facilitates adherence to social distancing measures, enabling individuals to conduct transactions safely and securely. The study also highlights limited alternatives to m-banking during this time. Additionally, the use of m-banking applications has been found to play a significant role in mitigating the spread of social contagion by providing convenient access to essential banking services. The implementation of stricter lockdown measures may further boost the usage of m-banking services (Girish and Manu 2020).

Another study conducted during the pandemic in India reveals that people have started accepting and trusting various m-banking platforms, using them for online bill payments and money transfers through UPI. The study concludes that m-banking advancements have made it convenient to avail all services at one's doorstep during the COVID-19 pandemic (Dubey et al. 2020).

3.2.3 M-banking &TAM

The rapid advancement of technology and the increasing popularity of e-commerce have sparked a significant body of research focusing on online banking, particularly m-banking. This extensive literature provides valuable insights into the factors influencing the adoption of m-banking, taking into account customers' varying perceptions of its value (Laukkanen 2007). Previous studies conducted by Nor and Pearson (2008), Nor et al. (2008), and Kim et al. (2008) have delved into the primary determinants of internet banking adoption, identifying risk and the complex nature of trust as crucial factors. Moreover, these studies have demonstrated the applicability of fundamental concepts such as social norms, perceived usefulness, perceived ease-of-use, attitude, and self-efficacy within the context of internet banking.

To address the acceptance of m-banking, Lee and Chung (2009) developed a model that draws on the Theory of Planned Behaviour (TPB) and the Technology Acceptance Model (TAM) literature. They introduced the concept of "perceived credibility" as a relevant factor in the realm of mobile commerce. Perceived credibility is closely related to, but distinct from, perceived risk and trust, encompassing an individual's conviction that utilizing m-banking does not entail security or privacy hazards.

In a pragmatic approach, Kim et al. (2008, 2009) and focused on establishing clients' initial trust in m-banking. Their research emphasized the significance of initial confidence as a predictor of m-banking adoption. Factors such as perceived competitive advantages over traditional banking networks, personal inclination towards embracing emerging technologies and forging new business partnerships, and belief in the robust security measures implemented by m-banking providers all contributed to users' initial confidence in m-banking.

Examining the impact of trust on m-banking adoption, Lee et al. (2007) incorporated the notion of perceived threat within the TAM framework. Their study explored specific dimensions of risk and trust related to confidence in banks, telephone companies, and wireless infrastructure. The findings underscored the direct influence of confidence and perceived usefulness on adoption behavior, while also highlighting the moderating role of confidence in the relationship between perceived damage and adoption behavior.

3.2.4 M-banking and Financial Inclusion in Rural India

According to Yu (2009), m-banking eliminates bank branch brick and mortar layout and increases branch density in rural India. It aided in removing supply-side barriers to financial inclusion (Sharma et al. 2014). M-banking in rural India permits the provision of finance-related services such as deposits, savings mobilisation, remittances, cash disbursements and repayments, bill payments, money transfers, and so on through banking facilitators or correspondents. Banking facilitators might boost their earning potential by conducting frequent transactions rather than small transactions using cell phones or portable devices. M-banking is helpful in handling large volumes of business transactions for underbanked and vulnerable rural clients. M-banking primarily encourages open m-banking, which minimises operational or transaction costs for both mobile clients and banks.

According to Mittal and Gupta (2021), m-banking is the greatest means to achieve the aim of financial inclusion, particularly for the under-banked or unbanked community. Financial inclusion strives to bring financial services to low-income persons within its scope. M-banking provides instant access to financial services and goods while also being cost-effective. It eliminates the need for a mortar and brick model. M-banking is a strategic endeavour for banks to make transactions with low-income customers. According to (Mas and Ivatury 2010), focusing solely on smaller banks and Micro Finance Institutions (MFIs) may result in greater distribution costs due to lower transaction rates and a potentially more scattered and remote position.

M-banking represents an innovative solution for providing cost-effective financial services in rural areas. It is possible to mitigate the disparity between mobile money and banking networks. Mobile financial services are accessible to geographically dispersed individuals who lack access to traditional banking facilities. The remodelled smartphone under the mobile money system serves as a repository for capital, however, it does not accrue interest at the rate offered by traditional bank savings accounts. The facilitation of microcredit disbursement and savings mobilisation is contingent upon the assistance provided by the mobile money infrastructure. The concept of financial inclusion alone cannot ensure the protection of the rural population utilising mobile money services (Banzal 2010). This area of study holds promise for further exploration by emerging researchers.

3.3 Research Methodology

The proposed study incorporates TAM attributes and investigates the role of these attributes in users' attitudes to use m-banking in India. Thus, the purpose of this article is to investigate the factors that influence non-adoption of m-banking by investigating user attitudes for non-adoption of m-banking, as well as demographic variables that influence the decision. This section also reviews the relevant literature on m-banking adoption and so proposes a conceptual structure implying a relationship between various variables and consumers' attitude.

Figure 3.1 show the model for the purpose of this study.

The variables so considered for the study are explained as follows:

3.3.1 Perceived Ease of Use (PEoU)

Abdullah et al. (2016) emphasize the concept of perceived ease of use, which relates to an individual's perception of the simplicity and user-friendliness of a particular technology. According to Gu et al. (2009), an integrated model has been proposed, indicating that perceived ease of use is a robust predictor of the adoption of m-banking. This finding suggests that the easier it is for individuals to use m-banking, the more likely they are to adopt it. As m-banking becomes increasingly prevalent

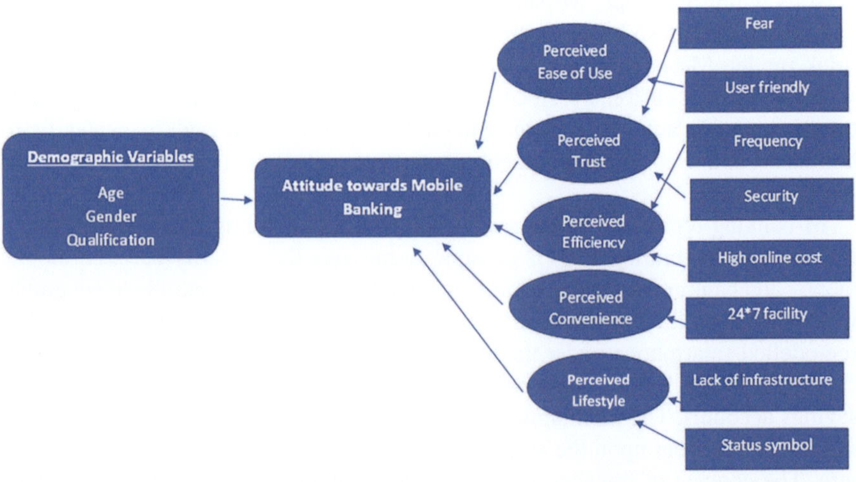

Fig. 3.1 Proposed hypothesis model. Source: Authors

in today's digital age, it is essential to understand the factors that influence its adoption. By recognizing the importance of perceived ease of use in m-banking adoption, organizations can design user-friendly m-banking applications to encourage greater adoption rates and enhance customer satisfaction. Suh and Han's (Suh and Han 2003) research found a significant correlation between perceived ease of use (PEoU) and attitudes towards Internet Banking. Lin (2011) argued that the ease of use provided by mobile devices and banking applications eliminates the need for users to exert significant effort in utilizing m-banking services, and the perceived ease of use (PEoU) plays a significant role in this context. Deb and David (2014) discovered that attitudes towards the perceived ease of use (PEoU) of m-banking significantly influenced its adoption in India. Similarly, Chawla and Joshi (2017) found that perceived ease of use (PEoU) was a critical factor in the adoption of various banking communication technologies, including the internet and cell phones. According to Davis (1989), the Technology Acceptance Model (TAM) defines perceived ease of use as an individual's perception of the speed and simplicity with which a technology can be used. Scholars have extensively studied this factor to understand the adoption of m-banking applications by consumers (Venkatesh and Davis 2000; Abdul-Hamid et al. 2019), although with relatively less focus on the Indian context. Therefore, based on the preceding discussion, the following hypothesis arises:

Hypothesis 1 (H_1): PEoU negatively impacts the attitude of consumer towards m-banking adoption.

3.3.2 *Perceived Trust*

Trust is defined as the level of confidence an individual has in an organization without experiencing the anticipated benefits of the intended activity. In the context of online banking services, trust is considered to be even more crucial compared to offline services due to the need for transactions to be secure, responsive, and confidential (Heijden et al. 2003). The world has recently faced an unprecedented challenge in the form of the novel coronavirus (COVID-19) pandemic, with efforts focused on controlling its spread (Alhumaid et al. 2021). As the infection rates soared, people naturally became anxious and concerned. Ahorsu et al. (2020) highlighted the fear of contacting individuals who might be infected with COVID-19, which unfortunately can amplify the impact of the disease itself. This fear, however, can serve as a motivating factor for users to adopt m-banking and develop trust in the system.

Agarwal et al. (2009) found that customer loyalty in e-banking is influenced by the sense of security and trust associated with the service. Hernandez and Mazzon (2007) conducted a study on the prevalence of online banking in Brazil and found that users' perceptions of their own security and privacy play a major role in their decision to switch from traditional banking methods. In their research, Gu et al.

(2009) introduced an integrated model that highlights a strong and positive association between the adoption of internet banking and trust. This finding is particularly relevant for understanding the behavior of young Indian consumers who engage with internet banking services (Bashir and Madhavaiah 2014). Additionally, the concepts of perceived risk and trust are interconnected and recognized as hindrances to the widespread adoption of mobile services (Lewis 2010). By addressing trust-related concerns and mitigating perceived risks, organizations can foster greater trust among users and promote the successful adoption of mobile services. Recognizing the significance of trust in both internet banking and mobile services is crucial for developing effective strategies and ensuring a positive user experience in the digital realm. In the context of mobile applications, trust assumes greater significance due to the inherent risks associated with wireless environments and potential security breaches (Riquelme and Rios 2010). As mobile technology continues to advance and play an increasingly integral role in our lives, understanding the factors that influence trust in mobile applications becomes crucial. By addressing trust-related concerns and implementing robust security measures, organizations can in still confidence in users and promote the widespread adoption of mobile services. Lin (2011) argued that m-banking users can develop knowledge-based trust if banks, telecom providers, and financial institutions are committed to delivering professional services while safeguarding customer data and transaction integrity. Kim et al. (2009) found that m-banking is perceived as more risky than traditional banking. The establishment of initial trust is considered a crucial factor in the adoption of m-banking (Nouri and Soltani 2019). However, the relationship between trust and the intention to adopt m-banking is not well-established (Koenig et al. 2010). Behboudi et al. (2013) suggested that the direct impact of m-banking's purpose depends on the levels of familiarity and perceived harm. Based on the aforementioned discussion, the following hypotheses are derived:

Hypothesis 2 (H_2): Trust perception has a detrimental effect on consumers' attitudes while adopting m-banking

3.3.3 Perceived Lifestyle Compatibility

Perceived lifestyle compatibility refers to the extent to which m-banking aligns with individuals' beliefs, perceptions, lifestyles, and preferences, as influenced by both external factors such as friends, peer circles, and superiors, as well as internal factors like family social norms that shape lifestyle and influence adoption behavior (Riquelme and Rios 2010). Deb and David (2014) found a positive relationship between social influence, including the approval of friends and family, and attitudes toward m-banking. Similarly, Bashir and Madhavaiah (2014) discovered that social influence positively impacted the decision to use internet banking. Building on the hypothesis of m-banking (Tan and Teo 2000), it is anticipated that individuals who use smartphones more frequently perceive m-banking as better fitting with their lifestyle, increasing their likelihood of adopting m-banking. Lin (2011) revealed that perceived compatibility is determined by how well m-banking aligns with individuals' ideas, perceptions, lifestyles, and preferences. Therefore, compatibility is a crucial factor in understanding the inclination of Indian customers to adopt m-banking (Singh and Srivastava 2014). Mohammadi (2015) further emphasized that reliability is a significant factor influencing consumers' perceptions of m-banking. Based on these discussions, the following hypotheses are proposed:

Hypothesis 3 (H_3): The perception of lifestyle compatibility has a detrimental effect on consumers' attitudes towards the adoption of m-banking

3.3.4 Perceived Efficiency

Perceived Performance refers to how much more value technology delivers than its predecessors, which can be measured in terms of efficacy, economic advantages, and enhanced standing (AlSoufi and Ali 2014; Estriegana et al. 2019). While many empirical studies do not explicitly discuss perceived output, it does manifest itself in the context of the quality depending on innovation indicated in the principle of diffusion of innovation. Consumer friendliness, relative benefit, dependability, observability, and trialability are creative attributes (Rogers 1995; Ahmad 2018). Customers regard mobile networks, which allow them to conduct banking transactions on the go and at any time, as an innovative mechanism that not only saves time and energy but also allows them to do additional financial transactions with less effort and expenditure. As a result, we propose the following hypotheses:

Hypothesis 4 (H_4): Consumer perception of efficiency has a negative effect on their decision to adopt m-banking

3.3.5 Perceived Convenience

The concept of perceived convenience pertains to the capacity to access financial services at any time of the day or night, throughout the week, in a manner that is considered to be superior to other banking channels. The scholarly literature has recognised comfort as a crucial determinant that shapes a user's inclination towards adopting technological innovations, such as ATM, SMS, m-banking, or Internet banking. In the specific context of m-banking, perceived convenience refers to the ability to access financial services at any time, seven days a week, in a manner that is perceived as more advantageous than other banking channels. The study conducted by Liao and Wong (2008) explored the impact of internet-based e-banking services on user experiences and highlighted the importance of factors such as simplicity and responsiveness to service requests. Agarwal et al. (2009) found that the level of ease experienced by consumers had a notable effect on their overall satisfaction with e-banking services in India. Several academic studies have indicated that the anticipated value of a programme has a favourable influence on one's attitude towards it. These studies include the works of Jackson and Leitch (1997), Jeung and Chihui (2009), and Davis (1989). Therefore, the perception of convenience is indicative of a positive disposition towards the utilisation of m-banking, leading to the formulation of the subsequent hypotheses:

Hypothesis 5 (H_5): The adoption of m-banking is negatively impacted by perceived convenience in the eyes of the consumer.

3.3.6 Demographic Attributes for M-banking Adoption

The present investigation incorporates demographic factors, including but not limited to gender, age, employment status, marital status, profession, educational attainment, and work experience, as variables that may influence the adoption of technological innovations. According to Wan's (2005) observations, there is a higher likelihood of males adopting m-banking technologies, which supports previous research findings. In a study conducted in Finland, Laukkanen and Pasanen (2008) found that there were differences in the usage of m-banking and other online banking services based on age and gender.

However, it was observed that factors such as age, employment status, occupation, and household size did not significantly distinguish between the groups. Previous studies have indicated that customer segments can be identified by considering factors related to product acceptance, which may vary based on demographic variables (Deb and David 2014; Mann and Sahni 2012). It can be argued that there exists a rational basis for perceiving that demographic factors may contribute to variations in cluster profiles, thereby leading to differences in technology adoption attitudes and objectives across various demographic groups. The study conducted by Flaviàn and colleagues in 2006 identified income, age, and gender as significant determinants affecting the uptake of online banking. The study conducted by Quazi and Talukder (2011) investigated the impact of demographic factors on the comprehension and adoption of technological innovations in Australian work environments. The researchers noted that individuals possessing higher levels of educational attainment exhibit a more positive attitude towards career progression, which in turn is reflected in their level of acceptance within the workplace. Correa et al. (2010) conducted a survey in the United States and discovered that there is a significant variation in the usage of social network instruments based on age and gender. Several studies, including those by Samudra and Phadtare (2012), Mattila et al. (2003), Karjaluoto et al. (2002), and Sathye (1999), have reported that individuals who are older, have received training, and possess greater wealth are more inclined to utilise internet banking services. According to research conducted by Akinci et al. (2004) and Kolodinsky et al. (2004), there is evidence to suggest that younger individuals exhibit a greater propensity to adopt technological advancements in comparison to their older counterparts. Jayawardhena (2000) observed that the acceptance of Internet banking is influenced by income and profession. According to Szopiński's (2016) research, there exists a significant relationship between the utilisation of online banking and demographic factors such as age, income, and educational attainment. Jha and Ye (2016) conducted a study on American adults using Facebook as a platform to investigate the variations in attitudes and usage patterns across different demographic categories such as social class, age, employment status, and income. Therefore, the aforementioned evidence leads to the extraction of the following hypotheses:

Hypothesis 6 (H_6): The attitudes of consumers towards using m-banking vary greatly by gender

Hypothesis 7 (H_7): Age of the respondent has a detrimental impact on consumer attitudes regarding the adoption of m-banking

Hypothesis 8 (H_8): Consumer attitude towards m-banking differs across qualification.

3.4 Data Analysis, Results and Discussion

The collection of data for the study was carried out with a method known as convenience sampling. The survey instrument was disseminated throughout the mountainous regions of Uttarakhand, encompassing the administrative divisions of the Garhwal locality. The regions encompassed Chamoli, Pauri, Tehri, Uttarkashi, and Rudraprayag districts. The survey participants were presented with a questionnaire to gather their perspectives and insights on the subject matter, with particular attention given to the unique characteristics and circumstances of rural areas. The survey instrument was designed to be easily comprehensible and straightforward in its structure and content. The study employed logistic regression to examine the association between the binary outcome variable and the predictor variable(s), regardless of whether they were categorical or continuous in nature, owing to the non-normal distribution of the outcome variable.

Initially, scholarly databases were consulted to establish the fundamental context of the research. The research utilised a field methodology to acquire original data, employing a questionnaire as the principal instrument for data gathering. The creation of the questionnaire was influenced by pertinent literature and constructs linked to the Technology Acceptance Model (TAM), in addition to the expert perspectives of professionals in the respective field. The Delphi process was employed to evaluate the validity of the questionnaire. The research aimed to gather the viewpoints of 30 academic professionals and specialists by means of a survey, and their assessments were taken into account with respect to the quality of the elements employed in calculating each factor as outlined in Table I.

The researchers employed a five-point Likert scale to assess the degree of agreement among participants in an online survey. Out of the total of 320 responses that were obtained, it was found that 48.8% of the participants self-identified as male, whereas 51.2% identified as female. Roughly 50% of the subjects belonged to the age group of 30 years or below. Based on the educational credentials of the subjects under consideration, it can be inferred that around 36.3% had completed their education up to the 10th grade, whereas only 15% had attained postgraduate degrees. The study revealed that a considerable percentage of the participants, precisely 43.8%, indicated that they have not made use of m-banking services. Nonetheless, a comparatively lesser proportion of 12.5% disclosed utilising it once in every three months, whereas 37% revealed employing it exclusively for the purpose of balance inquiries. Furthermore, a notable proportion of 31% of the participants have refrained from employing any m-banking facilities.

The reliability was checked for all the seven variables along with the demographic variables through correlation matrix using R. The test helped us identify the variables that are correlated with each other. Thus Fig. 3.2 concluded that perceived trust and perceived convenience are highly correlated with attitude.

Subsequently, the mean values of the seven constructs were calculated to investigate their impact on consumer attitudes towards m-banking adoption. To predict the likelihood of non-adoption, respondents' attitudes were categorized as either High Attitude or Low Attitude based on their scores. The data included median

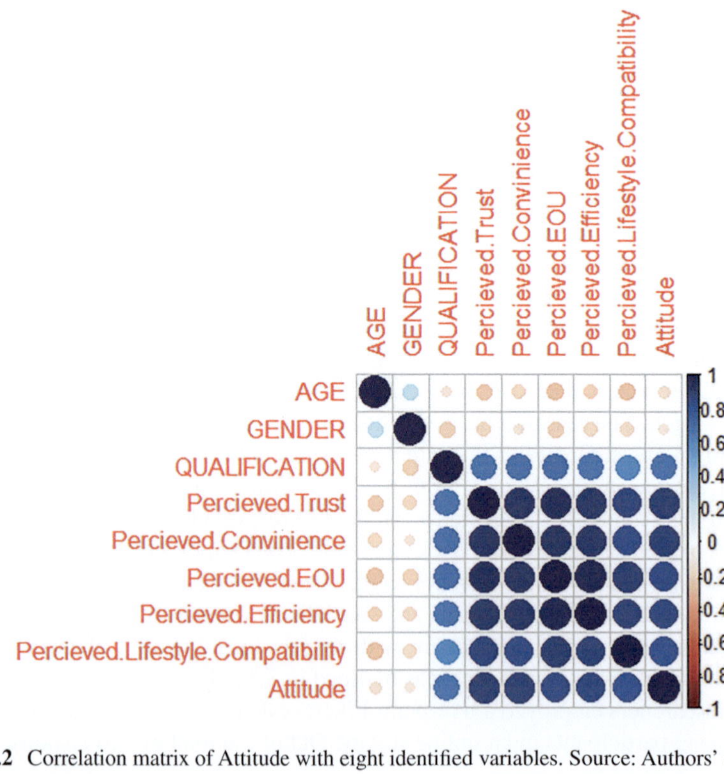

Fig. 3.2 Correlation matrix of Attitude with eight identified variables. Source: Authors' analysis

scores for Aim and Attitude predictors. No respondents obtained a median aim score of 51. Those who scored above 51 were classified as having a high inclination towards adopting m-banking, while those below 51 were deemed to have a low inclination. Out of 320 respondents, 128 exhibited a high intention to adopt m-banking, while 192 showed a weak attitude towards it.

Logistic regression analysis was employed to predict individuals' inclination toward adopting m-banking. The methodology utilized to develop a model for predicting consumer attitudes is elaborated in the subsequent section.

Let p denote the probability of a respondent having a positive attitude toward adopting m-banking, while 1 - p represents the probability of a respondent having a negative attitude (3.1). The odds ratio, p / (1 - p), indicates the likelihood of a respondent having a positive attitude toward m-banking.

In this study, the aim is to investigate the influence of various factors, such as trust, perceived ease of use, lifestyle compatibility, perceived productivity, and perceived comfort, on individuals' adoption attitudes toward m-banking. These factors are treated as independent variables. Additionally, three demographic variables - gender, age, and qualification - are included as predictors of adoption attitudes toward m-banking. The demographic variables are categorized as follows:

AGE: Users aged 30 years or below are coded as 0, while users above 30 years are coded as 1.
GENDER: Male is coded as 0, and Female is coded as 1.
QUALIFICATION: Users with qualifications up to tenth class are coded as 0, while those with qualifications above tenth class are coded as 1.

Hence, the logit regression model for m-banking attitude is expressed as:

$$Z = I_n \left[\frac{P_i}{(1?P_i)} \right] \tag{3.1}$$
$$= ?+?_1 X_1 + ?_2?X_2 + ?_3 X_3 + ?_4 X_4 + ?_5 X_5 + ?_6 X_6 + ?_7 X_7 + ?_8 X_8$$

Where,

X_1 = Perceived Trust
X_2 = Perceived Ease of Use (PEoU)
X_3 = Perceived Lifestyle Compatibility
X_4 = Perceived Efficiency
X_5 = Perceived Convenience
X_6 = Gender
X_7 = Age
X_8 = Qualification

The exponential Eq. (3.1) can be used to determine Pi, representing the probability of the ith respondent having a higher attitude towards m-banking. Specifically, Pi can be calculated as Pi = [Zi/ (1 + Zi)]. If the Pi probability of the ith respondent is greater than 0.5, it is expected that they will have a high inclination towards adopting m-banking. Conversely, a probability lower than 0.5 indicates a low inclination. A positive coefficient for a logistic regression variable suggests that an increase in the variable's value will result in higher odds in favor of a positive attitude towards m-banking. The significance of the Logistic Regression statistic lies in its ability to indicate a strong fit between the logistic regression equation and the data.

The Logistic Regression method will be employed to analyse attitude hypotheses spanning from H1 to H8. The present study employed logistic regression to analyse the attitude towards non-adoption of m-banking. The model included eight independent variables, which are elaborated upon as follows. The results are presented in Model of Fig. 3.3. The results suggest that the non-adoption of m-banking is positively influenced by perceived trust, perceived lifestyle compatibility, perceived efficiency, and perceived ease of use. This suggests the need to satisfy the four hypotheses (H1–H4). Upon scrutinising the coefficients associated with said variables, it has been determined that the perceived convenience variable holds no statistical significance, as its coefficient exhibits a negative value. The analysis of the demographic variables reveals that the age variable exhibits a negative coefficient and is deemed statistically insignificant. Thus, the hypothesis H6 is not substantiated. Gender and level of education are factors that positively influence the non-adoption of m-banking. Thus, it can be inferred that none of the factors hold statistical significance for conducting further analysis, given that their respective p-values exceed the threshold of 0.05 (Fig. 3.3).

The contrast between *null deviance* and *residual deviance* shows us the good-fit model (Manning 2007). The higher the pattern the greater the gap. Null deviance occurs if interception happens for no variables in the calculation whereas residual deviance occurs if all variables are taken into consideration. If the gap is large enough, it makes sense to call the model good fit. Thus, from our study there's a very small difference between the two, along with 8 degrees of freedom (df).

The lack of adoption of m-banking among rural individuals can be attributed to three demographic variables and five additional factors identified in the literature.

```
                              Estimate Std. Error z value Pr(>|z|)
(Intercept)                   -319.792 184145.777  -0.002    0.999
AGE                             -4.717 100251.120   0.000    1.000
GENDER                          30.364 108242.605   0.000    1.000
QUALIFICATION                   29.260 132222.502   0.000    1.000
Percieved.Trust                  9.777  16887.200   0.001    1.000
Percieved.Convinience           -7.432  33754.486   0.000    1.000
Percieved.EOU                    1.165  10481.646   0.000    1.000
Percieved.Efficiency             8.652  62375.725   0.000    1.000
Percieved.Lifestyle.Compatibility 11.983 43875.533  0.000    1.000

(Dispersion parameter for binomial family taken to be 1)

    Null deviance: 1.1090e+02  on 79  degrees of freedom
Residual deviance: 6.2693e-09  on 71  degrees of freedom
AIC: 18
```

Fig. 3.3 Estimated logistic regression model for attitude

Previous studies by Chawla and Joshi (2017), Kaushik and Rahman (2015), and Mohammadi (2015) suggest that perceived trust and perceived ease of use (PEoU) play a significant role in non-adoption. These findings are supported by Nouri and Soltani (2019) and Al-Somali et al. (2009). Akturan and Tezcan (2012) also found a direct impact of these variables on user attitudes. However, logistic regression and correlation matrix analyses conducted by Behl et al. and Das et al. indicate that the rural population lacks confidence and shows reluctance in adopting m-banking. The correlation matrix results show a strong positive association between perceived trust, perceived convenience, perceived ease of use, and attitude. However, logistic regression analysis suggests that perceived convenience is not statistically significant. Therefore, the non-adoption of m-banking in rural areas can primarily be attributed to perceived trust and perceived ease of use.

Heijden et al. (2003) proposed that risk and perceived ease of use (PEoU) are precursors to the online shopping mindset, with perceived trust being the most influential factor, followed by perceived ease of use, perceived performance, and perceived consistency with lifestyle. In the current study, anticipated convenience was found to be insignificant and remained unchanged in cases where m-banking was not adopted. There was a transition observed in all variables except for comfort, indicating that the factors impacting attitude and intention are consistent. Therefore, certain factors have an impact on both an individual's emotional state and objectives.

The study also revealed that demographic factors such as gender and educational attainment significantly influence attitudes. However, the impact of age on the non-adoption of m-banking was found to be statistically insignificant. The results indicate that user attitudes play a significant role in the adoption of m-banking. The decision to refrain from embracing m-banking is influenced not only by technology acceptance factors but also by attitudes towards m-banking.

3.5 Limitations and Conclusion of the Study

M-banking is a cost-effective and easy means of delivering financial facilities to rural Indian consumers especially given the current circumstance. However, lack of awareness in the communication technology adversely affects the usage of cell phones as means to access the online banking system. A larger pie of population uses mobile phones for mere purpose of communication. Most of the rural people possess scant knowledge about m-banking, its usage and implementation. It is noticeable that there are very few current mobile users access m-banking services even in the metropolitan areas as they are largely unaware of the benefits of m-banking. Arguably, rural areas do not receive sufficient consistency and reliability of cellular service which ultimately affects safety and reliability aspects which in turn are critical to a banking operation. These factors cement customers doubt on the credibility and safety of m-banking services.

The COVID-19 pandemic has had a significant impact on the use of mobile financial services, as many individuals have shifted to digital channels for their financial transactions. However, this transition has also brought to light several challenges that must be addressed to ensure the sustainability and effectiveness of mobile banking.

One of the main challenges is the lack of trust in mobile banking, which can be attributed to concerns regarding security and privacy. Users may have worries about the safety of their personal and financial information when utilizing mobile banking apps or platforms. To tackle this challenge, it is essential to implement robust security measures and educate users on how to protect their data.

Another challenge is the insufficient infrastructure and connectivity, particularly in rural or remote areas where access to reliable internet and mobile networks may be limited. This can hinder the adoption of mobile banking among specific populations, making it difficult for them to access financial services. Consequently, our research findings indicate that the adoption of mobile banking is influenced by factors such as perceived trust, perceived lifestyle compatibility, perceived efficiency, and ease of use. Therefore, it can be concluded that banking institutions can enhance the adoption of mobile platforms by focusing on these variables identified in our research.

Rural populations exhibit a proclivity towards accumulating modest savings, which they paradoxically tend to retain. M-banking services are currently limited and do not offer customers the option to make deposits. A potential limiting factor for the restricted utilisation of mobile services may be the requirement for individuals to physically visit a branch in order to deposit funds. The literature suggests that the quality of refund support in m-banking is inadequate (Noor 2011). The occurrence of frequent breakdowns in m-banking has resulted in challenges for consumers, as reported by Li (2013) and Ramdhony and Munien (2013). The utilisation of English as a standard language in m-banking applications poses a challenge to illiterate and rural consumers, thereby exacerbating their existing difficulties. This issue has been highlighted in previous research studies (Shareef et al. 2018). M-banking poses a significant challenge in terms of authentication, as highlighted by previous studies (Gupta and Mittal 2017; Nayak et al. 2014). It is imperative to raise awareness regarding the dependability of sources for downloading applications, particularly subsequent to the assessment of reviews, as suggested by Zhang et al. (2018). Vigilant scrutiny of user approval requests, implementation of mobile protection mechanisms, and regular security update enhancements are among the measures that can be taken to mitigate the challenges associated with m-banking and enhance the overall user experience of m-banking services.

According to scholarly research conducted by Mohan and Potnis (2015) and Behl et al., m-banking services are predominantly necessary for balance update inquiries among rural consumers in India. Service providers can promote the adoption of m-banking platforms to enable customers to conveniently monitor the current status of their financial transactions, both incoming and outgoing, thereby facilitating the integration of m-banking services. Organisations, including banks, may provide instruction to their employees on the usage of m-banking, as noted by

Wazid et al. and Singh et al. (2014). Users who possess the necessary skills to operate the programme may perceive it as user-friendly and exhibit a higher inclination towards its adoption.

Though numerous studies have been done on m-banking, this study is unique as it focuses on the Indian context wherein a substantial majority of the population resides in small towns and villages. It has contributed by providing information on how the rural consumers use this latest technology and thus can serve as a guide on how to penetrate this market further by the current providers. It should be noted that the government of India has though inaugurated the digital India initiative with an aim to seamlessly connect the country. The initiative[1] supports internet infrastructure in towns, public places, colleges, hospitals, and mobile Wi-Fi hotspots, etc. However, it can be established that a significant part of communication happens via cell phones. Nevertheless, the attitude of citizens of m-banking adoption differs because of socio-economic, political, and demographic variations.

Our study however suffers from limitation of restricted data due to ignorance of m-banking in rural areas. There is a scope of meta-analysis in the area of m-banking. The analysis may thus provide practical implications on the industry. A study on the perceptions of the users of m-banking in India may serve as an avenue for future research.

TAM and recent research on the implementation of technology have been included in the current analysis to describe consumer mind set and purpose towards implementation of m-banking. There are other technological implementation frameworks that can be evaluated, and a distinction can be made on the model's explanatory power. Also, the present research used cross-sectional sample where data was collected at a different time point. It may be viewed as a disadvantage taking into consideration the speed of the users' preference with regard to the usage and implementation of technologies. Users 'preference pattern was identified in 2014 by KPMG (US) where growing consumer tastes and technological statistics are among the key drivers of banks transformation programs (KPMG 2016). A retrospective analysis thus be necessary to ascertain the users' disposition and behavior towards m-banking. Future studies may also check if the variables already suggested in literature make a shift with time. Budding researchers may also check how culture influences the adoption of technologies in banking (Im et al. 2011).

In conclusion, m-banking has the potential to revolutionize the banking industry in India, particularly in rural areas where many individuals lack access to formal financial services. However, to encourage adoption of the technology, it is important to address the challenges facing m-banking adoption, including limited awareness and understanding of the technology among rural consumers, the limited availability and reliability of cellular service, and the limited services offered by m-banking. By addressing these challenges and working to build trust and provide real-time

[1] https://www.npci.org.in/PDF/npci/knowledge-center/Digital-Payment-Adoption-in-India-2020.pdf

updates on cash transactions, providers can encourage the adoption of m-banking among rural consumers and help to expand financial access and inclusion in India.

References

Abdul-Hamid IK, Shaikh AA, Boateng H, Hinson RE (2019) Customers' perceived risk and trust in using mobile money services—an empirical study of Ghana. Int J E-Bus Res (IJEBR) 15(1):1–19

Abdullah F, Ward R, Ahmed E (2016) Investigating the influence of the most commonly used external variables of TAM on students' Perceived Ease of Use (PEOU) and Perceived Usefulness (PU) of e-portfolios. Comput Hum Behav 63:75–90

Agarwal R, Rastogi S, Mehrotra A (2009) Customers perspectives regarding e-banking in an emerging economy. J Retail Consum Serv 16(5):340–351

Ahmad M (2018) Review of the technology acceptance model (TAM) in internet banking and mobile banking. Int J Inf Commun Technol Digital Convergence 3(1):23–41

Akinci S, Akoy S, Atilgan E (2004) Adoption of internet banking among sophisticated consumer segments in an advanced developing country. Int J Bank Mark 22(3):212–232

Akturan U, Tezcan N (2012) M-banking adoption of the youth market. Mark Intell Plan 30(4):444–459

Al-Somali SA, Gholami R, Clegg B (2009) An investigation into the acceptance of online banking in Saudi Arabia. Technovation 29(2):130–141

AlSoufi A, Ali H (2014) Customers perception of mbanking adoption in Kingdom of Bahrain: an empirical assessment of an extended tam model. arXiv Preprint arXiv:*1403.2828*

Banzal (2010) M-banking & M–commerce and related issues

Bashir I, Madhavaiah C (2014) Determinants of young consumers intention to use internet banking services in India. Vision 18(3):153–163

Behboudi M, Abedini Koshksaray A, Jalilvand Shirkhani Tabar M (2013) Mobile-banking adoption by Iranian bank clients. J Mark Manag 8(18):21–46

Chawla D, Joshi H (2017) High versus low consumer attitude and intention towards adoption of M-banking in India: an empirical study. Vision 21(4):410–424

Chircu AM, Mahajan V (2009) Perspective: revisiting the digital divide: an analysis of mobile technology depth and service breadth in the BRIC countries. J Prod Innov Manag 26(4):455–466

Correa T, Hinsley AW, De Zuniga HG (2010) Who interacts on the web? The intersection of users' personality and social media use. Comput Hum Behav 26(2):247–253

Davis FD (1989) Perceived usefulness, perceived ease of use, and user acceptance of information technology. MIS Q 13:319–340

Deb M, David EL (2014) An empirical examination of customers adoption of m-banking in India. Mark Intell Plan 32(4):475–494

Dubey V, Sonar R, Mohanty A (2020) FinTech, RegTech and contactless payments through the lens of COVID 19 times. Int J Adv Sci Technol 29(6):3727–3734

Estriegana R, Medina-Merodio JA, Barchino R (2019) Student acceptance of virtual laboratory and practical work: an extension of the technology acceptance model. Comput Educ 135:1–14

Gefen D, Karahanna E, Straub DW (2003) Trust and TAM in online shopping: an integrated model. MIS Q:51–90

Giovanis A, Athanasopoulou P, Assimakopoulos C, Sarmaniotis C (2019) Adoption of mobile banking services: a comparative analysis of four competing theoretical models. Int J Bank Mark 37(5):1165–1189

Girish V, Manu MY (2020) Impact of COVID-19 on mobile banking services. Stud Indian Place Names 40(70):2790–2797

Gopal M, Schnabl P (2022) The rise of finance companies and fintech lenders in small business lending. Rev Financ Stud 35(11):4859–4901
Gu J-C, Lee S-C, Suh Y-H (2009) Determinants of behavioral intention to m-banking. Expert Syst Appl 36:11605–11616
Gupta S, Abbas AF, Srivastava R (2022) Technology Acceptance Model (TAM): a bibliometric analysis from inception. J Telecommun Digit Econ 10(3):77–106
Heijden VH, Verhagen T, Creemers M (2003) Understanding online purchase intentions: contribution from technology and trust perspectives. Eur J Inf Syst 12(1):41–48
Hernandez JM, Mazzon JA (2007) Adoption of internet banking: proposition and implementation of an integrated methodology approach. Int J Bank Mark 25(2):72–88
Im I, Hong S, Kang MS (2011) An international comparison of technology adoption: testing the UTAUT model. Inf Manag 48(1):1–8
Internet in India (2016) An IAMAI & KANTAR IMRB Report. http://bestmediainfo.com/wp-content/uploads/2017/03/Internet-in-India-2016.pdf
Jackson CM, Leitch RA (1997) Toward an understanding of the behavioral intention to use an information system. Decis Sci 28(2):357–389
Jayawardhena C (2000) Changes in the banking sector—the case of internet banking in the UK. Int Res Electron Netw Appl Policy 10(1):19–30
Jeung-tai ET, Chihui C (2009, June) Perceived innovativeness, perceived convenience and TAM: effects on mobile knowledge management. In: 2009 Third International Conference on Multimedia and Ubiquitous Engineering. IEEE, pp 413–420
Jha S, Ye C (2016) The impact of demographic variables on perception of importance and continued usage of Facebook in the US. Glob Bus Rev 17(1):1–15
Karjaluoto H, Mattila M, Pento T (2002) Factors underlying attitude formation towards online banking in Finland. Int J Bank Mark 20(6):261–272
Kaushik AK, Rahman Z (2015) Innovation adoption across self-service banking technologies in India. Int J Bank Mark 33(2):96–121
Kim D (2006) The effects of trust-assuring arguments on consumer trust in Internet stores: application of Toulmin's model of argumentation, vol 17. Inf Syst Res, pp 286–300
Kim D, Ferrin D, Rao H (2008) A trust-based consumer decision-making model in electronic commerce: the role of trust, perceived risk, and their antecedents. Decis Support Syst 44(2):544–564
Kim G, Shin B, Lee H (2009) Understanding dynamics between initial trust and usage intentions of m-banking. Inf Syst J 19(3):283–311
Koenig-Lewis N, Palmer A, Moll A (2010) Predicting young consumers' take up of mobile banking services. Int J Bank Mark 28(5):410–432
Kolodinsky JM, Hogarth JM, Hilgert MA (2004) The adoption of electronic banking technologies by US consumers. Int J Bank Mark 22(4):238–259
KPMG (2016) Digital banking. KPMG
Kumbhar V (2011) Financial inclusion through m-banking services: scope and problems in India. MPRA
Laukkanen T (2007) Internet vs m-banking: comparing customer value perceptions. Bus Process Manag J 13:788–797
Laukkanen T, Pasanen M (2008) M-banking innovators and early adopters: How they differ from other online users? J Financ Serv Mark 13(2):86–94
Lee K, Chung N (2009) Understanding factors affecting trust in and satisfaction with m-banking in Korea: a modified DeLone and McLean's model perspective. Interact Comput 21:385–392
Lee K, Lee H, Kim S (2007) Factors influencing the adoption behavior of m-banking: a South Korean perspective. J Internet Bank Commer
Lewis NK (2010) Predicting young consumers' take up of m-banking services. Int J Bank Mark 28(5):410–432
Li F (2013) Why users adopt mobile banking service: an empirical study. In: 2013 10th International Conference on Service Systems and Service Management. IEEE, pp 490–493

Liao Z, Wong WK (2008) The determinants of customer interactions with internet enabled e-banking services. J Oper Res Soc 59(9):1201–1210
Lin (2011) An empirical investigation of m-banking adoption: the effect of innovation attributes and knowledge-based trust. Elsevier, pp 252–260
Mago S, Chitokwindo S (2014) The impact of mobile banking on financial inclusion in Zimbabwe: a case for Masvingo Province. Mediterr J Soc Sci 5(9):221
Malaquias RF (2016) An empirical study on trust in m-banking: a developing country perspective. Comput Hum Behav:453–461
Malaquias F, Malaquias R, Hwang Y (2018) Understanding the determinants of mobile banking adoption: a longitudinal study in Brazil. Electron Commer Res Appl 30:1–7
Mann BJ, Sahni SK (2012) Profiling adopter categories of internet banking in India: an empirical study. Vision 16(4):283–295
Manning C (2007) Logistic regression (with R). Changes 4:1–15
Mas I, Ivatury G (2010) The early experience with branchless banking. CGAP
Mattila M, Karjaluoto H, Pento T (2003) Internet banking adoption among mature customers: early majority or laggards? J Serv Mark 17(5):514–528
Mittal G, Gupta P (2021) A bibliometric review of Mobile Banking adoption literature. In: Proceedings of the International Conference on Advances in Management Practices (ICAMP 2021)
Mohammadi H (2015) A study of m-banking usage in Iran. Int J Bank Mark 33(6):733–759
Mohan L, Potnis D (2015) Mobile banking for the unbanked poor without mobile phones: comparing three innovative mobile banking services in India. In: 2015 48th Hawaii International Conference on System Sciences. IEEE, pp 2168–2176
Nair AB, Prabhu KS, Aditya BR, Durgalashmi CV, Prabhu AS (2021) Study on the usage of mobile banking application during COVID-19 pandemic. Webology 18(2):190–207
Nayak N, Nath V, Goel N (2014) A study of adoption behaviour of mobile banking services by Indian consumers. Int J Res Eng Technol 2(3):2347–4599
Nor K, Pearson J (2008) An exploratory study into the adoption of internet banking in a developing country: Malaysia. J Internet Commer:29–67
Nor K, Shanab E, Pearson JM (2008) Internet banking acceptance in Malaysia based on the theory of reasoned action. J Inf Syst Technol Manag 51:3–14
Nouri BA, Soltani M (2019) A hybrid model of factors affecting adoption of M-banking technology between customers of iranian banks. Serbian J Manag:49–75
Quazi A, Talukder M (2011) Demographic determinants of adoption of technological innovation. J Comput Inf Syst 52(1):34–47
Ramdhony D, Munien S (2013) An investigation on mobile banking adoption and usage: a case study of Mauritius. WORLD 3(3):197–217
Riquelme HE, Rios RE (2010) The moderating effect of gender in the adoption of m-banking. Int J Bank Mark 28(5):328–341
Rogers EM (1995) Diffusion of innovations, 4th ed
Samudra MS, Phadtare M (2012) Factors influencing the adoption of m-banking with special reference to Pune City. ASCI J Manag 42(1):51–65
Sathye M (1999) Adoption of internet banking by Australian consumers: an empirical investigation. Int J Bank Mark 17(7):324–334
Shareef MA, Baabdullah A, Dutta S, Kumar V, Dwivedi YK (2018) Consumer adoption of mobile banking services: An empirical examination of factors according to adoption stages. J Retail Consum Serv 43:54–67
Sharma RK, Jain V, Gupta S (2014) Financial inclusion in rural oman: a demand and supply analysis. Int J Manag Int Bus Stud 4(3):285–296
Singh R, Malik G (2019) Impact of digitalization on Indian rural banking customer: with reference to payment systems. Emerg Econ Stud 5(1):31–41
Singh S, Srivastava RK (2014) Factors influencing the adoption of m-banking in India. Int J E-Serv Mobile Appl 6(4):1–15

Suh B, Han I (2003) Effect of trust on customer acceptance of Internet banking. Electron Commer Res Appl:247–263
Sulaiman A, Jaafar NI, Mohezar S (2007) An overview of mobile banking adoption among the urban community. Int J Mob Commun 5(2):157–168
Szopiński TS (2016) Factors affecting the adoption of online banking in Poland. J Bus Res 69(11):4763–4768
Tan M, Teo TS (2000) Factors influencing the adoption of internet banking. J Assoc Inf Syst 1(1):1–42
Taskinsoy J (2020) A move towards a cashless society accelerates with the novel coronavirus induced global lockdown. Available at SSRN 3747750
Tiwari R, Buse S, Herstatt C (2006, July) Mobile banking as business strategy: impact of mobile technologies on customer behaviour and its implications for banks. In: 2006 Technology Management for the Global Future-PICMET 2006 Conference, vol 4. IEEE, pp 1935–1946
Venkatesh V, Davis FD (2000) A theoretical extension of the Technology Acceptance Model: four longitudinal studies. Manag Sci 46:186–204
Wan WW-L (2005) Customers adoption of banking channels in Hong Kong. Int J Bank Mark 23(3):255–272
Yu S (2009) Factors influencing the use of m-banking: the case of SMS-based M-banking. Sci J King Faisal Univ
Zhang T, Lu C, Kizildag M (2018) Banking “on-the-go”: examining consumers’ adoption of mobile banking services. Int J Qual Serv Sci 10(3):279–295

Chapter 4
Integration of Artificial Intelligence Technology in Islamic Financial Risk Management for Sustainable Development

Early Ridho Kismawadi, James Aditchere, and P. C. Libeesh

4.1 Introduction

In the financial sector, the incorporation of Artificial Intelligence (AI) technology into financial risk management has generated considerable interest. It has been demonstrated that the implementation of AI technology improves the efficacy and accuracy of risk management processes, resulting in enhanced decision-making and financial stability (I. Lee 2021; Mallikarjuna and Rao 2019; Schätter et al. 2019). In the context of Islamic finance, where Shariah compliance is a fundamental requirement, the incorporation of AI technology into financial risk management can promote transparency, consistency, and Shariah compliance (Morgan et al. 2023). The application of AI technology in the Islamic finance sector can also contribute to the achievement of sustainable development objectives. Sustainable development necessitates a strong, resilient, and inclusive financial system in addition to promoting environmental sustainability, social justice, and economic growth. The incorporation of AI technology into financial risk management can assist in overcoming

E. R. Kismawadi (✉)
Faculty of Islamic Economics and Business, Department of Islamic Banking, IAIN Langsa, Langsa, Aceh, Indonesia
e-mail: Kismawadi@iainlangsa.ac.id

J. Aditchere
Department of Supply Chain and Information Systems, Kwame Nkrumah University of Science and Technology, Kumasi, Ghana
e-mail: Jamesaditchere@carisca.knust.edu.gh

P. C. Libeesh
NSB Academy, Business School, Bangalore, India
e-mail: libeesh@nsb.edu.in

M. Irfan et al. (eds.), *Applications of Block Chain technology and Artificial Intelligence*, Financial Mathematics and Fintech,
https://doi.org/10.1007/978-3-031-47324-1_4

these obstacles by enhancing the efficacy of risk management processes, minimizing potential systemic risks, and expanding financial inclusion.

This paper investigates the potential of artificial intelligence (AI) technology to enhance financial risk management in Islamic finance, focusing on its implications for sustainable development. This paper is a literature review on the incorporation of artificial intelligence technology into Islamic finance and financial risk management. In addition, this paper will investigate the potential benefits and difficulties of integrating AI technology into financial risk management in the Islamic finance sector. The conclusion of the paper will include recommendations for policymakers and practitioners on how to effectively integrate AI technology into Islamic financial risk management to promote sustainable development. This article contributes to the literature on the implementation of artificial intelligence technology in Islamic financial risk management and its implications for sustainable development. Relevant are the findings of this study for policymakers, regulators, practitioners, and academics interested in the incorporation of AI technology in financial risk management and sustainable development in the Islamic finance industry.

This research has the potential to make a substantial impact on the growth of an Islamic financial sector that is more efficient, inclusive, and sustainable. To preserve the financial stability of businesses and prevent losses brought on by financial hazards, financial risk management is crucial in the financial sector. It has been demonstrated that integrating AI technology into financial risk management improves the efficiency and accuracy of decision-making, enhancing the financial performance of an organization.

Following sharia law is crucial in the Islamic finance industry (Bhatti 2019). The application of AI technology in financial risk management can support sharia compliance. As a result, there will be more consistency and openness in decision-making, which will boost investor confidence in the business. In addition, the major goal of the Islamic financial sector is financial inclusion in order to advance sustainability and economic equality. By increasing the efficiency and precision of financial decision-making, the adoption of AI technology can contribute to the growth of financial inclusion. Additionally, through boosting financial stability, lowering systemic risks, and promoting financial inclusion, the application of AI technology can aid in the achievement of sustainable development goals. Sustainable development in the context of Islamic finance also refers to the accomplishment of sharia-based environmental, social, and economic objectives. It is essential that this research significantly contributes to the growth of a more efficient, inclusive, and sustainable Islamic banking industry through the application of AI technology in financial risk management. As a result, the Islamic finance sector as well as society at large may greatly benefit from this research.

4.2 Artificial Intelligence Technology in Islamic Financial Risk Management

In the financial sector, the incorporation of Artificial Intelligence (AI) technology into financial risk management has become a popular topic of discussion (Bussmann et al. 2021; Harymawan et al. 2023; Hemphill and Kelley 2021; Larkin et al. 2022; Li et al. 2022a; Milojević and Redzepagic 2021; Morgan et al. 2023; Weber et al. 2023). It has been demonstrated that the application of AI technology improves the efficacy and precision of risk management procedures. In the context of Islamic finance, where Shariah compliance is a fundamental requirement, the incorporation of AI technology into financial risk management can promote transparency, consistency, and Shariah compliance. The application of AI technology in the Islamic finance sector can also contribute to the achievement of sustainable development objectives (Braganza et al. 2021; Gupta et al. 2023; Leal Filho et al. 2022; Liengpunsakul 2021).

In the financial industry, financial risk management is essential for ensuring the financial stability of businesses and preventing losses caused by financial risks (Coombs 2022; Fraczkiewicz-Wronka et al. 2021; Harb et al. 2022; Nguyen and Hoang Dinh 2021; Zhang et al. 2021). It has been demonstrated that the incorporation of AI technology into financial risk management enhances the efficacy and precision of decision making, thereby enhancing the financial performance of organizations. However, Shariah principles must be adhered to in the Islamic finance sector. Consequently, the use of AI technology can assist in ensuring compliance with sharia principles in financial risk management, which will increase transparency and consistency in decision-making, thereby boosting investor confidence in the company.

The incorporation of AI technology can provide the financial sector with numerous benefits, particularly in terms of financial risk management. In the context of Islamic finance, where Sharia compliance is a fundamental requirement, AI technology can aid in promoting transparency, consistency, and Sharia compliance. For instance, AI technology can be used to conduct automated audits of financial statements and identify possible Shariah violations. In addition, the use of AI technology in the Islamic financial sector can increase the efficiency of operational processes and financial transactions. AI can be used to monitor fund movements, prices, and investment decisions, for instance. Through the automation of business processes such as credit checking, risk management, and the collection of accounts receivable, AI technology can also help reduce operational costs and boost productivity. Long-term, the incorporation of AI technology into financial risk management can boost public confidence in the Islamic finance sector and bolster its position on the global market.

Moreover, the incorporation of AI technologies can contribute to the expansion of financial inclusion by enhancing the efficacy and precision of financial decision-making. In the context of Islamic finance, sustainable development also includes the achievement of environmental, social, and economic objectives founded on sharia.

Utilizing AI can aid in attaining sustainable development objectives by enhancing financial stability, decreasing systemic risks, and expanding financial inclusion. The incorporation of AI technology into Islamic finance can increase financial inclusion by enhancing the efficacy and precision of financial decisions. Using AI technology, for instance, the credit decision-making process can be enhanced by identifying credit risk more precisely and calculating the credit repayment capabilities of prospective debtors more efficiently. Moreover, the incorporation of AI technology can improve financial access for individuals who are not served by conventional financial institutions, such as through smartphone-accessible digital financial platforms. In this instance, AI technology can expedite the identity verification and loan decision-making processes, making it simpler for individuals in need of financing to acquire financing. Thus, the incorporation of AI technology can enhance financial inclusion and contribute to the industry's sustainable growth.

In addition, the incorporation of AI technology into financial risk management can assist in optimizing capital allocation and mitigating investment loss risk. The incorporation of AI technology in the Islamic finance industry can aid in the administration of Sharia-compliant investment portfolios, such as investments in the halal sector or those that do not contradict religious teachings. By reducing the risk of fraud and financial crime, the use of AI technology can also enhance the security and privacy of financial transactions. It is possible to detect suspicious activity in financial transactions, identify security risks in the financial system, and prevent deception using AI technology. Therefore, the incorporation of AI technology into financial risk management can assist in overcoming these obstacles and accelerating the achievement of sustainable development objectives in the Islamic finance industry. However, keep in mind that AI technology is not an ideal solution and should be used with caution. Adherence to Shariah principles and ethical values should always be a top priority in the Islamic finance industry, and the use of AI technology should always be strictly regulated and monitored.

Utilizing AI for financial risk management in the Islamic finance sector can have numerous benefits. Initially, AI technology can enhance the efficacy and precision of risk management procedures, thereby facilitating improved and more timely decision making. Second, AI technology can assist with ensuring Shariah compliance in financial risk management, boosting transparency and consistency, and boosting investor confidence. Thirdly, the incorporation of AI technology into financial risk management can facilitate the achievement of sustainable development objectives by enhancing financial stability, reducing systemic risk, and expanding financial inclusion.

Financial risk management is significantly facilitated by artificial intelligence technology (Bussmann et al. 2021; Milojević and Redzepagic 2021; Nimmy et al. 2022), especially in the Islamic finance sector. First, the incorporation of AI technologies can significantly enhance risk management processes, allowing for more informed and timely decisions. For instance, artificial intelligence technology can be used to foresee credit risk and estimate creditors' repayment capabilities, making it simpler for Islamic banks to grant loans to customers. Second, the application of AI technology to financial risk management can help ensure Shariah compliance. In the Islamic finance industry, Sharia principles should be applied to all business processes, including risk management (Coombs 2022; Nguyen and Hoang Dinh 2021; Rusydiana 2021; Zhang et al. 2021). AI technology can aid in monitoring the risk management procedure and identifying potential Shariah violations. This increases the industry's transparency and consistency, which in turn boosts investor confidence. Thirdly, AI technology can assist the Islamic financial sector in achieving its sustainable development objectives. By precisely and exhaustively analyzing data, AI technology can improve financial stability and reduce systemic risk. In terms of financial inclusion, AI technology can enhance the effectiveness and precision of financial decision-making, allowing more people to obtain access to essential financial services.

However, there are obstacles to integrating AI technology into financial risk management in the Islamic finance industry. First, there are important data security and privacy concerns to address. Second, there are fears that the replacement of human labor with AI technology could reduce employment. Thirdly, there are concerns about the impact of AI technology on corporate governance and regulatory functions. Concerns regarding the incorporation of AI technology into financial risk management in the Islamic finance industry include the potential for AI technology to replace human labor, resulting in a loss of employment. For instance, AI technology can be used to perform human-performed tasks, such as risk analysis and decision making, thereby reducing the need for human labor. In contrast, the incorporation of AI technologies can generate new employment opportunities associated with their development and operation. To overcome these obstacles and capitalize on the available opportunities, it is necessary to conduct extensive research and develop appropriate strategies for integrating AI technology into financial risk management in the Islamic finance industry.

The potential impact of AI technology on corporate governance and the function of regulators is an additional concern related to the use of AI technology in financial risk management in the Islamic finance industry. For instance, the use of AI technology can grant capital owners and company managers a great deal of power, thereby upsetting the equilibrium between capital owners, company managers, and regulatory supervisors. In the event of a system error or failure, the use of AI technology can also make determining responsibility and accountability more difficult. Consequently, it is essential to carefully consider how AI technology can be integrated in accordance with Sharia principles and applicable regulations.

4.3 Application of AI Technology in Financial Fraud Identification and Handling

The application of artificial intelligence (AI) technology in the identification and resolution of financial misconduct is crucial in the financial sector, particularly in the Islamic finance sector, which emphasizes justice and adherence to Sharia principles. The focus of this discussion is on how the incorporation of AI technology into financial risk management can aid in the identification of fraud patterns, the analysis of transaction data in real-time, and the detection of suspicious behavior that can harm financial institutions and their consumers (Azrai Azaimi Ambrose and Abdullah Asuhaimi 2021; Grira and Labidi 2021; Rashid et al. 2023; Rickinghall 2022).

Using AI-based behavioral modeling algorithms, for instance, financial institutions can automatically monitor and detect fraudulent activities such as money laundering, unlawful transactions, and Sharia-compliant activities. For instance, AI systems can analyze atypical transaction patterns, suspicious transaction amounts, and spending patterns that do not correspond to past consumer behavior. When the system detects any indication of fraud, financial institutions can promptly block transactions, request additional verification, or notify the appropriate authorities.

The application of AI technology to the detection and resolution of financial fraud not only contributes to the preservation of financial stability, but also reduces losses caused by fraudulent actions. With AI's ability to gather and analyze massive amounts of data, as well as detect complex patterns undetectable by humans, financial institutions can prevent fraud and protect customer interests with proactive measures.

For instance, an Islamic bank could use AI technology to identify suspicious electronic banking transactions. If the system detects a transaction that violates Sharia principles, such as the use of non-halal funds or other violations of Islamic finance rules, the bank can promptly stop the transaction and conduct further inquiries. Thus, the deployment of AI technology aids in ensuring Shariah compliance, preserving the bank's reputation, and fostering customer trust.

With the implementation of AI technology in the identification and management of financial fraud, the Islamic finance sector can face the increasingly complex and sophisticated challenges of fraud with greater efficiency (Al-Araj et al. 2022; Richet 2022; Rodrigues et al. 2022; Singhal et al. 2022). Utilizing AI technology can aid financial institutions in securing the financial system, safeguarding customer funds, and preserving public confidence in the Islamic finance sector.

In addition, the use of AI technology in the detection and management of financial misconduct can help financial institutions reduce the cost and time required for manual investigations. The manual process of analyzing transactions and identifying fraud patterns on a large scale can be extremely time and resource intensive. Using AI technology, however, the process can be completed automatically and efficiently.

For instance, an Islamic financial institution could integrate an AI monitoring system with a customer transaction database. Real-time analysis and comparison of transaction patterns with previously identified fraud patterns. If the system identifies a match or indication of fraud, an automated report can be generated and the investigation team can take appropriate action. This not only accelerates the process of detecting and responding to fraud, but also reduces the risk of financial losses resulting from fraud.

In addition, the incorporation of AI technology can aid in obtaining a deeper understanding and insight into financial fraud trends. By conducting a comprehensive analysis of transaction data, AI systems are able to identify new and evolving fraud patterns. This data can be used to enhance security systems and create more effective prevention strategies. In addition, the results of the data analysis can contribute to efforts to update Islamic finance regulations and policies in order to anticipate emergent fraud threats.

By incorporating AI technology into the identification and resolution of financial fraud, the Islamic finance sector can enhance consumer safety and confidence. In addition, the application of AI technology conveys that financial institutions are committed to safeguarding customer funds and interests through the use of cutting-edge technology solutions and are serious about combating fraud threats.

The Islamic finance industry derives substantial benefits from the deployment of AI technology in the detection and management of financial fraud. Using sophisticated behavioral modeling and data analysis algorithms, financial institutions can identify fraud patterns, prevent financial losses, ensure Shariah compliance, and increase public trust. By continuing to innovate and adapt to technological advancements, the Islamic finance sector will be better able to combat fraud and safeguard customers and the financial system as a whole.

4.4 Improving Customer Experience Through Personalization and Better Financial Services

The incorporation of AI technology into the Islamic finance industry can not only improve financial stability, but also enhance the customer experience through personalization and the provision of superior financial services. By utilizing AI technology, financial institutions can provide services that are more personalized and tailored to the specific requirements of each client (Bawack et al. 2021; J.-S. Chen et al. 2021; Hoyer et al. 2020; Mostafa and Kasamani 2022; Sujata et al. 2019).

For instance, Islamic financial institutions can utilize chatbots or virtual assistants enabled by technology for natural language processing. These chatbots are capable of interacting with consumers via digital platforms and delivering prompt, accurate assistance. Customers may, for instance, inquire about financial products or services, request transaction-related information, or seek advice regarding their financial plans. With AI's ability to comprehend and respond to human language, chatbots can provide pertinent responses and efficiently resolve customer issues.

In addition, the incorporation of AI technology permits financial institutions to generate investment recommendations that are more aligned with the financial profiles and objectives of their customers. By analyzing historical data and using artificial intelligence to model customer behavior, financial institutions can gain a deeper understanding of customers' investment preferences and risk tolerance. A customer with a conservative risk profile, for instance, may be better adapted to invest in stable, low-risk financial instruments, whereas a customer with an aggressive risk profile may be more interested in instruments with the potential for high returns but also greater risk. Financial institutions can assist clients in administering and optimizing their investment portfolios by providing recommendations that correspond to customer profiles.

This enhancement of the consumer experience is also consistent with the objective of financial inclusion in the Islamic finance industry. By offering customized and pertinent financial services, financial institutions can increase public access to and participation in a Sharia-compliant financial system. The use of AI technology to improve financial services can help reduce financial disparities and expand individuals' access to services that meet their specific requirements.

Islamic financial institutions must invest in appropriate AI technologies, such as sophisticated data analytics systems and dependable IT infrastructure, in order to enhance customer experience through personalization and better financial services. In addition, the key to success is training and developing employees who are proficient in utilizing AI technology. Thus, the incorporation of AI technology in the Islamic finance sector can provide significant benefits for enhancing the customer experience and attaining the sector's broader financial inclusion objectives. To achieve this, financial institutions must establish partnerships with the leading AI technology providers and collaborate with other financial industry stakeholders. Cooperation between financial institutions, regulators, and technology providers will expedite the adoption and integration of artificial intelligence technology in financial risk management.

In order to improve customer experience through personalization and better financial services, Islamic financial institutions must also guarantee Shariah compliance in the use of AI technology. This can be accomplished by incorporating scholars and Sharia experts into the development and implementation of AI technology. By collaborating with them, financial institutions can ensure that solutions adhere to Shariah principles and instill consumer confidence in adopting AI technology.

In addition, it is essential to increase public understanding and awareness of the benefits and ramifications of utilizing AI technology. Financial institutions can involve customers in the development process and educate them on AI technology, data security, and privacy policies. Thus, clients will be more assured and motivated to utilize the offered services.

The incorporation of AI technology into the Islamic finance industry can provide substantial benefits for enhancing the customer experience through personalization and the provision of superior financial services. By utilizing AI technology, financial institutions are able to provide more personalized services, rapidly analyze data, make relevant recommendations, and increase public participation in the Sharia-based financial system. To accomplish this, however, investments in technology infrastructure, cooperation between financial institutions and technology providers, adherence to Sharia principles, and a greater understanding and awareness of AI technology will be required.

4.5 The Effect of AI Technology Integration in Financial Risk Management on Financial Stability

The incorporation of artificial intelligence (AI) technology into financial risk management has a significant impact on the sector's financial stability. The application of AI technology in financial risk management can enhance the ability to make accurate decisions and identify risks, thereby reducing potential systemic risks and protecting the overall stability of the financial system (Sikorski 2021; Thiruppathi and Kirubhashankar 2023; Vučinić and Luburić 2022).

For instance, AI technology can be used to analyze market data, investor behavior, and relevant economic factors in real-time when making investment decisions. By utilizing sophisticated algorithms and data processing, AI systems are able to provide more accurate information and empower decision makers to make more prudent investment decisions. This reduces the possibility that uncontrolled investments will undermine financial stability.

Moreover, AI technology can enhance the identification of systemic hazards. AI systems can continuously monitor and analyze data to identify patterns and tendencies that indicate systemic risk. By obtaining faster and more accurate insights into these risks, appropriate preventative and mitigating measures can be implemented to preserve financial stability. For instance, AI systems can monitor the relationships between financial institutions and identify potential interconnection risks that could impact the entire system.

In addition, the application of AI technology can improve risk comprehension and transparency. By analyzing vast amounts of data, AI can uncover intricate relationships and identify hazards that humans may overlook. This information can be used to help stakeholders gain a greater understanding of the risks and take the necessary precautions.

Using AI technology, for instance, an Islamic bank can analyze consumer data and financial transactions in accordance with Sharia principles. The bank can identify and manage Sharia compliance-related risks by leveraging AI's ability to recognize patterns that are not in accordance with Sharia principles. This aids Islamic institutions in maintaining financial stability and gaining the public's trust.

Therefore, the use of AI technology in financial risk management can have a positive effect on the financial stability of the Islamic finance industry. Through its ability to make better decisions, identify accurate risks, and increase transparency, AI technology can aid in maintaining the integrity and viability of the Islamic finance sector by safeguarding system stability and minimizing system risks. For instance, Islamic banks that employ AI technology in their financial risk management can reduce the risk of Sharia-violating misappropriation of funds, fraud, and Sharia-contravention violations, which threaten financial stability. Through more precise and systematic analysis, AI technology can assist in identifying potential infractions and implementing the necessary preventative measures.

In addition, the use of AI technology in financial risk management can strengthen the Islamic finance sector's supervision and regulation system. By exhaustively analyzing data and detecting anomalies, AI technology can assist supervisory and regulatory authorities in identifying practices that have the potential to violate rules or threaten the financial system's stability. This allows authorities to maintain the integrity of the Islamic financial sector in a more efficient manner.

As an illustration, an Islamic financial institution could use AI technology to monitor transactions in real time. AI is capable of analyzing suspicious transaction patterns, such as an atypical increase in transaction volume or payment patterns that do not correspond to consumer preferences. By rapidly identifying potential risks, these financial institutions can take immediate action, such as blocking transactions or conducting further investigations, to protect financial stability and maintain customer confidence.

The incorporation of AI technology into financial risk management in the Islamic finance industry has a substantial positive impact on financial stability. By enhancing intelligent decision-making, precise risk identification, preventing violations of Sharia principles, and bolstering supervision and regulation, AI technology can mitigate potential systemic risks and safeguard the overall stability of the financial system. This is advantageous for all parties, including Islamic banks, customers, communities, and authorities, in the development of a robust, trustworthy, and sustainable Islamic finance industry.

4.6 Challenges and Constraints of AI Technology Integration in Financial Risk Management in the Islamic Finance Sector

The incorporation of artificial intelligence (AI) technology into financial risk management in the Islamic finance sector is inextricably linked to a number of obstacles and restrictions. The availability and quality of data required to train and operate AI systems is a significant obstacle. Data pertaining to Shariah principles, such as transactions, financial products, and Islamic law compliance, should be readily accessible in the context of Islamic finance. However, the data is frequently unstructured, dispersed, or poorly documented, making it challenging to implement AI technologies that require high-quality data.

In addition, information security is a significant obstacle in the Islamic finance industry's integration of AI technology (Ahmad et al. 2021; Andreev et al. 2021; Cappellozza et al. 2022; Li et al. 2022b). AI relies on sensitive and confidential data that must be adequately protected against security threats like cyberattacks and data theft. Successful implementation of AI technology in financial risk management requires a robust and stringent security system, as well as adherence to applicable privacy and data protection regulations.

In addition to regulations and legal frameworks, integrating AI technology into financial risk management in the Islamic finance sector is hampered by these factors. The application of AI technology must adhere to applicable regulations and

Shariah standards, as well as Islamic finance's underlying principles. Regulators must ensure that the use of AI technology in financial risk management does not violate Shariah principles and that customer interests and the stability of the financial sector as a whole are prioritized.

Lastly, user confidence is a crucial factor in the adoption of AI technology in the Islamic finance industry. Users must be confident that the use of AI technology will not compromise their privacy or be used for nefarious purposes. It is essential for financial institutions and other stakeholders to provide explicit explanations of the utility, benefits, and limitations of artificial intelligence (AI) technology in financial risk management, as well as to communicate data protection efforts.

As an illustration, an Islamic bank that wishes to integrate AI technology into its financial risk management must overcome the difficulty of obtaining Shariah-compliant transaction and financial product data. In addition, the bank must employ a skilled security team to ensure that the AI systems it employs are protected from security threats and data abuse. Moreover, the bank must coordinate with regulatory authorities and ensure compliance with all applicable regulations.

Collaboration between the financial industry, regulators, and other stakeholders is essential in the face of these challenges. Together, they must establish a clear regulatory framework and facilitate the data exchange required for the implementation of AI technologies in financial risk management. Moreover, investments in a secure and resilient technology infrastructure are necessary for the successful integration of AI technologies.

As a concrete measure, Islamic banks can form alliances with technology service providers that specialize in developing AI solutions for the financial industry. This can help them obtain access to ready-to-use AI technologies and receive expert assistance in overcoming technical and security obstacles. In addition, Islamic institutions can involve customers and the general public in the development and implementation of AI technology. Transparent explanations of the benefits, uses, and preservation of data will aid in establishing trust and decreasing resistance to AI technology.

Incorporating AI technology into financial risk management in the Islamic finance sector is a complex process fraught with obstacles and limitations. However, with good cooperation between the financial industry, regulators, and other stakeholders, as well as the adoption of concrete measures such as collaboration with technology service providers and customer engagement, the use of AI technology can provide significant benefits for increasing effectiveness, transparency, and Shariah compliance in financial risk management in the Islamic finance sector.

4.7 Effective Integration of Artificial Intelligence Technology

Several recommendations can be considered by policymakers and practitioners to achieve effective integration between AI technology and financial risk management in the Islamic finance industry. First, it is essential to establish a defined regulatory framework to facilitate the use of AI technology in financial risk management in accordance with Shari'ah principles. Clear regulations will provide direction for usage, data protection, and compliance. Additionally, collaboration between the financial industry, regulators, and other interested parties should be encouraged. Using dialogue and collaborative forums, problems can be resolved collectively. Prioritize the investment in a secure and resilient technology infrastructure. A robust infrastructure will assure the data protection, system security, and data availability necessary for AI technology usage. Through training, seminars, and educational programs, efforts should be made to increase understanding and awareness of the benefits and repercussions of using AI technology. By adhering to these recommendations, effective integration between AI technology and financial risk management can be achieved in the Islamic finance sector, bringing significant benefits to the Islamic finance industry and society as a whole.

Another suggestion is to enhance cooperation between Islamic financial institutions, academic institutions, and the technology sector. Integrating AI technology into financial risk management requires an in-depth understanding of Shariah principles and the particular requirements of the Islamic finance industry. Islamic finance-related research and development can be conducted if Islamic financial institutions and academic institutions with expertise in AI technology form close partnerships. In addition, collaboration with the technology industry developing AI solutions is essential for the development and implementation of Islamic finance-appropriate solutions. Through this collaboration, knowledge transfer, innovation, and technology adoption in financial risk management in the Islamic finance sector can be enhanced.

In addition, it is essential to promote the incremental adoption of AI technology by beginning with specific and quantifiable applications. It enables Islamic financial institutions to test and evaluate the efficacy of AI technology in financial risk management while minimizing unintended risks and consequences. Using this phased approach, Islamic financial institutions can identify and resolve potential issues, as well as gain a deeper understanding of the application of AI technology within the context of Islamic finance.

In addition to these recommendations, it is essential to adopt an inclusive approach to the incorporation of AI technology in Islamic financial risk management. This involves empowering communities and ensuring that AI technology is accessible and utilized by a broad range of stakeholders, such as micro and small finance institutions and financially underserved communities. In the context of Islamic finance, financial inclusion is a key objective, and artificial intelligence (AI) technology can be used to attain broader and more sustainable inclusion.

By implementing these recommendations, the incorporation of AI technology into financial risk management in the Islamic finance sector can be accomplished with greater efficiency and intent. This will improve efficiency, the integrity of decision-making, and financial stability as a whole. In addition, with the correct application of AI technology, the Islamic finance sector can contribute to achieving environmental, social, and economic sustainable development objectives.

4.8 Application of AI Technology in Sharia Risk Management

Application of AI technology to Islamic risk management is a significant development in the Islamic finance industry (Begoli et al. 2019; Bradley 2020; H. Lee 2020; Zhou et al. 2019). In this context, AI technology can be utilized to identify and manage Shariah compliance-related hazards. For instance, AI technology can be used to analyze financial contract documents and extract Shariah-compliant information, such as riba (interest), gharar (uncertainty), and maysir (speculation) regulations. AI algorithms can be programmed to autonomously evaluate a contract's adherence to Shariah principles, identify potential violations, and suggest corrective action. In addition, AI technology can be used to monitor financial activities in real time in order to identify transactions that violate Sharia principles, such as usury transactions and investments in haram sectors. Financial institutions can identify suspicious patterns and take the necessary precautions by employing AI-based data analysis and behavioral modeling.

AI technology can also be applied to Sharia supervision and auditing, in addition to risk management. Using natural language processing techniques, for instance, AI systems can analyze fatwa documents, academic research, and Islamic financial literature to identify inconsistencies between business policies and practices and Sharia principles. This will promote greater Sharia compliance and transparency in the Islamic financial sector.

The implementation of AI technology in Sharia risk management not only aids financial institutions in ensuring compliance, but also improves the supervisory process's efficacy and precision. With the implementation of AI technology, financial institutions can reduce the risk of data breaches, enhance their reputations, and increase consumer trust. The incorporation of AI technology into Islamic risk management has the potential to significantly enhance compliance and supervision within the Islamic finance industry. However, financial institutions must ensure that the algorithms and models they employ are compliant with Shariah regulations and fatwas. In addition, human supervision and control are required to maintain accuracy and ethics in the context of Sharia when utilizing AI technology.

With the incorporation of AI technology into Sharia-compliant risk management, financial institutions can provide greater advantages in terms of Sharia-compliant risk management. AI technology can assist in identifying potential Shariah-related risks, enabling financial institutions to take the necessary steps to mitigate these risks. For instance, Islamic financial institutions can use AI technology to develop predictive models that can identify Islamic financing-related risks, such as liquidity risk, financing risk to haram sectors, and credit risk. With more precise and real-time analysis, financial institutions can proactively take the required measures to mitigate these risks, such as diversifying financing or conducting more stringent monitoring of financing quality.

In addition, AI technology can be utilized to improve decision-making in Sharia risk management (Dara et al. 2022; Di Vaio et al. 2022; Han et al. 2023; Senoner et al. 2022; Valle-Cruz et al. 2022). Algorithms based on artificial intelligence can assist in assessing the financial products and services offered by financial institutions and assuring their compliance with Shariah principles. Thus, financial institutions are able to ensure that the products and services they provide to consumers adhere to Sharia law. In addition to risk management and decision making, AI technology can also be used to increase Sharia risk management's transparency and accountability. By utilizing blockchain technology powered by artificial intelligence, financial institutions can precisely record and track Islamic-compliant financial transactions, thereby ensuring data integrity and facilitating more thorough audits.

The application of AI technology in Shariah risk management presents Islamic financial institutions with an excellent opportunity to strengthen supervision, enhance compliance with Shariah principles, and develop customer trust. It is essential to note, however, that the successful integration of AI technology within the Sharia context also depends on an appropriate regulatory framework, effective supervision, and a comprehensive comprehension of Sharia principles.

4.9 Islamic Financial Risk Management for Sustainable Development

In the Islamic finance industry, risk management plays a crucial role in promoting sustainable developmen (H. Chen and Zhao 2022; Hummel et al. 2021; Inshakova et al. 2021; Nobanee et al. 2021; Popkova and Sergi 2021; Vagin et al. 2022) t. It entails identifying, evaluating, and mitigating risks while ensuring Shariah compliance. This strategy aims to attain financial stability, moral conduct, and social responsibility within the Islamic financial system.

One of the most important aspects of Islamic financial risk management is the incorporation of moral and social considerations into investment decisions. Investing in activities regarded harmful or incompatible with Sharia principles, such as gambling, alcohol, or interest-based transactions, is prohibited by Islamic finance. This procedure ensures that investments are directed toward socially responsible projects and industries that promote sustainable development. For instance, Islamic banks may prioritize financing renewable energy projects, the development of ecological infrastructure, or social impact initiatives that promote education or alleviate poverty.

The emphasis on profit-and-loss sharing (PLS) arrangements is an additional essential aspect of Islamic financial risk management. PLS contracts, such as Mudarabah (profit sharing) and Musyarakah (partnership), are preferable to interest-based loans in Islamic finance. These contracts allocate risk and return among the parties, align their interests, and promote an equitable distribution of wealth. This strategy promotes more equitable and inclusive economic growth by ensuring that risks are shared among stakeholders and by promoting a participatory framework for sustainable development.

In addition, Islamic financial risk management incorporates risk-sharing mechanisms such as Takaful (Islamic insurance) to mitigate ambiguity and prevent potential losses. Takaful is based on the principles of mutual aid and solidarity, in which participants aggregate their contributions to provide protection from risk. The system fosters community resiliency by encouraging shared responsibility and providing assistance to individuals and businesses in times of financial hardship.

Collaboration among various stakeholders, including regulators, financial institutions, academics, and investors, is necessary for the effective implementation of Islamic financial risk management practices. Regulators play a crucial role in establishing robust frameworks and standards that ensure Shariah compliance and advance sustainable development objectives. In order to enhance the efficacy, transparency, and precision of risk assessment and mitigation, financial institutions must implement risk management strategies and innovative technologies, such as artificial intelligence and blockchain.

In addition, governance and accountability play a significant role in Islamic financial risk management. Transparency and accountability are essential for maintaining the Islamic financial system's integrity and fostering stakeholder confidence. There are in place effective risk monitoring and management frameworks, including independent Shariah boards, risk committees, and internal control mechanisms.

In addition, Islamic financial institutions can utilize technological advancements to enhance their risk management procedures. Integration of artificial intelligence (AI) and big data analytics, for instance, can facilitate real-time risk assessment, early detection of potential hazards, and more precise decision-making. Algorithms propelled by AI can analyze vast quantities of data to identify patterns, forecast market trends, and enhance risk modeling. This technology-based approach enables proactive risk management and enables institutions to respond swiftly to emergent risks, thereby contributing to the overall stability and sustainability of the Islamic finance industry.

Effective Islamic financial risk management has significant implications for sustainable development. It ensures that financial activities are consistent with ethical principles and social responsibility, resulting in investments that benefit society and the environment. Islamic finance can help reduce income inequality and promote inclusive economic development by integrating risk-sharing mechanisms and promoting equitable wealth distribution. In addition, a focus on ethical considerations and responsible investment contributes to the attainment of the United Nations Sustainable Development Goals (SDGs), such as poverty alleviation, environmental sustainability, and social welfare.

Incorporating ethical considerations, governance frameworks, risk-sharing mechanisms, and technological advancements, Islamic finance risk management plays an essential role in promoting sustainable development. Islamic finance can contribute to a more resilient and equitable financial system that promotes social progress and environmental preservation by adhering to Sharia principles and concentrating on long-term sustainability. Continued efforts to enhance risk management practices and collaboration among stakeholders will strengthen the global impact of Islamic finance on sustainable development.

4.10 Conclusion

The incorporation of AI technology in financial risk management has the potential to bring numerous benefits to the Islamic financial industry, such as increased efficiency and precision, adherence to Sharia principles, increased transparency and consistency, and promotion of financial inclusion and sustainable development. However, ethical concerns must take precedence in the application of AI technology. Future research should investigate the application of AI technology in Islamic finance in greater depth, particularly in the areas of risk management, investment portfolio management, and financial inclusion. In addition, the impact of AI technology on the development, stability, and long-term viability of the Islamic financial sector must be assessed. To maximize the benefits of AI technology in the Islamic finance industry, well-designed policies, regulations, and implementation standards are essential.

Governments and regulators must ensure that the financial sector's use of artificial intelligence technology does not compromise consumer privacy or increase cybersecurity risks. Companies utilizing AI technology must respect consumer privacy and implement best data protection practices. To this end, researchers should continue to examine the potential applications of AI technology in the financial sector, including the Islamic finance sector, and its effects on decision-making, adherence to Sharia principles, financial inclusion, and attainment of sustainable development goals. The research could cast light on how AI technology can be utilized to enhance performance and Shariah compliance in the Islamic finance industry, as well as provide guidance for future policy development.

References

Ahmad T, Zhang D, Huang C, Zhang H, Dai N, Song Y, Chen H (2021) Artificial intelligence in sustainable energy industry: status Quo, challenges and opportunities. J Clean Prod 289. https://doi.org/10.1016/j.jclepro.2021.125834

Al-Araj R, Haddad H, Shehadeh M, Hasan E, Nawaiseh MY (2022) The effect of artificial intelligence on service quality and customer satisfaction in Jordanian Banking Sector. WSEAS Trans Bus Econ 19:1929–1947. https://doi.org/10.37394/23207.2022.19.173

Andreev AI, Kazanin AG, Zayed NM (2021) The relationship of science and personnel policy development in the problem field of the digital economy. Acad Strateg Manag J 20:1–13. https://www.scopus.com/inward/record.uri?eid=2-s2.0-85102747423&partnerID=40&md5=8673d592297ece427085b75f21d2f873

Azrai Azaimi Ambrose AH, Abdullah Asuhaimi F (2021) Cash waqf risk management and perpetuity restriction conundrum. ISRA Int J Islam Financ 13(2):162–176. https://doi.org/10.1108/IJIF-12-2019-0187

Bawack RE, Wamba SF, Carillo KDA (2021) Exploring the role of personality, trust, and privacy in customer experience performance during voice shopping: Evidence from SEM and fuzzy set qualitative comparative analysis. Int J Inf Manag 58. https://doi.org/10.1016/j.ijinfomgt.2021.102309

Begoli E, Bhattacharya T, Kusnezov D (2019) The need for uncertainty quantification in machine-assisted medical decision making. Nat Mach Intell 1(1):20–23. https://doi.org/10.1038/s42256-018-0004-1

Bhatti M (2019) Resolving Islamic finance disputes through arbitration in the Middle East. In: The growth of islamic finance and banking: innovation, governance and risk mitigation (pp 254–261). https://www.scopus.com/inward/record.uri?eid=2-s2.0-85116589970&partnerID=40&md5=36cd31a25f661dd2da390db9348af563

Bradley P (2020) Risk management standards and the active management of malicious intent in artificial superintelligence. AI Soc 35(2):319–328. https://doi.org/10.1007/s00146-019-00890-2

Braganza A, Chen W, Canhoto A, Sap S (2021) Productive employment and decent work: the impact of AI adoption on psychological contracts, job engagement and employee trust. J Bus Res 131:485–494. https://doi.org/10.1016/j.jbusres.2020.08.018

Bussmann N, Giudici P, Marinelli D, Papenbrock J (2021) Explainable machine learning in credit risk management. Comput Econ 57(1):203–216. https://doi.org/10.1007/s10614-020-10042-0

Cappellozza A, Moraes GHSM, Perez G, Simões AL (2022) Antecedent factors of violation of information security rules. RAUSP Manag J 57(1):85–103. https://doi.org/10.1108/RAUSP-02-2021-0022

Chen H, Zhao X (2022) Green financial risk management based on intelligence service. J Clean Prod 364. https://doi.org/10.1016/j.jclepro.2022.132617

Chen J-S, Le T-T-Y, Florence D (2021) Usability and responsiveness of artificial intelligence chatbot on online customer experience in e-retailing. Int J Retail Distrib Manag 49(11):1512–1531. https://doi.org/10.1108/IJRDM-08-2020-0312

Coombs N (2022) Narrating imagined crises: how central bank storytelling exerts infrastructural power. Econ Soc 51(4):679–702. https://doi.org/10.1080/03085147.2022.2117313

Dara S, Dhamercherla S, Jadav SS, Babu CM, Ahsan MJ (2022) Machine learning in drug discovery: a review. Artif Intell Rev 55(3):1947–1999. https://doi.org/10.1007/s10462-021-10058-4

Di Vaio A, Hassan R, Alavoine C (2022) Data intelligence and analytics: a bibliometric analysis of human–artificial intelligence in public sector decision-making effectiveness. Technol Forecast Soc Chang 174. https://doi.org/10.1016/j.techfore.2021.121201

Fraczkiewicz-Wronka A, Ingram T, Szymaniec-Mlicka K, Tworek P (2021) Risk management and financial stability in the polish public hospitals: the moderating effect of the stakeholders' engagement in the decision-making. Risks 9(5). https://doi.org/10.3390/risks9050087

Grira J, Labidi C (2021) Banks, funds, and risks in islamic finance: literature and future research avenues. Financ Res Lett 41. https://doi.org/10.1016/j.frl.2020.101815

Gupta BB, Gaurav A, Panigrahi PK, Arya V (2023) Analysis of artificial intelligence-based technologies and approaches on sustainable entrepreneurship. Technol Forecast Soc Chang 186. https://doi.org/10.1016/j.techfore.2022.122152

Han H, Shiwakoti RK, Jarvis R, Mordi C, Botchie D (2023) Accounting and auditing with blockchain technology and artificial Intelligence: a literature review. Int J Account Inf Syst 48. https://doi.org/10.1016/j.accinf.2022.100598

Harb E, El Khoury R, Mansour N, Daou R (2022) Risk management and bank performance: evidence from the MENA region. J Financ Report Account. https://doi.org/10.1108/JFRA-07-2021-0189

Harymawan I, Anridho N, Minanurohman A, Ningsih S, Kamarudin KA, Raharjo Y (2023) Do more masculine-faced CEOs reflect more tax avoidance? Evidence from Indonesia. Cogent Bus Manag 10(1). https://doi.org/10.1080/23311975.2023.2171644

Hemphill TA, Kelley KJ (2021) Artificial intelligence and the fifth phase of political risk management: an application to regulatory expropriation. Thunderbird Int Bus Rev 63(5):585–595. https://doi.org/10.1002/tie.22222

Hoyer WD, Kroschke M, Schmitt B, Kraume K, Shankar V (2020) Transforming the customer experience through new technologies. J Interact Mark 51:57–71. https://doi.org/10.1016/j.intmar.2020.04.001

Hummel K, Laun U, Krauss A (2021) Management of environmental and social risks and topics in the banking sector - an empirical investigation. Br Account Rev 53(1). https://doi.org/10.1016/j.bar.2020.100921

Inshakova AO, Sozinova AA, Litvinova TN (2021) Corporate fight against the covid-19 risks based on technologies of industry 4.0 as a new direction of social responsibility. Risks 9(12). https://doi.org/10.3390/risks9120212

Larkin C, Drummond Otten C, Árvai J (2022) Paging Dr. JARVIS! Will people accept advice from artificial intelligence for consequential risk management decisions? J Risk Res 25(4):407–422. https://doi.org/10.1080/13669877.2021.1958047

Leal Filho W, Yang P, Eustachio JHPP, Azul AM, Gellers JC, Gielczyk A, Dinis MAP, Kozlova V (2022) Deploying digitalisation and artificial intelligence in sustainable development research. Environ Dev Sustain. https://doi.org/10.1007/s10668-022-02252-3

Lee H (2020) Role of artificial intelligence and enterprise risk management to promote corporate entrepreneurship and business performance: evidence from korean banking sector. J Intell Fuzzy Syst 39(4):5369–5386. https://doi.org/10.3233/JIFS-189022

Lee I (2021) Cybersecurity: risk management framework and investment cost analysis. Bus Horiz 64(5):659–671. https://doi.org/10.1016/j.bushor.2021.02.022

Li W, Paraschiv F, Sermpinis G (2022a) A data-driven explainable case-based reasoning approach for financial risk detection. Quant Finance 22(12):2257–2274. https://doi.org/10.1080/14697688.2022.2118071

Li T, Xu W, Wang W, Zhang X (2022b) Obstacle detection in a field environment based on a convolutional neural network security. Enterp Inf Syst 16(3):472–493. https://doi.org/10.1080/17517575.2020.1797180

Liengpunsakul S (2021) Artificial intelligence and sustainable development in China. Chinese Econ 54(4):235–248. https://doi.org/10.1080/10971475.2020.1857062

Mallikarjuna M, Rao RP (2019) Evaluation of forecasting methods from selected stock market returns. Financ Innov 5(1). https://doi.org/10.1186/s40854-019-0157-x

Milojević N, Redzepagic S (2021) Prospects of artificial intelligence and machine learning application in banking risk management. J Cent Bank Theory Pract 10(3):41–57. https://doi.org/10.2478/jcbtp-2021-0023

Morgan TR, Roath AS, Glenn Richey R (2023) How risk, transparency, and knowledge influence the adaptability and flexibility dimensions of the responsiveness view. J Bus Res 158. https://doi.org/10.1016/j.jbusres.2022.113641

Mostafa RB, Kasamani T (2022) Antecedents and consequences of chatbot initial trust. Eur J Mark 56(6):1748–1771. https://doi.org/10.1108/EJM-02-2020-0084

Nguyen LTM, Hoang Dinh P (2021) Ex-ante risk management and financial stability during the COVID-19 pandemic: a study of Vietnamese firms. China Financ Rev Int 11(3):349–371. https://doi.org/10.1108/CFRI-12-2020-0177

Nimmy SF, Hussain OK, Chakrabortty RK, Hussain FK, Saberi M (2022) Explainability in supply chain operational risk management: a systematic literature review. Knowl-Based Syst 235. https://doi.org/10.1016/j.knosys.2021.107587

Nobanee H, Alhajjar M, Abushairah G, Al Harbi S (2021) Review reputational risk and sustainability: a bibliometric analysis of relevant literature. Risks 9(7). https://doi.org/10.3390/risks9070134

Popkova EG, Sergi BS (2021) Dataset modelling of the financial risk management of social entrepreneurship in emerging economies. Risks 9(12). https://doi.org/10.3390/risks9120211

Rashid A, Akmal M, Shah SMAR (2023) Corporate governance and risk management in Islamic and convectional financial institutions: explaining the role of institutional quality. J Islam Account Bus Res. https://doi.org/10.1108/JIABR-12-2021-0317

Richet J-L (2022) How cybercriminal communities grow and change: an investigation of ad-fraud communities. Technol Forecast Soc Chang 174. https://doi.org/10.1016/j.techfore.2021.121282

Rickinghall M (2022) Impact of fintech on Islamic bank performance in Malaysia: descriptive study on fintech. Front Artif Intell Appl 352:93–102. https://doi.org/10.3233/FAIA220088

Rodrigues VF, Policarpo LM, da Silveira DE, da Rosa Righi R, da Costa CA, Barbosa JLV, Antunes RS, Scorsatto R, Arcot T (2022) Fraud detection and prevention in e-commerce: a systematic literature review. Electron Commer Res Appl 56. https://doi.org/10.1016/j.elerap.2022.101207
Rusydiana AS (2021) Bibliometric analysis of journals, authors, and topics related to COVID-19 and Islamic finance listed in the Dimensions database by Biblioshiny. Sci Ed 8(1):72–78. https://doi.org/10.6087/kcse.232
Schätter F, Hansen O, Wiens M, Schultmann F (2019) A decision support methodology for a disaster-caused business continuity management. Decis Support Syst 118:10–20. https://doi.org/10.1016/j.dss.2018.12.006
Senoner J, Netland T, Feuerriegel S (2022) Using explainable artificial intelligence to improve process quality: evidence from semiconductor manufacturing. Manag Sci 68(8):5704–5723. https://doi.org/10.1287/mnsc.2021.4190
Sikorski M (2021) Digital innovations and smart solutions for society and economy: pros and cons. Found Manag 13(1):103–116. https://doi.org/10.2478/fman-2021-0008
Singhal A, Dube P, Jain VK (2022) Modelling barriers of artificial intelligence in banking sectors using total interpretive structural modelling. Int J Appl Decis Sci 15(3):311–335. https://doi.org/10.1504/IJADS.2022.122649
Sujata J, Aniket D, Mahasingh M (2019) Artificial intelligence tools for enhancing customer experience. Int J Recent Technol Eng 8(2 Special Issue 3):700–706. https://doi.org/10.35940/ijrte.B1130.0782S319
Thiruppathi A, Kirubhashankar CK (2023) New ranking of generalized quadrilateral shape fuzzy number using centroid technique. Intell Autom Soft Comput 36(2):2253–2266. https://doi.org/10.32604/iasc.2023.033870
Vagin SG, Kostyukova EI, Spiridonova NE, Vorozheykina TM (2022) Financial risk management based on corporate social responsibility in the interests of sustainable development. Risks 10(2). https://doi.org/10.3390/risks10020035
Valle-Cruz D, Fernandez-Cortez V, Gil-Garcia JR (2022) From E-budgeting to smart budgeting: exploring the potential of artificial intelligence in government decision-making for resource allocation. Gov Inf Q 39(2). https://doi.org/10.1016/j.giq.2021.101644
Vučinić M, Luburić R (2022) Fintech, risk-based thinking and cyber risk. J Cent Bank Theory Pract 11(2):27–53. https://doi.org/10.2478/jcbtp-2022-0012
Weber P, Carl KV, Hinz O (2023) Applications of explainable artificial intelligence in finance—a systematic review of finance, information systems, and computer science literature. Manag Rev Q. https://doi.org/10.1007/s11301-023-00320-0
Zhang Y-J, Bouri E, Gupta R, Ma S-J (2021) Risk spillover between Bitcoin and conventional financial markets: An expectile-based approach. N Am J Econ Financ 55. https://doi.org/10.1016/j.najef.2020.101296
Zhou H, Sun G, Fu S, Liu J, Zhou X, Zhou J (2019) A big data mining approach of PSO-Based BP neural network for financial risk management with IoT. IEEE Access 7:154035–154043. https://doi.org/10.1109/ACCESS.2019.2948949

Chapter 5
Blockchain and IOT Devices: A Futuristic Approach for Digital and Smart Supply Chain

Bindiya Jain, Indrajit Ghosal, Akshita Chotia, and P. G. S. Amila Jayarathne

5.1 Introduction

Block is records of recent transactions and blockchain is a database to store data (in form of block) in encrypted format. In blockchain system information is in link manner, so we can easily rectify our previous work and ready for next. Blocks are part of network, IOT devices, and supply chain management system are used to keep record of database. SCM is the centralize database structure where all notes or systems are connected with each other. IBM unable us and share access IOT device data and without control of Central and manage. Chang et al. (2022) Blockchain needs real data like company info, demand, supply, material, investment, destruction, customer demand, supply material and logistics services etc. Supply chain management with real time data generate and decision-making power for new future inside.

According to Chang et al. (2022) and Raghavendra and Amalanathan (2023) Amazon Web Services is a biggest example of SCM in today's scenario. Supply chain management is a global and all manufacturer, suppliers, retailers, and logistic

---

B. Jain
Poornima University, Jaipur, Rajasthan, India
e-mail: bindiya.jain@poornima.edu.in

I. Ghosal (✉)
Brainware University, Barasat, Kolkata, India

A. Chotia
Amity University, Noida, India

P. G. S. Amila Jayarathne
Sri Lanka Institute of Information Technology (Sabatical), University of Sri, Colombo, Sri Lanka
e-mail: jayewardenepura.amilaj@sjp.ac.lk

M. Irfan et al. (eds.), *Applications of Block Chain technology and Artificial Intelligence*, Financial Mathematics and Fintech,
https://doi.org/10.1007/978-3-031-47324-1_5

companies and of course customer are connected with each other in the medium of internet. Traditional supply chain management system always needs controls, rules and regulations, always try to stop the increase of price in market. But in today's scenario track ability and transparency is needed in each stage. In supply chain management track and trace method have the ability to identify present and past location of products in shared immutable ledger (Sodhi et al. 2022; Panigrahi et al. 2023) We can track and trace complex history of products and future followers. The mean challenge is visibility gap, inaccurate supply demand prediction, manual error, and compliance. These all challenges can be overcome if you use blockchain technology and IOT device. Pal (2023) have found blockchain is used to build application for multiple parties can transaction directly by individual network with shared immutable ledgers, crypto graphically recorded transactions with no single network owner. Smart and Digital supply chain management system work with internet of things to automatically tracking the condition of product, quality control, transportation method, product authenticity and ethical supply chain management.

Blockchain and supply chain management integration define by Queiroz et al. (2020) that supply chain management is a global and all manufacturer, suppliers, retailers, and logistic companies and of cause customer are connected with each other in the medium of internet. Traditional supply chain management system always needs controls, rules and regulations, always try to stop the increase of price in market. In this study researcher aims to find out the trackability and transparency is needed of each stage in blockchain technology. A Study done by Sangeetha et al. (2020) In supply chain management track and trace method have the ability to identify present and past location of products in all inventory. We can track and trace complex history of products and future followers. The mean challenge is visibility gap, inaccurate supply demand prediction, manual error, and compliance. These all challenges can be overcome if you use blockchain technology and IOT device. Blockchain is used to build application for multiple parties can transaction directly by individual network with shared immutable ledgers, cryptographically recorded transactions with no single network owner. Smart and Digital supply chain management system work with internet of things to automatically tracking the condition of product, quality control, transportations method, product authenticity and ethical supply chain management.

5.2 Theoretic Contextual

5.2.1 Supply Chain Managing

Awasthi and Kansra (2023) Blockchain is a distributed digital ledger that fully tracks transactions. It has several attributes, including subsidiarity, accuracy, security, and privacy. The method and uses of blockchain in healthcare, as well as how and where blockchain benefits the healthcare sector, are covered in the current study. According to Raghavendra and Amalanathan (2023), inventory management is essential for sustaining the effectiveness of supply chain management. Since fruits and vegetables are perishable by nature, inventory management is necessary to prevent loss and wastage due to overstocking and stock out situations.

5.2.2 Blockchain Technology

According to Chang et al. (2022) research on focusing to improve blockchain security and provide solution to find effectiveness and scalability to overcome the challenges. Cluster index and blockchain both on renewable energy to solve various challenges. These blockchain technology for supply chain management study indicate that traceability is a key characteristic. Exploring BCT's potential in environmental sciences and agriculture could be promising areas for future research. Belotti et al. (2019) and Prasad and Ghosal (2022) predicted director to consumer brands through Artificial Neural Network. Another research by Kumar and Singh (2023) found Smart contracts play a significant role in streamlining trade and business between identified and anonymous parties by removing the requirement for an intermediary to conduct the transaction and the fees and time delays that go along with it. By maintaining the encrypted records of transactions between participants, a corporation may quickly increase transparency and foster customer trust. Maintaining the effectiveness of supply chain management requires effective inventory management told by Raghavendra and Amalanathan (2023).

Reported Sodhi et al. (2022) said why emerging supply chain technologies firstly upset Block Chain, IOT, and Artificial Intelligent. The specific characteristics, profits, and tasks related with each of these technologies on the stock chain context. It may also discuss previous studies, case examples, and frameworks related to the adoption and implementation of these technologies in supply chain management. The authors establish a foundation of knowledge and theoretical understanding to investigate how managers perceive these technologies and the factors influencing their adoption experiences. This informs the subsequent survey conducted with supply chain managers, allowing for a comparison of their perceptions and priorities regarding areas, affordances, and limits crosswise different emerging skills. Sawwalakhe et al. (2023) focuses on the many applications and historical advancements of machine learning (ML) and artificial intelligence (AI) in the financial sectors, particularly from the perspective of Indian capital markets. Unique technology like AI and ML can be applied in many different fields. P. Kumar et al. (2023) The latest algorithms using AI, ML, and other cutting-edge technologies, combined with high-tech computer capacity, enable the new quantitative methodologies to offer amazing possibilities. The limitations imposed by human resources and the peculiarities of the financial industries, however, make it difficult to apply the most recent tools and processes. Pal (2023) Due to their widespread use in a variety of sectors, artificial intelligence (AI) and machine learning (ML) have integrated themselves into our daily lives. Technology called artificial intelligence enables a machine to act just like a person. A branch of artificial intelligence called machine learning enables a machine to automatically learn from the previous data and make more accurate predictions about future events. Our lives run online today, especially with the COVID-19 pandemic, and practically all of our job is done online. The research done by Sarker et al. (2022) have found techniques, applications, and research challenges towards automation, intelligent, and smart systems. Ai-based modelling: Artificial Intelligence concept is great emitter of co2 in our environment. When we discuss Internet is always useful for transform communication connectivity knowledge accessibility social and digital technology. The energy needed for a single internet search and CO2 emission raise up drastically. Emission of greenhouse gas which further lead to Global warming. The escalating the use of IoT devices with additional sensor and communication consume use amount of energy and high carbon emission the future challenge of IOT. The Kamath and Habeeb (2023) study on the changing nature of money shows the biggest changes. The globe is approval how simple it is to get high-speed internet services across all spectrums, and as a result, fiat cash is changing to virtual currency and cryptocurrencies. The trading of cryptocurrencies is largely driven by professional and institutional investors who trade in big sums of money and engage in speculation.

According to Sangeetha et al. (2020) proved that SCM system productivity and accountability is high when we bind with Block chain and IOT device technology increase trackability and trackability. Any drastic remove, fraud activities, money laundering, forget- checks, misrepresentation of goods and services are handled by 107 devices. In this paper researcher wants to explore modern methodologies of supply management system integrate with Blockchain and IOT device services. Ramachandran et al. (2018) give the ring the application of SCM with blockchain and cryptocurrency domain use to transforming data in the financial industry. Distributed ledger, large-scale IOT devices, decentralized system work on the applications of Blockchain technology. These technology changes have many challenges and opportunities to do more. In this article, we explore the blockchain's potential for IoT applications and look at the difficulties in developing blockchain-based IoT applications.

Argade et al. (2023) Understanding the function of cryptocurrencies and their impacts on the global and Indian digital financial (DiFi) ecosystem is the goal of this study. It has sparked innovative disruptive technology and hastened financial system reform. Rakhami et al. (2021) said supply chain system is a strategic business program. If it has absence of mutual trust there are many barriers of implementation blockchain Technology widely uses of cryptocurrency transaction and efficient trust in IOT domain. With all these models simplifies data sharing and reduce computational, storage and latency requirement while increase the security and visibility. Argade et al. (2023) Understanding the function of cryptocurrencies and their impacts on the global and Indian digital financial (DiFi) ecosystem is the goal of this study. It has sparked disruptive technical advancement and expedited banking system upheaval. According to N. Kulkarni et al. increased life expectancy, the shift from nuclear to single-parent families, and the absence of a strong social security system. This chapter looks at the relationship between technology's importance in financial advice with a focus on personal financial management.

Yadav et al. (2023) The study is a continuation of the ongoing debate about Bitcoin as a stock market and cryptocurrency asset diversifier. Cryptocurrencies have been examined in certain research as a diversification asset, while others have disputed this. In the area of multi-tier and multi-site trackability Agrawal et al. (2021) has done research on the production for transparency and quality insurance. This education examines and process a blockchain traceability frameworks for the contact of supply chain partners, and related networks architecture. The applicability of the development blockchain by testing two parameters. A distributed ledger can be used to store and authenticate supply chain transactions as part of the proposed system-based trust among supply chain partners. A unique opportunity, flexibility, and authority are provided by blockchain-based transactions to all parties involved in supply networks for trace-back, creature supply curtailment, and sustainability (Panigrahi et al. 2023). Current industries have a sense of place in the market and importance. The global digital uprising has an impact on supply, assembly, and procurement. Nowadays, machine learning is a prominent topic among academics and business experts. According to Verma et al. (2023), when new knowledge and technology advance, criminals use numerous cyber centres to further their cybercrime. Banks and other financial organisations are implementing artificial intelligence to reduce cybercrime and other dangers. Artificial intelligence technology provide a range of options that enable banks to increase prosperity and wealth. According to several academic works, cybercrime costs worldwide were estimated to reach $450 million.

5.2.2.1 Conceptual Diagram

How blockchain system is work with ownership? (Fig. 5.1)

5.3 Methodology

Blockchain and IOT both technologies have huge challenge for supply chain management. Because technical it is complexity, interchange, regular uncertainty and complaints are occurred in organization. Blockchain technology is immutable (Kumar and Singh 2023; Minoli and Occhiogrosso 2018; Chang et al. 2022). As we seen the example of AWS (huge supply chain management system), If other companies also take administrative control by themselves. In blockchain technology all blocks are individual and immutable. Besides that, each multi-nation company have their own supply chain management system with blockchain technology with IOT devices (Sodhi et al. 2022; Schwartz et al. 2020) Blockchains stand for cutting-edge technology that has the potential to change current business paradigms. In this situation, considering its impact on the actual world, blockchain technology is

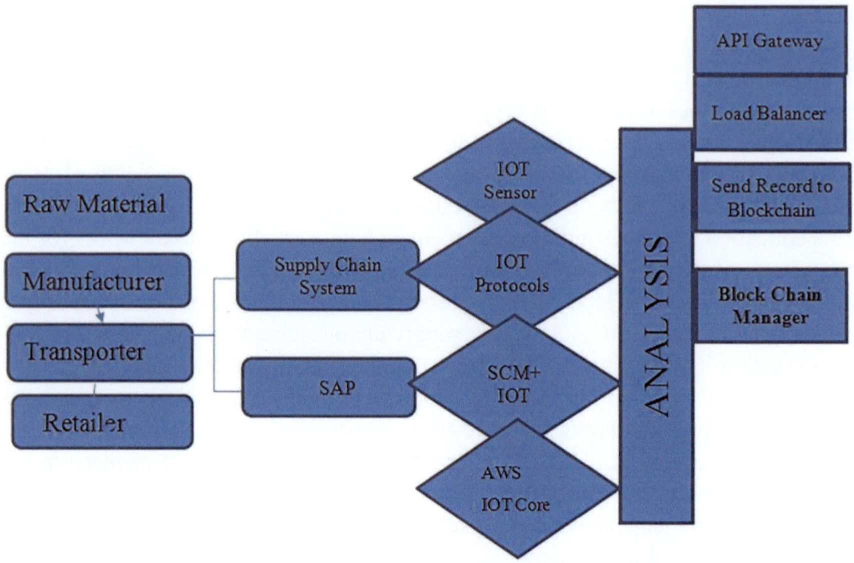

Fig. 5.1 Conceptual Diagram 1 source By author Block Chain, and IOT devices complete Supply Chain System

revolutionizing product safety because of progressions in traceability. Result, product source will advance life excellence and minor the cost of the healthcare system. Applications for blockchain are already affecting social change, environmental sustainability, and even environmental sustainability. For instance, there is reproach on this subject regarding the ecological costs of blockchain contacts connected to the waste energy used in the network. From the perspective of community transformation, blockchains are already changing the interactions between clients and businesses, in addition to those between businesses without a middleman to verify the transaction like with bitcoins and smart contracts (Raghavendra and Amalanathan 2023; Sodhi et al. 2022). The results showed that, despite the fact that blockchain-SCM requests are still in here start, about industries has collected knowledge these is relevant to the integration of blockchain such as the electric power industry with IoT.

Blockchain and SCM (Wang and Su 2020; Sodhi et al. 2022) Blockchain technology offers a decentralized, immutable record that can improve supply chain security, traceability, and transparency. Blockchain technology offers a decentralized, unchangeable record that can improve supply chain security, traceability, and transparency. It enables all participants in the supply chain network to have a synchronized view of transactions, making it difficult to tamper with or manipulate data. This can help address issues such as counterfeiting, product provenance, and trust among stakeholders. Smart contracts on the blockchain can automate and enforce contractual agreements, reducing administrative overhead and streamlining processes.

IoT and SCM (Pilkington 2016; Michael et al. 2018) contains involving carnal devices or objects for the internet, enabling those to collect are discussion data. In supply chain management, IoT devices can embedded in products, vehicles, warehouses, and other relevant points to monitor and track their movements, conditions, and environmental factors in real-time. This data can provide valuable insights into inventory management, logistics optimization, demand forecasting, and quality control. By combining IoT with SCM, businesses can achieve greater visibility, efficiency, and responsiveness throughout the supply chain. When combining blockchain, (SCM), and the (IOT), several methods and approaches can be employed to leverage the synergies between these technologies.

Shared immutable ledger Implementation (Michael et al. 2018; Belotti et al. 2019) Blockchain, as a form of shared immutable ledger, provides a decentralized and immutable ledger for recording transactions. Organizations can implement blockchain networks to create a shared and tamper-resistant record of supply chain data. This can be done through private, permissioned blockchains, where trusted participants have access and control over the network, or through public blockchains, where transactions are transparent to all network participants.

Smart Contracts and Automation (Yaga et al. 2019; Wang and Su 2020) Canny contracts are self-execution agreements coded on the blockchain. They can automate certain aspects of supply chain processes, such as payment settlements, inventory management, and quality control. By defining and enforcing predefined rules and conditions, smart contracts reduce the need for intermediaries and enhance the efficiency and accuracy of transactions.

IoT Sensors and Data Integration (Sangeetha et al. 2020; Sodhi et al. 2022) IoT devices equipped with sensors can collect and transmit data in real-time from various points in the supply chain. This data can include information about product place, fever, moisture, and other related parameters. Integrating IoT-generated data with blockchain allows for real-time tracking, monitoring, and verification of supply chain activities, enhancing transparency and traceability. Supply Chain Traceability and Provenance (Michael et al. 2018; Verma et al. 2023) Blockchain can be used to establish end-to-end traceability and provenance of goods during the supply chain. By recording step of the product's journey on the blockchain, including origin, manufacturing processes, transportation, and storage conditions, stakeholders can verify authenticity, identify bottlenecks, and address issues such as counterfeiting or product recalls more effectively.

Data Privacy and Access Control (Pal 2023; Panigrahi et al. 2023) Blockchain provides a transparent and immutable ledger, but not all supply chain data needs to be visible to all participants. Implementing privacy-enhancing techniques, such as encryption or zero-knowledge proofs, allows for selective sharing of information while preserving data confidentiality. Contact Blockchain technology offers a decentralized, immutable record that can improve supply chain security, traceability, and transparency. Blockchain technology offers a decentralized, unchangeable record that can improve supply chain security, traceability, and transparency. Mechanisms can also be established to ensure that only authorized parties can access specific data on the blockchain. Sarker (2022) told Interoperability and

Standards Establishing interoperability between different blockchain platforms, SCM systems, and IoT devices Schwartz et al. (2020) is crucial for seamless data exchange and collaboration across supply chain networks. Developing and adopting industry-wide standards, protocols, and APIs enable smooth integration and communication between disparate systems, enhancing data integrity and streamlining processes.

Collaborations with other organizations often start by conducting pilot projects to test and validate the feasibility and benefits of combining blockchain, SCM, and IoT in their specific supply chain context. This research can involve selected suppliers, partners, or specific segments of the supply chain. Collaborations between industry players, technology providers, and research institutions can also drive innovation and foster knowledge sharing in the application of these technologies.

It is important to note that specific methods and approaches employed will be contingent on the unique requirements, resources, goals each organization and (Michael et al. 2018; Panigrahi et al. 2023)

5.3.1 Result and Analysis

5.3.1.1 Block Chain Technology

Types of Business:

Business to Customer (Fig. 5.2)

Block chain is shared, immutable ledger which record each process and transaction, tracking, assets in a business network. These assets are tangible and intangible to tracked and trace in block chain network to reduce cast, and risk. Block chain is faster and more exact and Because it provides immediate, full, and transparent information kept on an immutable ledger, it is perfect for information delivery. Because it provides immediate, full, and transparent information kept on an immutable ledger, it is perfect for information delivery. These immutable ledgers are also accessible by permission only. Block chain network track order, payment, account, production, supply, return products, fail delivery, and disaster management are involved and all details are available each ands with full security, efficiently and great confidence. A Chain of block those are contain record of transactions and all block are connected with each other. The transaction store in block in block chain secured because of cryptography. Network participants have private key which is called personal digital signature (Ra et al. 2023; Agrawal et al. 2021) Distributed ledger technology for transaction in cryptocurrency feel with strong and storing method. Blockchain technology basically consistent, immutable and linear transactions between network. It is a transparent system and validity in the whole process of transaction.

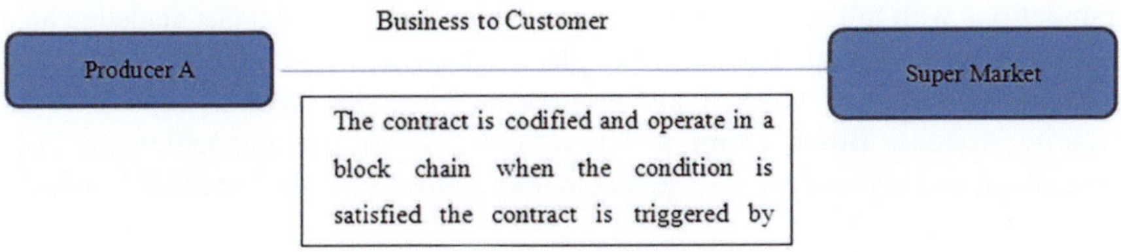

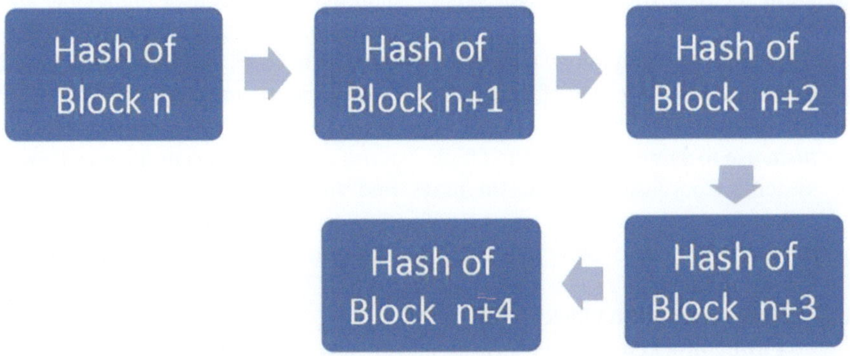

Fig. 5.2 Design of Block in Block chain. Source by own

5.3.1.2 Key Elements of Block Chain

Circulate ledger method All network and contributor can access circulated immutable record of transactions ledger. Transactions can be recorded only ones. **Immutable records** Means no changes are applicable in existing recorded transactions. If we need to change any error occurred then new transactions duty be added to reverse the error. Both the dealings are visible. **Smart contacts** They are basically called set of rules which is used in block chain to spread transactions and executed automatically. it is defined by corporate bones transfer, include terms, travel and insurance etc.

5.3.1.3 How Block Chain Works

It is a recorded as of block of data Products those are in block chain all tangible and intangible transaction can be recorded the data in block. All the information like who, what, when, where, how much, conditions, temperature, and all shipping details are maintain in block. **Each block is connected to each other** All transactions with full information are store in block and these blocks made in chain and move over from one place to other place. Block are transmitted exact time and sequence and alter of sequence and inserted between new block is impossible for security reasons. **Block chain is irreversible** Blocks are temper-evident, key strengthen, and immutable, build ledger to transactions and other members can trust.

5.3.1.4 Block Chain Characteristics

(a) **Data security** Block chain for supply chain management uses immutable and shared account those are access by authorized members. The block chain distributed ledger technology data make a structure in block and each block stored by transactions or group of transaction. Each block is connected with each other by cryptography chain. All transactions store in block with consensus mechanism use to agreement, trust and high security across the decentralized network system. Consensus mechanism plays and importance role of security by encrypting and automated group verification. Block chain network is label like public, private, and allow to participate, and permission to how we can access through network.

(b) **Private and public block chain**

Public blockchain This network allows anyone to join and participate validate transactions and find consensus mechanism.

private blockchain Private network use by confirm members access privilege only permit, and known organizations to join. The term selective endorsement means private block chain in e permissioned networks achieve consensus through a process.

5.3.1.5 Block Chain Storage

This means file create a safe copy and brake into multiple part. These parts are called **shared** and the process is call "<u>Shareding</u>". This shareding process is useful when transaction is occurred and we want to prevent data loss. These shared are all in encrypted with private key and it is impossible to view the content by any node in the network. Permission less block can access without restriction. Permission block change set of users who are granted identities to access certificate.

5.3.1.6 Vulnerability in Block Chain System

Code Manipulation code manipulation means exchange important data through code exploitation, and stolen key. BitfineX is worlds to largest dollar-based exchange of bitcoin and digital community with deep liquidity with dollar/bitcoin currency paid. Bitfine X Hong Kong based firm also in big risk because of stolen private key and personal digital signature. Employee computer hacked the laundering strategy time the theft did not send funds directly to address, own by exchange this is called employee computer hacked.

5.3.1.7 Supply Chain Management

(SCM) Supply chain management Raghavendra and Amalanathan (2023) is a network-based work where each part work independently. Means all part work independent but interlinked with chain. The production process of any good from the natural resources, transferring, storing and delivering to the finished product to customer are part of supplies chain system. All people, activities, resources, and information involved in supply chain system. It is basically management of all the part of the system. So that all activities are able to run smoothly and having a communication with each other. In each part of the systems having a full security, proper control and follow all protocols then supply chain management system create excellent benchmark in society. Managed and control supply chain management system have any benefits in terms of cost. Costing from raw material to finished goods can be possible very high, if we are controlling them. In supply chain management raw material cost, then intermediate product, storage, transport, security of products, disaster management, finish goods warehouse, packing and packaging all the handle by independently. Proper manage in supply chain management systems also helpful for less expense of warehouse. Quick delivery of product from manufacturing unit to retailer are help to reduce that cost. Better time management and analysis of previous data and time is also cost saving method. We can analysis and predict the future use of products. Our previous experience also helpful to make cost reduce. Supply chain management case the term

"Right time, right product to right customer".

Supply chain management system can split into three different categories product, information and finance (Fig. 5.3). Production flow, time, products deliver and quality assurance is a part of supply chain system in every business. The flow of information of sending and receiving sales and purchase order, update delivery status, return notes and products are also needed to update timely and securely. The

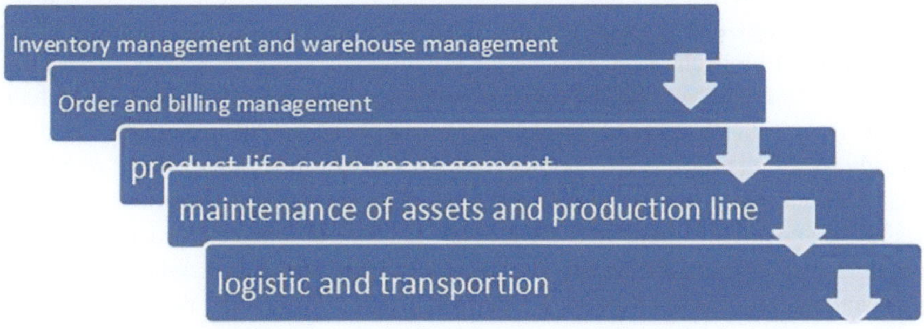

Fig. 5.3 Mechanism of supply chain management system. Source by author

finance flow like payment, billing, credit notes, debit note, purchase return and sales returns notes are updated in time necessary to proper management. Supply management software is tool to help simplify and improve the accuracy in each activity. This supply chain software keep organized and efficiency. It helped bring down the cost in term of production, purchasing raw and other accessories. Minimized risk and improve customer service.

5.4 Major Findings

Blockchain and IOT both technologies have huge challenge for supply chain management. Because technical it is complexity, interchange, regular uncertainty and complaints are occurred in organization. Blockchain technology is immutable. As we seen the example of AWS (huge supply chain management system), If other companies also take administrative control by themselves. In blockchain technology all blocks are individual and immutable. Besides that, each multi-nation company have their own supply chain management system with blockchain technology with IOT devices. These concepts work very fast and all vulnerability of blockchain system is stop.

5.4.1 Blockchain Technology and (SCM) Supply Chain Connections

Utilising blockchain technology in an SCM environment is anticipated to result in disruptive developments for many business sectors. Traditional relationship structures are thus already altering, mostly due to the transactions. According to the outcomes so far, it is realistic to assume that the adoption of blockchain-SCM integration will grow in both theoretical and real-world settings. According to the study's findings, there are industries with a lot of expertise integrating systems, including the electric power industry, intelligent transportation systems, healthcare systems, and Internet of Things applications. These sectors' use cases for blockchain technology can serve as models for other industries. In this literature on the integration of blockchain-SCM developing nations is seriously lacking. Decision-makers, consultants, and practitioners interested in learning more about blockchain applications in the SCM field. These context, supply chains all around the world are already experiencing the effects of various paradigm-disrupting effects of blockchains. Blockchains are changing the supply chain tracking enhancement, real-time visibility and greatly benefit from it. The addition of blockchain technology to decentralized operations, as seen in the electric power sector, where consumer transactions are handled through smart contracts, eliminates the need for an intermediary to govern them.

This study has significant management implications, decision-makers, consultants, and practitioners interested in learning more about blockchain applications in the SCM field. Supply chains all around the world are already experiencing the effects of various pattern-disrupting effects of blockchains. Blockchains are changing the supply chain tracking enhancement follow-up principles. Real-time visibility is therefore something that SCM may greatly benefits. The addition of blockchain technology to decentralized operations, as seen in the electric power sector, where business to consumer transactions is handled through smart contracts, eliminates the need for an intermediary to govern them. This issue is a significant interruption to this established industry that could spread to others. According to a methodological observing platform, due to the novelty, the result, techniques reporting operations research and analytical modelling like mathematical optimization, simulation, decision analysis, queueing theory, and multiple-criteria decision analysis, among others represent pertinent fields and opportunities for further research on the integration of blockchain and SCM. Surveys can be used as a tool to acquire relevant understanding advancements.

The researcher work in the blockchain-SCM interaction within well-established SCM sectors, like distribution, product traceability, intelligent transportation systems etc. In traditional SCM This study made significant contributions by summarizing the existing blockchain-SCM classic issues.

5.4.1.1 Implications of the study

The convergence of blockchain, SCM, and IoT is likely to bring several benefits in the future such as **Enhanced Transparency** Blockchain's transparent and auditable nature, coupled with IoT's real-time data, can provide end-to-end visibility into supply chain operations, allowing consumers and businesses to verify the origin, authenticity, and quality of products. **Improved Efficiency** Smart contracts and automation enabled by blockchain can streamline supply chain processes, reducing paperwork, intermediaries, and manual errors. IoT devices can optimize inventory levels, predict maintenance needs, and enable just-in-time production. **Trust and Collaboration** Blockchain's distributed ledger fosters trust among supply chain participants, as all parties can access and contribute to a shared, tamper-resistant record of transactions. This can promote collaboration and cooperation among stakeholders. **Proactive Risk Management** IoT sensors can detect anomalies, environmental changes, or potential disruptions in real-time, allowing proactive risk mitigation strategies. Blockchain's immutable record can ensure the accuracy and integrity of data related to compliance, certifications, and regulatory requirements. **Sustainability and Ethical Sourcing** The combination of blockchain, SCM, and IoT can support sustainable and ethical supply chains by enabling the tracking of materials, certifications, and responsible sourcing practices. This transparency empowers consumers to make informed choices and encourages companies to adopt more responsible practices.

Overall, the integration of blockchain, SCM, and IoT holds the potential to revolutionize supply chain operations, making them more transparent, efficient, and resilient. However, widespread adoption and addressing challenges like scalability, interoperability, and data privacy will be key for realizing the full potential of these technologies in the future. **Interoperability** In the future, efforts to establish interoperability standards and protocols will likely increase. This will enable seamless integration and communication between different blockchain platforms, IoT devices, and supply chain systems, promoting widespread adoption and collaboration. **Scalability** As the adoption of blockchain, SCM, and IoT grows, addressing scalability concerns will be crucial. Developing solutions that can handle the increasing volume of transactions, data, and devices while maintaining performance and efficiency will be a key focus. **Data Analytics and AI Integration** combination of blockchain, SCM, and IoT can generate massive amounts of data. In the future, advancements in data analytics and artificial intelligence (AI) will play a significant role in extracting actionable insights from this data. AI algorithms can identify patterns, optimize supply chain operations, and enable predictive analytics for enhanced decision-making. **Sustainability and Circular Economy** integration of blockchain, SCM, and IoT can contribute to the development of sustainable source chains and the circular economy. By tracking and verifying the origin, composition, and lifecycle of products, these technologies can support initiatives such as recycling, waste reduction, and responsible sourcing.

5.5 Conclusion

The combination of these technologies' suggestions several benefits. It improves transparency by providing end-to-end distinguishability and proof of product origin, authenticity, and quality. It improves efficiency by industrializing processes, optimizing inventory management, and enabling predictive analytics. It fosters trust and collaboration among supply chain participants through the use of distributed ledgers. Additionally, it supports sustainability initiatives by tracking responsible sourcing practices and promoting the round cheap.

In mixture of blockchain, source chain management (SCM), and the IOT holds significant promise of the future the supply chains. These technologies together, organizations can achieve enhanced transparency, traceability, efficiency, and trust throughout their supply chain operations. While there are tasks to address, such as adoption barriers, technical complexities, and data privacy concerns, ongoing research, technological advancements, and relationship among stakeholders can help overcome these limitations. Blockchain provides a decentralized and immutable ledger that enables secure and transparent recording of transactions. Smart contracts automate and enforce contractual agreements, reducing administrative overhead and streamlining processes. The IoT, with its network of connected devices and sensors, allows real-time data collection and monitoring at various points in the source chain.

In summary, a convergence of Blockchain, SCM, and IoT has the latent to revolutionize supply chain operations, making them more transparent, efficient, and resilient in the future. Organizations that hold and harness these technologies stand to gain a competitive edge in an increasingly interconnected and data-driven global marketplace.

5.5.1 *Limitations and Future Scope*

Adoption Challenges while the potential benefits are significant, widespread adoption of blockchain, SCM, and IoT in supply chain operations may face challenges. Industries and organizations will need to overcome barriers such as initial investment costs, technological complexity, resistance to change, and the need for collaboration among stakeholders. **Technical Complexity i**mplementing and managing blockchain, SCM, and IoT systems can be technically complex. Issues related to data integration, system compatibility, security, and privacy need to be carefully addressed to ensure seamless operation and protect sensitive information. **Data Privacy and Security t**he use of IoT devices and the storage of supply chain data on a distributed ledger raise concerns about data privacy and security. Ensuring robust encryption, access control mechanisms, and data governance frameworks will be essential to mitigate risks and maintain stakeholder trust. **Regulatory and Legal Frameworks t**he intersection of blockchain, SCM, and IoT may require new regulatory frameworks to address legal aspects related to data ownership, liability, and compliance. Administrations and regulatory bodies need to stay updated with technological advancements to establish appropriate policies and standards. **Infrastructure and Connectivity.** For the effective implementation of IoT devices and blockchain networks, reliable infrastructure and connectivity are crucial. In regions with limited access to internet connectivity or reliable power supply, deploying and maintaining these technologies can be challenging.

While these limitations exist, ongoing research, technological advancements, and collaborative efforts among industry stakeholders can help overcome them and unlock the full potential of Blockchain, SCM, IOT in the future of source chains.

References

Agrawal TK, Kumar V, Pal R, Wang L, Chen Y (2021) Blockchain-based framework for supply chain traceability: a case example of textile and clothing industry. Comput Ind Eng 154:107130

Argade NU, Shukla P, Singh MP (2023) Emergence of crypto currency and the digital financial landscape: a critique. In: Advanced machine learning algorithms for complex financial applications, pp 132–155

Awasthi V, Kansra P (2023) Blockchain in the healthcare industry: process and applications. In: Advanced machine learning algorithms for complex financial applications. IGI Global, pp 81–93. Blockchain was first conceptualized in 2008

Belotti M, Božić N, Pujolle G, Secci S (2019) A vademecum on blockchain technologies: when, which, and how. IEEE Commun Surv Tutor 21(4):3796–3838
Chang A, El-Rayes N, Shi J (2022) Blockchain technology for supply chain management: a comprehensive review. FinTech 1(2):191–205
Kamath RR, Habeeb M (2023) Cryptocurrency and the Indian Monetary System. In: Advanced machine learning algorithms for complex financial applications. IGI Global, pp 31–42
Kumar S, Singh R (2023) Blockchain and smart contracts for secure and sustainable development. In: Advanced machine learning algorithms for complex financial applications, pp 18–30
Michael J, Cohn ALAN, Butcher JR (2018) Blockchain technology. The Journal 1(7):1–11
Minoli D, Occhiogrosso B (2018) Blockchain mechanisms for IoT security. Internet Things 1:1–13
Pal T (2023) The exploratory study of machine learning on applications, challenges, and uses in the financial sector. In: Advanced machine learning algorithms for complex financial applications, pp 156–165
Panigrahi RR, Dash M, Shaikh ZH, Irfan M (2023) Review of machine learning techniques in the supply chain management of Indian industry: a future research agenda. In: Advanced machine learning algorithms for complex financial applications, pp 199–219
Pilkington M (2016) Blockchain technology: principles and applications. In: Research handbook on digital transformations. Edward Elgar Publishing, pp 225–253
Prasad B, Ghosal I (2022) Forecasting buying intention through artificial neural network: an algorithmic solution on direct-to-consumer brands. FIIB Bus Rev 11(4):405–421
Queiroz MM, Telles R, Bonilla SH (2020) Blockchain and supply chain management integration: a systematic review of the literature. Supply Chain Manag 25(2):241–254
Ra PK, Kumar S, Singh V (2023) Data modeling in finance challenges. In: Advanced machine learning algorithms for complex financial applications, pp 183–198
Raghavendra AN, Amalanathan S (2023) Role of AI in the Inventory Management of Agri-Fresh Produce at HOPCOMS. In: Advanced machine learning algorithms for complex financial applications, pp 115–131
Sangeetha AS, Shunmugan S, Murugan G (2020, October) Blockchain for IoT enabled supply chain management - a systematic review. In: 2020 fourth international conference on I-SMAC (IoT in social, mobile, analytics and cloud) (I-SMAC). IEEE, pp 48–52
Sarker IH (2022) Techniques, applications, and research challenges towards automation, intelligent, and smart systems. Ai-based modelling. SN Comput Sci 3(2):158
Sawwalakhe R, Arora S, Singh TP (2023) Opportunities and challenges for artificial intelligence and machine learning applications in the finance sector. In: Advanced machine learning algorithms for complex financial applications, pp 1–17
Schwartz J et al (2020) Green artificial intelligence. Commun ACM 63(12):54–63
Sodhi MS, Seyedghorban Z, Tahernejad H, Samson D (2022) Why emerging supply chain technologies initially disappoint: blockchain, IoT, and AI. Prod Oper Manag 31(6):2517–2537
Verma D, Kansra P, Kaur S (2023) A theoretical perspective of artificial intelligence in hostility of cyber threats in the banking sector. In: Advanced machine learning algorithms for complex financial applications, pp 43–54
Wang Q, Su M (2020) Integrating blockchain technology into the energy sector—from theory of blockchain to research and application of energy blockchain. Comput Sci Rev 37:100275
Yadav MP, Sharma S, Parmar B (2023) Does the cryptocurrency index provide diversification opportunities with MSCI World Index and MSCI Emerging Markets Index Cryptocurrency and Portfolio Diversificaiton. In: Advanced machine learning algorithms for complex financial applications, pp 94–114
Yaga D, Mell P, Roby N, Scarfone K (2019) Blockchain technology overview. arXiv preprint arXiv:1906.11078

Chapter 6
Artificial Intelligence and Blockchain Technology in Banking: Bibliometric Analysis

Monika Sirothiya, Nitendra Tiwari, Parvez A. Khan, and Raditya Sukmana

6.1 Introduction

In the wake of the Industrial Revolution, new financial technologies emerged, and these are known collectively as FinTech. It's just an integration of technical developments and banking services. Banking 1.0 is rooted in the past and relies on tried-and-true banking practices, while Banking 4.0 utilizes cutting-edge technology across all departments of a bank, including artificial intelligence. To keep up with the times and their rivals, banks have been adopting innovative technological solutions. For instance, in the 1960s, customers of Barclays Bank were among the first to use ATMs, a product of Banking 2.0. Banking 4.0 refers to the era of significant technological advancements in artificial intelligence that have allowed for cheaper data processing, storage, and speedier networking from 2017 onward. The age of Banking 4.0 is infused with machine learning, automation technologies, the Internet of Things (IoT), artificial intelligence (AI), and blockchain, etc., because of the unrelenting product and process innovation and transformation. A recent study estimates that by 2022, AI will bring at least USD 1 trillion per year in value to the international banking sector (Consultants 2022). At a CAGR (compound annual

M. Sirothiya (✉)
Research Rescue Consultancy, Bhopal, Madhya Pradesh, India

N. Tiwari
TCS Bengaluru, Bengaluru, India

P. A. Khan
Department of Commerce, Sophia Evening College, Bangalore, Karnataka, India

R. Sukmana
Faculty of Economics and Business, Department of Islamic Economics, Universitas Airlangga, Surabaya, Indonesia
e-mail: raditya-s@feb.unair.ac.id

M. Irfan et al. (eds.), *Applications of Block Chain technology and Artificial Intelligence*, Financial Mathematics and Fintech,
https://doi.org/10.1007/978-3-031-47324-1_6

growth rate) of 6%, the global financial services market is forecast to rise to USD 28.529 trillion by 2025–2030. This is largely attributable to the widespread implementation of AI in the reorganization of banking operations, especially in the wake of COVID-19 (Ross 2015).

Artificial Intelligence refers to the application of advanced technologies and algorithms that enable computers and systems to perform tasks typically requiring human intelligence. In the context of banking, AI involves the use of machine learning, natural language processing, and other AI techniques to analyze vast amounts of data, make predictions, automate processes, and provide personalized services. Business Insider found that 80% of financial institutions are aware of AI's potential benefits. AI is being rapidly adopted by the banking and financial sector because of the benefits it may bring in terms of efficiency, customer service, productivity, and operational expenses (Singh 2023). Robotics, economic modelling, finite element analysis, modelling of interstate conflict, and missing data estimation are just a few of the areas where these techniques have been successfully implemented (Panigrahi et al. 2023). By leveraging AI, banks aim to enhance customer experiences, streamline operations, improve efficiency, and make data-driven decisions, ultimately transforming the way banking services are delivered.

Blockchain is another technology that has altered the face of banking in addition to AI. This term is used to describe a digital ledger that is not controlled by any one entity (Ozili 2019). Transactions and other activity are recorded chronologically and made public. Blockchain was designed to be a distributed, secure, immutable ledger for keeping track of financial transactions with no single point of failure. Blockchain technology has evolved from its inception in 1991 as a research project (Investopedia 2021). This evolution has resulted in the creation of cryptocurrency (i.e., digital, or virtual money), Decentralised Finance (DeFi), non-fungible tokens (NFTs), and smart contracts. Blockchains have the potential to revolutionize the financial industry by enabling decentralized, more efficient, transparent, and secure systems. Consequently, fundamentally altering the way banks operate (Bunea et al. 2016). Due to improved data quality, transparency, and internal controls, a recent analysis by Accenture (2017) predicted that blockchain-based systems might deliver a possible 70% savings on central financial reporting.

The strength of these two technologies may decide the extent and breadth of Banking 4.0. The discussion of the compatibility of AI with BT requires a firm grounding in the nature of both technologies. Instead of relying on only one technology, banks may now provide superior service thanks to the synergy between AI and blockchain (Vedapradha and Ravi 2021). Christine Lagarde, (IMF's Managing Director) stressed on three themes: virtual/digital currencies, new models of financial intermediation and AI & blockchain integration are critical while framing banking policy for the year 2040. The adoption of blockchain and AI in banking is not only transforming traditional banking processes but also opening up opportunities for financial inclusion and innovation. Academic literature serves a crucial function in recording the progress, obstacles, and ramifications of integrating AI and BT within the banking industry.

The purpose of this research is to conduct a bibliometric analysis of the literature concerning AI and blockchain technology in the financial services industry. Opportunities for these technologies and the topics that have received the greatest press can be gleaned from the analysis. This study will first categorize papers by country and subject area to better understand publication patterns. Analysis of the clusters is then conducted based on the keywords and terms from the titles and abstracts of the reviewed publications.

6.2 Literature Review

6.2.1 Application of AI in Banking

Artificial intelligence will inevitably find its way into the financial sector. To eliminate potential risks to the efficiency of company operations, artificial intelligence (AI) is being used to replace human labour in the financial sectors. This AI is built entirely on the machine learning programming method. There was a decrease in fraud, more efficiency, improved dependability and accuracy, accelerated service, and fewer customer complaints at banks that implemented AI technologies (As indicated in Fig. 6.1).

Chatbots: Chatbot technology, which utilizes a database of commonly asked questions to hold natural conversations with and swiftly solve problems for customers, is one of the most unique and exciting applications of artificial intelligence. More and more financial institutions are using chatbots to interact with their customers and address their problems without any human assistance (Huang and Lee 2022).

Sentiments analysis: The primary goal of financial institutions is to design and deliver goods and services that meet the needs and preferences of their clients. Sentiment analysis technology powered by artificial intelligence can anticipate customers' feelings, opinions, and reactions across several mediums, including email, social media, and surveys. This technology is meant to foresee consumers' tastes.

Robo-advisors refer to digital financial advisory platforms that utilize algorithms to customize investment portfolios for investors with minimal or no human intervention. The AI service in question exhibits a greater degree of anthropomorphic qualities, is oriented towards providing recommendations, and is regarded as possessing a higher level of worth (Irfan et al. 2023).

Automation: Artificial intelligence (AI) technology is used in the banking industry in a variety of ways, including digital computers that can count currency precisely and quickly without human participation. This automated technology helps banks do more business every day, and it also makes counting money easier on workers and less prone to arithmetic error.

Fraud detection: Financial institutions face a higher susceptibility to fraudulent activities owing to the magnitude of their financial transactions and intricacy of their

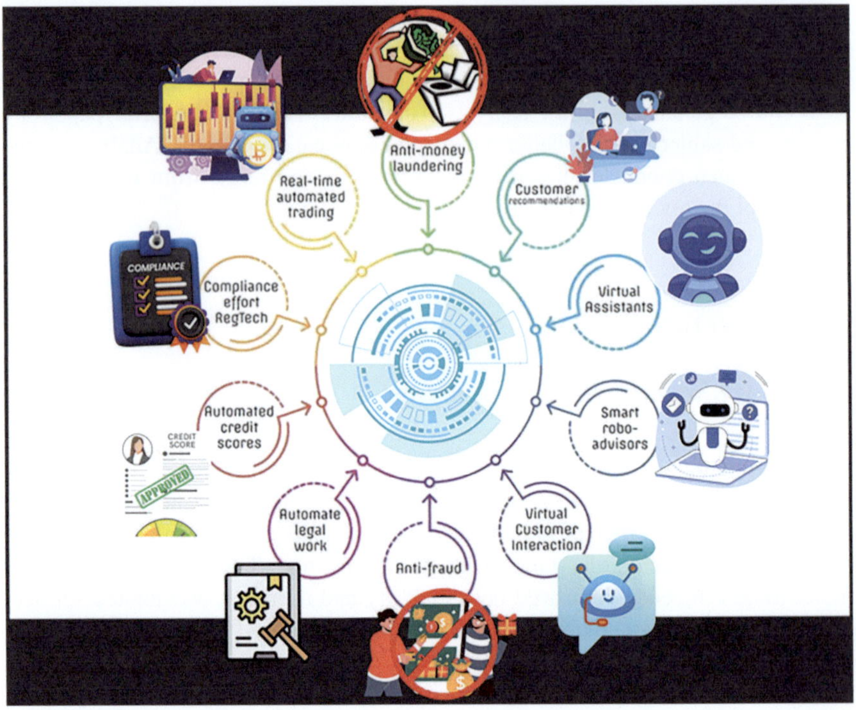

Fig. 6.1 Applications of Artificial Intelligence in Banking sector

operational procedures. Artificial intelligence employs mathematical computations and intricate algorithms to oversee the conduct of both customers and personnel through the implementation of unsupervised learning programs. Therefore, the implementation of AI technology can facilitate fraud prevention (Verma 2022).

6.2.2 Application of Blockchain in Banking

These days, blockchains are used for everything from settling financial assets to conducting economic transactions to predicting the market and providing services to businesses. Significant progress has been made in the financial markets as a result of blockchain implementation, with benefits including improved efficiency and operational performance in digital payment and settlement, derivatives and securities trading, commercial banking processes, loan management, and digital currency exchanges (Gao et al. 2018; Jayasuriya Daluwathumullagamage and Sims 2021).

Over-the-counter markets, proxy voting, automated compliance, asset rehypothecation, and contingent convertible bonds are some more examples of financial applications (McWaters et al. 2016). Blockchain applications like centralized

Fig. 6.2 Applications of Blockchain technology in Banking

administration, reporting, compliance, and company operations can reduce expenses (Accenture 2017) (Fig. 6.2).

Digital currencies/Cryptocurrencies: With enhanced monetary norms, banks can benefit from Blockchain. Banks will want to process and settle monetary trades faster and safer using digital currency. Bitcoin and Ethereum are examples of the distributed public blockchain network that provides cryptocurrency with its distinctive qualities i.e., its trustworthiness, immutability/transparency, decentralization, and cheap transaction costs.

Decentralized finance (DeFi) is utilized for the purposes of borrowing and lending. The technology's robust verification capabilities have the potential to mitigate the likelihood of loan defaults. The international bank BNP Paribas (Paris, France) is developing a blockchain platform to allow regular investors to lend money to businesses via mini bonds (Rizzo 2016).

Customer KYC: The implementation of blockchain technology can enhance the banks know-your-customer (KYC) and anti-money-laundering (AML) defenses by verifying the authenticity of potential customers and ensuring that they are not fraudulent or deceitful actors. Blockchain KYC technology can save banks \$160 M yearly (Thomson Reuter 2022). OnRamp, Bloom's newest Know Your Customer and Anti-Money Laundering (KYC and AML) compliance solution, includes features like sanction and politically exposed person (PEP) screening and ID verification.

Smart contracts are blockchain-based applications that are coded using computer programming. The contract's clauses are encoded within the procedure and are executed automatically upon the occurrence of specific contingencies, without the need for a central intermediary. The utilization of smart contracts for routine and recurring transactions such as supply and lease agreements, letters of credit, and royalties, which are uncomplicated and straightforward to manage, is a logical course of action (Cheng et al. 2018).

Clearance and Settlements Systems: The implementation of distributed ledger technology has the potential to facilitate direct settlement of transactions and enhance transaction tracking capabilities beyond the current standards set by SWIFT (Society for Worldwide Interbank Financial Communication). The utilization of blockchain technology enables the execution of "atomic" transactions, wherein transactions are promptly cleared and settled upon payment. The Corda platform facilitates decentralized transactions among participants, thereby eliminating the requirement for central authorities and enabling a seamless commerce experience.

Securities: Tokenizing conventional securities like equities, bonds, and alternatives, then storing them on public blockchains, is one way that blockchain technology could help create more streamlined, interoperable financial markets. Securities transactions conducted via blockchain might reduce annual global trade processing expenses by $17 billion to $24 billion (CB Insights 2022).

Fraud Reduction: Bank databases, exchanges, and clearing houses are particularly vulnerable to hacking and security risks due to their centralized nature. Blockchains' decentralized structure would inhibit such assaults, and their real-time payment execution would allow fraud detection and prevention (Lo et al. 2017; Bonneau et al. 2014). Through a trail of transactions and cryptocurrency, blockchain network transactions could help detect fraudsters. Regulators might use blockchain data to detect fraud and unlawful activity.

Decentralized autonomous organizations (DAO) are an additional significant innovation enabled by blockchain technology. A DAO is an organization without a chief executive officer, managers, employees, or office facilities. A smart contract is created and executed based on the computer code it contains.

6.2.3 AI and Blockchain: Decentralized Intelligent Banks

The amalgamation of Blockchain and AI facilitates the resolution of issues pertaining to precision, delay, concentration, safeguarding, and confidentiality. The implementation of Blockchain technology has the potential to address concerns related to privacy and reliability. Singh et al. (2020) assert that artificial intelligence possesses the capability to construct a sophisticated analytical algorithm on the Blockchain for the purpose of safeguarding against cybersecurity breaches. The implementation of blockchain technology ensures the encryption of data, rendering it highly secure and thereby facilitating its utilization in artificial intelligence systems.

Over time, artificial intelligence aids in the identification, interpretation, and acknowledgement of legal precedents and databases, promoting autonomous cooperation. The public nature of data in Blockchain and its record-keeping capabilities enable the provision of decentralized AI platforms. These platforms can offer data and computer power, while also ensuring transparency in AI decision-making processes. According to Dinh and Thai (2018), AI has the potential to facilitate the design and scalability of Blockchain, as well as automate and optimize its processes to enhance efficiency.

6.3 Data and Methodology

The current research applied Bibliometric analysis as it is a prevalent and precise technique utilized for evaluating large amounts of data (Donthu et al. 2021). The initial stage in performing bibliometric analysis and our investigation was conducted using Scopus Index, which is a widely utilized database among researchers globally. TITLE-ABS-KEY "blockchain", "artificial intelligence", and "banking" were combined in advanced search operation results into 100 documents available on Scopus database. The study has limited documents from the year 2018–2023.

The outcomes obtained have been retrieved in CSV format for the purpose of being analyzed and visualized through VOSviewer, in order to identify and examine the patterns in the bibliometric structure. The software known as "Visualization of Similarity" or "VOS" was developed to enable users to visually comprehend the relationships between various entities on a map (Van Eck and Waltman 2014).

6.4 Results and Discussions

The selected research articles are described, the sample statistics are presented, keywords are analyzed, and citations are analyzed in this section.

6.4.1 Descriptive Analysis

Total number of papers and increasing trend presented in graph 1, indicates that number of papers on blockchain and artificial intelligence are increasing. The trend shows the highest number in the year 2022. The notable surge in the quantity of publications can be attributed to the heightened interest of the Banking industry in the adoption of blockchain and artificial intelligence technology.

Displayed statistics indicate that based on category of documents published by type the maximum percentage is occupied by Conference papers (44%), followed by articles (28%). The other document type includes book chapters, reviews and books (Fig. 6.3).

6.4.2 Country Wise Analysis of Documents

The world map in Fig. 6.4 shows the top ten countries that constitute 73% of total publications. The knowledge generation in the field of blockchain and artificial intelligence in banking sector is dominated by India (28) and Russia Federation (10). United states (8), China (5), United Arab Emirates (5), United Kingdom (5), Australia, Malaysia and Turkey have 4 documents each.

6.4.3 Citation Analysis

Citation analysis is a methodology employed to ascertain the quantity of citations that an article receives from other scholarly articles (Kataria et al. 2020) (Table 6.1).

6.4.4 Bibliometric Analysis of Themes Using Keyword Analysis

When performing the co-occurrence mapping, all of the terms that aided in the full counting technique were taken into account. The analysis in the study was constrained to get a more precise outcome. A minimum of three occurrences of a keyword was applied as a limiting factor, resulting in 50 out of a total of 509 keywords (Fig. 6.5). For its two-part keyword and co-word analysis, the study amassed all the sample articles' keywords. The connections between the keywords used in each discipline and time were first shown using graphical co-word networks. Second, the

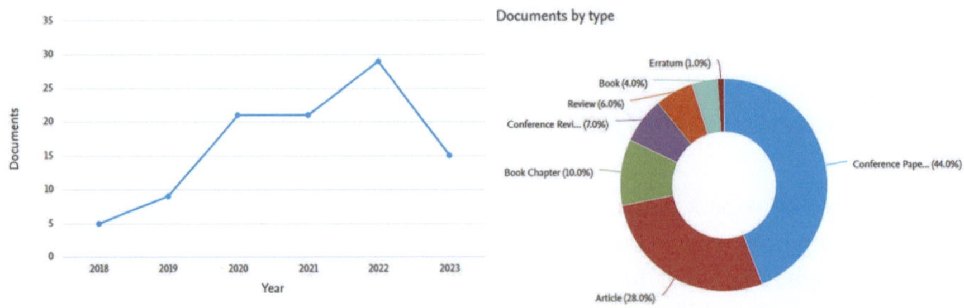

Fig. 6.3 Documents by year and type

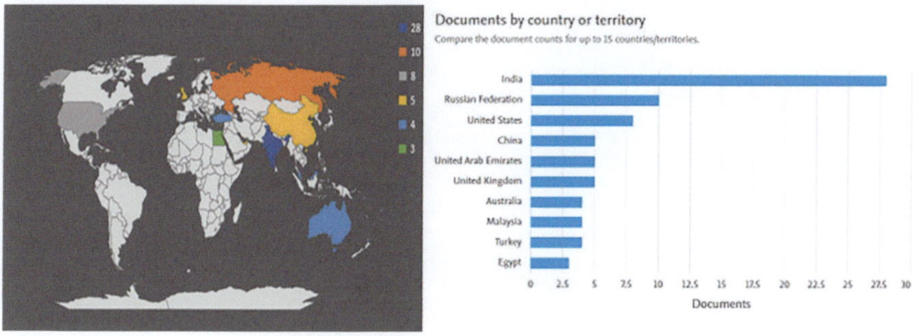

Fig. 6.4 Documents by country

Table 6.1 Top 5 most active country, sources, and authors

Subject	Count of citations
Country:	
United States	44
Iran	33
Russian Federation	28
India	25
United Arab Emirates	14
Sources:	
ACM International conference proceeding series	28
International journal of electrical and computer engineering	22
International journal of scientific and technology research	11
Journal of the knowledge economy	11
Proceedings of the 3rd international conference on intelligent sustainable systems, ICISS 2020	9
Author:	
Xu, R., Lin, X., Dong, Q., & Chen, Y. (2018)	31
Gejke, C. (2018)	22
Naimi-Sadigh, A., Asgari, T., & Rabiei, M. (2021).	14
Rabbani, M. R., & Khan, S. (2020).	11
Mehrotra, A. (2023).	10

study grouped the keywords that appeared several times into themes to better understand the shifts in research interests over time.

Within the network, individual terms are visually represented by circular shapes, with the size of each circle being proportionate to the frequency of its appearance across publications. The visualization depicts distinct clusters of terms, each denoted by a unique color. The thickness of the curved lines represents the intensity of the links between subject areas or keywords, while the length of the curved lines refers to the frequency with which terms appear. The interconnection between different topics are shown by the clusters.

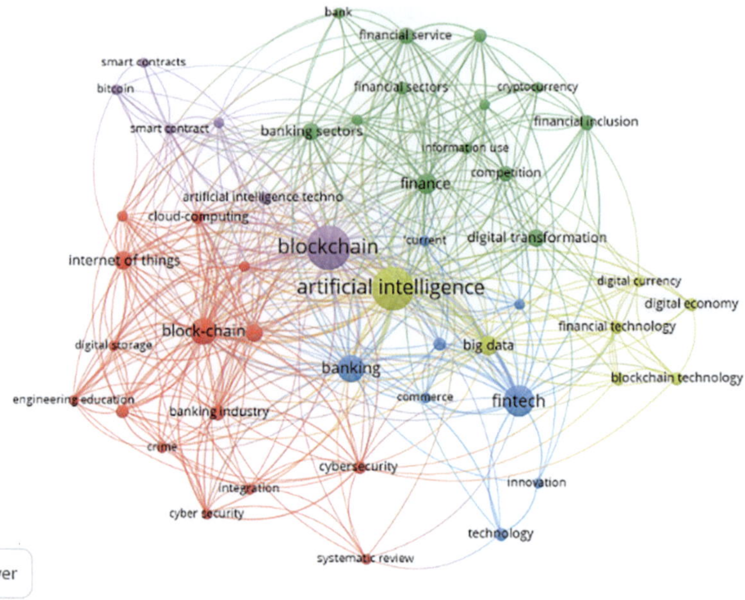

Fig. 6.5 Network visualization: circle representation

Four clusters out of five, red, yellow, purple, and blue ones, are considered for explaining in detail for the current study.

The grouping denoted by **purple** represents a conglomeration of technologies including blockchains, smart contracts, distributed ledgers, and artificial intelligence. Blockchain is a form of distributed ledger. Distributed ledgers rely on autonomous computer systems, commonly known as nodes, to document, disseminate, and harmonize transactions within their corresponding digital ledgers, as opposed to maintaining centralized data as observed in conventional ledgers. Bitcoin frequently appears as an alternative to traditional currencies. To facilitate international transactions, blockchain technology was initially created.

Following the advent of Ethereum, blockchain technology has evolved into a computational network that encompasses decentralized applications, smart contracts, and other features. Cryptocurrency and token systems serve as fundamental units for each platform (Pothavarjula and Sirisha 2022). Smart contracts enable the automatic execution of transactions upon the fulfillment of predetermined conditions, while also providing comprehensive control over the contractual agreements.

To automatically execute a transaction upon satisfying the predetermined requirements for the conclusion and full control of the contracts, a digital protocol for transferring information using a mathematical algorithm is used; this is what is known as a "smart contract "(Saleh and Awny 2020). Authors also discussed that convergence of artificial intelligence and blockchain technology will increase competition, digital revolution and increase the efficiencies in banking activities (Golubev et al. 2020).

The cluster represented by **blue** color; visualized keywords used by researchers are mainly revolve around innovation in banking specially Fintech. Fintech conveys the integration of technology with financial services. The authors proposed that blockchain based financial technologies overcomes the transitions issues in Fintech payment systems (including cryptocurrencies), credit markets (including peer-to-peer lending), and insurance systems (Rjoub et al. 2023).

The **yellow** cluster area focuses on artificial intelligence technology, big data, digital economy, and digital currency. Artificial intelligence (AI) technologies are primarily utilized by financial institutions to assess vulnerabilities, detect, and mitigate payment fraud, enhance anti-money laundering measures, and conduct additional administrative know-your-client verifications (Swain and Gochhait 2022). The literatures also highlight how the use of AI in data analysis helps avoid many risks when concluding transactions by considering customer characteristics like conscientiousness, law-abidingness, efficiency, emotional stability, and reliability (Kirillova et al. 2021). Moreover, AI technologies, neural networks, and other innovations help improve the skills of bank employees (Mamela et al. 2020).

Red area highlighted keywords such as cybersecurity, internet of things, cloud computing, and digital storage. The cluster deliberated on the way the amalgamation of artificial intelligence and blockchain technology is contributing to the enhancement of security, efficiency, and productivity of banking applications. Cybercrime refers to the act of gaining unauthorized access to computer networks with the intention of obtaining data, causing damage to the operating system, hardware, or programs through an attack. The Internet of Things (IoT) refers to a systematic arrangement of interconnected devices that are susceptible to malware attacks and other forms of security breaches as a result of their interconnectivity to the Internet. The utilization of Artificial Intelligence (AI) in countering Advanced Persistent Threats (APTs) is a pivotal aspect of safeguarding data against various forms of disruptive activities. Sridevi and Kumar et al. (2019) have reported that artificial intelligence is utilized for detecting errors in both network and application systems. The banking industry worldwide is confronted with a significant threat in the form of credit card fraud. Cloud computing enables banks to reduce their infrastructure expenses and securely store their data (Btoush et al. (2021).

Figure 6.6 displays the breadth and depth of investigation into the subject at hand. The greater the concentration of colors, the greater the number of studies being conducted on the subject. There is a lot of talk about blockchains, AI, Fintech, the IoT, banking, finance, big data, smart contracts, digital storage, and cybersecurity.

Each keyword is analyzed by the software, which counts links, total link strengths, and keyword co-occurrences. The number of occurrences represents the total number of times the search term appears in the articles. Table 6.2 displays the most often occurring keywords.

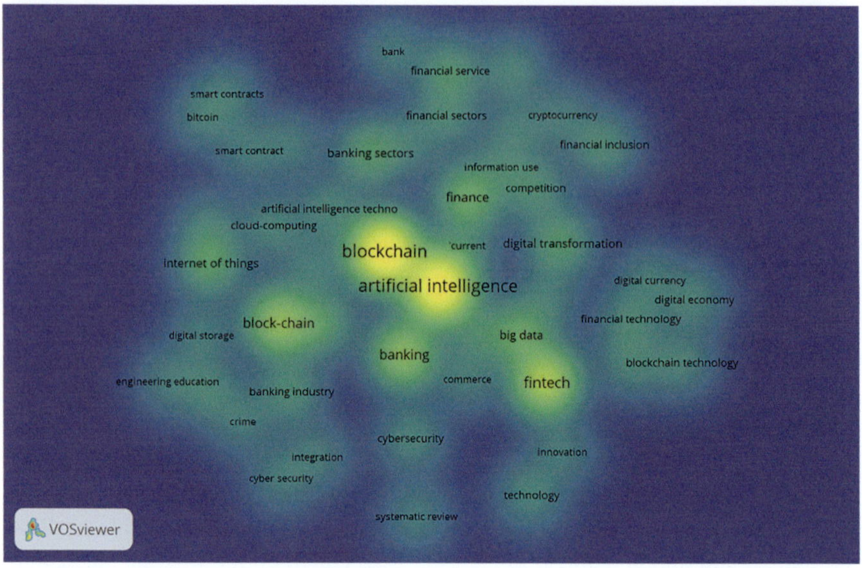

Fig. 6.6 Density visualization

Table 6.2 Most occurred keywords

Keywords	Cluster number	Links	Total link strength	Occurrences
Blockchain	5 (purple)	43	170	43
Artificial intelligence	3 (yellow)	43	168	42
Fintech	4 (blue)	23	75	22
Banking	4 (blue)	31	80	16
Internet of things	1 (red)	25	49	8

6.5 Discussion

The global market for blockchain in banking and financial services is estimated to be worth USD 87470.03 million by 2030 (Vantage 2023) from an initial value of USD 3305.67 million in 2022. The expansion of the blockchain technology market is being propelled by the increasing adoption of distributed ledger technologies (DLT) in the banking and financial services industry for transparency and accountability in financial transactions. Xu et al. (2020) propose a Blockchain Integrated Time banking (BIT) system as a means of protecting a decentralized community exchange economy. To improve the scalability, efficiency, and security of blockchain networks, it is necessary to synchronize blockchain with AI's data analysis, pattern recognition, and automation capabilities (Saha 2023).

The adoption of blockchain and AI technologies in the banking sector is revolutionizing the industry by offering enhanced security, efficiency, and transparency. Blockchain, a decentralized and immutable ledger, provides a secure and tamper-proof platform for recording transactions and verifying data integrity. When combined with AI, it enables banks to automate processes, streamline operations, and reduce costs. AI algorithms can analyze the vast amount of data stored on the blockchain, extracting valuable insights for risk assessment, fraud detection, and customer behavior analysis. Moreover, the integration of blockchain and AI promotes faster and more secure cross-border transactions, eliminates intermediaries, and enhances regulatory compliance. Overall, the synergy between blockchain and AI is reshaping the banking landscape, making it more resilient, customer-centric, and technologically advanced. Financial risks (including credit, market, and liquidity) and non-financial risks (such cybersecurity, fraud, and financial crime) are all being managed by banks with the use of these technologies.

Clearly, current applications of AI and blockchain in banking and finance are simply the tip of the iceberg, and soon the power of technology to give better experiences, cut costs, reduce risks, enhance revenues, and possibly even proceed to more accurate forecasts will become a reality. It's important to note that while AI and blockchain offer numerous benefits to the banking sector, there are also challenges and considerations, such as data privacy, regulatory compliance, scalability, and integration with existing systems. Banks must carefully evaluate and implement these technologies while adhering to relevant regulations and ensuring the security of customer data.

6.6 Conclusion

This research gives a bibliometric examination of the literature on blockchain and AI in the financial services industry. Consequences for regulatory compliance, fraud detection, and customer credit scores all improve when blockchain technology and AI are combined. Banking sectors are adopting these technologies for security, digital storage, fraud prevention, decentralization, fast payment, risk management and reducing infrastructure cost. Along with this Banking 5.0, brings a cultural transition to customers, the banking industry has been shifting from people-centric to customer-centric for years. It is required for financial institutions and Banks to apply innovative technologies in their business models and more attention to default preventions for providing good experience to customers.

The present research employs a bibliometric analysis to investigate the literature on the topic of "blockchain and artificial intelligence." This study utilized the Scopus database to analyze scholarly literature pertaining to the intersection of blockchain and artificial intelligence within the Banking sector. The timeframe for the analysis spanned from 2018 to 2023. Future research endeavors may employ diverse data sources, including SpringerLink and the Web of Science. The study analyzed 100 documents to identify trends in publications, document types, major countries, authors based on citations, and themes related to blockchain and artificial intelligence in banking. This was accomplished through the use of keyword network map analysis, facilitated by Vos Viewer software. Future research endeavors may be broadened by venturing into alternative domains or examining diverse facets through the utilization of varied keywords.

References

Accenture (2017) Banking on blockchain, a value analysis for investment banks; Technical report; Accenture Consulting, Dublin

Bonneau J, Narayanan A, Miller A, Clark J, Kroll JA, Felten EW (2014) Mixcoin: anonymity for bitcoin with accountable mixes. Springer, Berlin and Heidelberg, pp 486–504

Btoush E, Zhou X, Gururaian R, Chan KC, Tao X (2021, October) A survey on credit card fraud detection techniques in banking industry for cyber security. In: 2021 8th international conference on behavioral and social computing (BESC). IEEE, pp 1–7

Bunea A, Otoiu C, Roşu GC (2016) Blockchain: practical implications for accounting, auditing, and corporate governance. Audit Financ 14(142):690–696

CB Insights (2022) How blockchain is disrupting banking services. CB Insights. https://www.cbinsights.com/research/blockchain-disruptingbanking/#:~:text=Blockchain%20tech%20removes%20the%20middleman,in%20global%20trade%20processing%20costs

Cheng JC, Lee NY, Chi C, Chen YH (2018, April) Blockchain and smart contract for digital certificate. In: 2018 IEEE international conference on applied system invention (ICASI). IEEE, pp 1046–1051

Consultants M (2022) Benefits of artificial intelligence in the banking sector. Millinium Consultants, Kuala Lumpur. https://www.millenniumci.com/benefits-of-artificial-intelligence-in-the-banking-sector. Accessed on 11 September 2022

Dinh TN, Thai MT (2018) AI and blockchain: a disruptive integration. Computer 51(9):48–53. https://doi.org/10.1109/MC.2018.3620971

Donthu N, Kumar S, Mukherjee D, Pandey N, Lim WM (2021) How to conduct a bibliometric analysis: an overview and guidelines. J Bus Res 133:285–296

Gao J, Zhang Y, Liu Z (2018) Blockchain technology in financial markets: a comprehensive review. J Financ Innov 6(2):45–56

Gejke C (2018) A new season in the risk landscape: Connecting the advancement in technology with changes in customer behaviour to enhance the way risk is measured and managed. J Risk Manag Financ Inst 11(2):148–155

Golubev A, Ryabov O, Zolotarev A (2020, September) Digital transformation of the banking system of Russia with the introduction of blockchain and artificial intelligence technologies. In IOP conference series: materials science and engineering, vol 940(1). IOP Publishing, p 012041

Huang SY, Lee C-J (2022) Predicting continuance intention to fintech chatbot. Comput Hum Behav 129:107027. https://doi.org/10.1016/j.chb.2021.107027

Investopedia (2021) What is blockchain? [online] Available at: https://www.Investopedia.com/terms/b/blockchain.asp

Irfan M, Elhoseny M, Kassim S, Metawa N (2023) Advanced machine learning algorithms for complex financial applications [online]. Available at https://www.igi-global.com/book/advanced-machine-learning-algorithms-complex/290048#table-of-contents

Jayasuriya Daluwathumullagamage D, Sims A (2021) Fantastic beasts: blockchain based banking. J Risk Financ Manag 14(4):170

Kataria A, Kumar S, Sureka R, Gupta B (2020) Forty years of employee relations–the international journal: a bibliometric overview. Empl Relat Int J 42(6):1205–1230. https://doi.org/10.1108/ER-10-2019-0410

Kirillova EA, Bogdan VV, Blinkova EV, Zulfugarzade T, Yunusova KV (2021) The main features of the use of digital technologies in the financial and banking sector. Management

Kumar S, Tomar S, Verma D (2019) Women's financial planning for retirement. Int J Bank Mark 37(1):120–141

Lo SK, Xu X, Chiam YK, Qinghua L (2017) Evaluating suitability of applying blockchain. In: Paper presented at the IEEE International Conference on Engineering of Complex Computer Systems, Fukuoka, Japan, November 5–8, pp 158–161

Mamela TL, Sukdeo N, Mukwakungu SC (2020, August) Adapting to artificial intelligence through workforce re-skilling within the banking sector in South Africa. In: 2020 international conference on artificial intelligence, big data, computing and data communication systems (icABCD). IEEE, pp 1–9

McWaters R, Kerner S, Janoska P (2016) Blockchain technology in the insurance sector. Geneva Pap Risk Insurance-Issues Pract 41(4):579–598

Mehrotra A (2023) FinTech driven financial inclusion-the hype and the reality of missed targets. Int J Public Sect Perform Manag 11(2):165–176

Naimi-Sadigh A, Asgari T, Rabiei M (2021) Digital transformation in the value chain disruption of banking services. J Knowl Econ:1–31

Ozili PK (2019) Impact of digital finance on financial inclusion and stability. Borsa Istanbul Rev 19(3):167–176

Panigrahi RR, Dash M, Shaikh ZH, Irfan M (2023) Review of machine learning techniques in the supply chain management of indian industry: a future research agenda. In: Advanced machine learning algorithms for complex financial applications. IGI Global, pp 199–219

Pothavarjula PP, Sirisha B (2022, March) An investigation of decentralized ledger applications using Ethereum in a blockchain network. In 2022 9th International Conference on Computing for Sustainable Global Development (INDIACom). IEEE, pp 524–529

Rabbani MR, Khan S (2020) Agility and fintech is the future of Islamic finance: a study from Islamic banks in Bahrain. Int J Sci Technol Res 9(3):6955–6957

Rizzo P (2016) French Bank BNP is testing blockchain for mini-bonds. Available online: http://www.coindesk.com/french-bank-bnp-testing-blockchain-mini-bonds/. Accessed on 12 January 2017

Rjoub H, Adebayo TS, Kirikkaleli D (2023) Blockchain technology-based FinTech banking sector involvement using adaptive neuro-fuzzy-based K-nearest neighbors algorithm. Financ Innov 9(1):65

Ross S. What percentage of the global economy is comprised of the financial services sector. Investopedia 2015, 5, 2015. Available online: https://www.investopedia.com/ask/answers/030515/what-percentage-global-economy-comprised-financial-servicessector.asp. Accessed on 11 September 2022

Saha P (2023) When AI meets blockchain, revolutionizing smart contracts and digital payments. Available at https://www.financialexpress.com/business/blockchain-when-ai-meets-blockchain-revolutionising-smart-contracts-and-digital-payments-3109288/

Saleh A, Awny MM (2020) Digital transformation strategy framework. In: 29th international conference of the international association for management of technology: towards the digital world and industry X. 0, IAMOT

Singh S (2023) AI in banking – how artificial intelligence is used in banks. https://appinventiv.com/blog/ai-in-banking/

Singh S, Sharma PK, Yoon B, Shojafar M, Cho GH, Ra IH (2020) Convergence of blockchain and artificial intelligence in IoT network for the sustainable smart city. Sustain Cities Soc 63:102364. https://doi.org/10.1016/j.scs.2020.102364

Swain S, Gochhait S (2022, October) ABCD technology-AI, Blockchain, Cloud computing and Data security in Islamic banking sector. In 2022 international conference on sustainable islamic business and finance (SIBF). IEEE, pp 58–62

Thomson Reuter (2022) How KYC and AML Compliance can be improved with Blockchain Technology. Retrieved from https://www.thomsonreuters.com/en/resources/articles/2022/03/how-kyc-and-aml-compliance-can-be-improved-withblockchain-technology/

Van Eck NJ, Waltman L (2014) Visualizing bibliometric networks. In: Measuring scholarly impact: methods and practice. Springer International Publishing, Cham, pp 285–320

Vantage (2023) Global blockchain in banking and financial services market. Retrieved from https://www.vantagemarketresearch.com/industry-report/blockchain-in-banking-and-financial-services-market-1460#:~:text=Global%20Blockchain%20In%20Banking%20And,50.60%25%20over%20the%20forecast%20period

Vedapradha R, Ravi V (2021) A review on AI and blockchain based approaches for banking industry. Procedia Comput Sci 175:72–80

Verma J (2022) Application of machine learning for fraud detection—a decision support system in the insurance sector. In: Big data analytics in the insurance market. Emerald Publishing Limited, Bingley, pp 251–262

Xu R, Lin X, Dong Q, Chen Y (2018, November) Constructing trustworthy and safe communities on a blockchain-enabled social credits system. In: Proceedings of the 15th EAI international conference on mobile and ubiquitous systems: computing, networking and services, pp 449–453

Xu R, Zhai Z, Chen Y, Lum JK (2020, September) BIT: a blockchain integrated time banking system for community exchange economy. In: 2020 IEEE International Smart Cities Conference (ISC2). IEEE, pp 1–8

Chapter 7
Transforming Finance: Exploring the Potential of Decentralized Business Models Enabled by Blockchain Technology

Mini Jain, Hari Prapan Sharma, and Iqwal Thonse Hawaldar

7.1 Introduction

Increasing trading opportunities and minimizing exchange costs are frequently made possible through intermediaries. In business, intermediates frequently assist parties in finding one another, developing a relationship of trust, and completing a transaction (Jøsang et al. 2007). The ability of the parties to a transaction to connect, negotiate, and enforce a contract may be hindered in the absence of an intermediary. However, intermediaries frequently have a lot of influence over how economic transactions are structured and can use that influence to promote their interests, which raises queries regarding monopoly status (Rajan and Zingales 2001). How human cultures treat dominance in financial transactions, there is a conflict amid the need for efficient transactions and concerns over monopoly power. Large financial institutions that enable and control financial transactions are at the center of this tension in financial systems (Panigrahi et al. 2023).

Financial institutions have long been instrumental in easing the burden of exchange costs by organizing and arranging economic transactions (Braun and Gabor 2020). Financial institutions minimize exchange costs by connecting traders and encouraging confidence (Rusmita et al. 2023). As we move toward a digital economy, E-Currency, and financial technology (FinTech) are beginning to take over the role that large financial institutions used to play.

Digital technology can sometimes minimize exchange costs, expand the range of transactions, empower peer-to-peer trade, and ignite new Fin-Tech innovation waves. Although financial institutions are no longer necessary, intermediaries still

M. Jain · H. P. Sharma (✉)
Institute of Business Management, GLA University, Mathura, India

I. T. Hawaldar
Department of Accounting and Finance, Kingdom University, Riffa, Bahrain

M. Irfan et al. (eds.), *Applications of Block Chain technology and Artificial Intelligence*, Financial Mathematics and Fintech,
https://doi.org/10.1007/978-3-031-47324-1_7

exist. Frequently, one intermediary is replaced by another e.g., Financial Institution, Technology company (Avgouleas and Kiayias 2019). The progression towards decentralization and de-intermediation is gaining momentum, and blockchain-based decentralized finance is poised to be the next significant development in this evolution. A new paradigm with an emphasis on decentralization and intermediation is made possible by recent developments in blockchain technology. Enabling peer-to-peer transactions through decentralized belief, decentralized platforms, and blockchain technology can do away with the importance of middlemen in financial transactions (Javaid et al. 2021). Blockchain technology can thereby significantly rise the volume and effectiveness of peer-to-peer transactions, making business models that were before impractical conceivable. Blockchain technology will enable financial services that are more decentralized, inventive, beyond national boundaries, interoperable, and transparent.

Comparing this innovative paradigm to one based on Transaction Cost Economics (TCE) (Ahluwalia et al. 2020). TCE, on the other hand, places a strong emphasis on opportunism. However, this innovative model is built on distributed trust, which is a type of trust that "flows laterally across individuals" without prior trust associations. Due to the validity, immutability, and verifiability of transactions recorded on a blockchain, blockchain technology can foster decentralized trust. They are safeguarded by cutting-edge cryptography and validated by decentralized consensus.

Thus, the blockchain serves as a solitary source of actuality for all parties involved in a trade, improving the efficiency of transactions between buyers and sellers (Lahkani et al. 2020; Awasthi and Kansra 2023). Second, while TCE recognizes the importance of tiers and intermediaries in minimizing exchange costs, this innovative model emphasizes minimizing exchange costs through decentralization and non-intermediation. Blockchain technology can increase transaction opportunities by creatively linking peers directly to peers while lowering costs related to searches, contracts, and execution. This is made possible by decentralization and the elimination of middlemen.

Platforms and Trusts that are Decentralized blockchain technology enable innovators and Entrepreneurs to be aware of the possibility of developing a free-standing financial system with no contribution from financial institutions. This will minimize exchange costs, broaden financial inclusion, improve open access, encourage invention without restriction, and offer up innovative business prospects (Mhlanga 2020). Although the action is still in its early phases, it shows how blockchain technology can be used to create new business models that are based on decentralization and cutting out middlemen. Should this movement gain momentum, there is a possibility of disrupting existing industries and unlocking novel opportunities for innovation and entrepreneurship.

7.2 The Potential of Decentralized Finance

7.2.1 Decentralization

Within a centralized financial system, financial institutions assume critical functions as intermediaries and overseers of financial transactions (Kumar et al. 2022). Financial transactions are made more effective and seamless by intermediaries, who also help to lower transaction costs. Financial institutions, however, have the potential to expand and control economic activity as significant mediators that promote financial transactions. Due to their centralized control, financial giants such as the State Bank of India, Square Google Pay, and PayPal can amass disproportionate amounts of market power and wealth. In contrast, rather than centralized organizations, decentralized peer-to-peer networks promote financial transactions in decentralized financial systems. Through diminishing the influence of central institutions, decentralized networks have the capability to lower transaction costs and generate network effects, all while avoiding the burdensome costs associated with monopolies (Singh and Tarkar 2022). As decentralized peer-to-peer networks become more common, no one can gain sufficient monopoly power to control the network and prevent others from using it, and network effects benefit everyone. and more prospects for transactions.

7.2.2 Innovation

Decentralized finance encourages combinatorial and permissionless innovation. Open innovation and experimentation can be encouraged by centralized platforms, but the platform's owner frequently has the power to restrict access and withdraw permissions in order to enforce governance rules (Athanere and Thakur 2022). Consequently, in situations where the platform owner unilaterally alters the terms, third-party developers often face the risk of having their access to the hosting platform terminated.

The majority of platform owners are helpful and friendly to third-party developers, but sometimes they make decisions unilaterally that are detrimental to them. Decentralized platforms, on the other hand, don't have administrators, enabling free access and permissionless innovation (Hsieh and Vergne 2023). This indicates that programmers are unrestricted in their ability to design and test new apps (Javaid et al. 2022a). Decentralized platforms empower developers by providing them with assured access and enabling them to enhance decentralized finance through permissionless innovation, fostering natural and unforeseen advancements. New financial technologies will lay the groundwork for future innovation in a decentralized financial environment, allowing for new partnerships and products (Feng et al. 2022). The decentralized financial ecosystem fosters a climate of permissionless innovation and open sourcing, which facilitates the development of novel combinations and advancements. Prominent examples of decentralized financial platforms and applications like Bitcoin, Tether, Libra, and Ethereum are widely utilized, making their underlying technology accessible through permissive open-source licenses. This accessibility allows individuals to leverage the technology and create new applications, fostering a collaborative environment for novel advancements. You could create your program (Amershi et al. 2019). To prevent others from utilizing their valuable technology, centralized financial services and platforms often take good attention to the protection of their intellectual property through copyrights, patents, and trademarks. Combinatorial innovation can speed up financial innovation, increase market competitiveness, and produce newer, improved, and more affordable financial services.

7.2.3 Interoperability

Interoperability can be enhanced through decentralized finance. Due to the siloed nature of traditional finance, there are many trade restrictions. Due to the need for many financial institutions to keep their books of accounts, it may be impossible for one financial service to work with another. Transferring money between silos can be time-consuming and expensive as a result. In contrast, decentralized finance relies on open standards and public blockchains, which significantly improve the interoperability among different businesses (Sharma and Chaturvedi 2021). High interoperability enables value and financial capital to flow seamlessly across various services and countries, resulting in the creation of an Internet of Value.

While projects developed on a shared public blockchain benefit from substantial interoperability, decentralized finance still faces challenges in achieving complete interoperability due to issues stemming from blockchain incompatibility (Wang et al. 2023). Innovators and Entrepreneurs are exploring two possible options to accomplish full interoperability. One possible approach is to promote the establishment of a prominent platform and encourage all projects to adopt it. Currently, Ethereum stands out as the most widely adopted decentralized financial platform, offering extensive interoperability for all projects built upon it. 64.89% of all publicly financed projects, both inside and outside of decentralized finance, are constructed on Ethereum (Metcalfe 2020). Relying on a single blockchain may not be conducive to supporting diverse projects in the future, thus making it undesirable to rely on one dominant platform for achieving interoperability. Enhancing blockchain interoperability is a preferable course of action. As a result, projects can be developed on many blockchains and still be fully interoperable. Numerous projects, including Cosmos and Polkadot, are presently trying to integrate various blockchains for complete interoperability (Fig. 7.1).

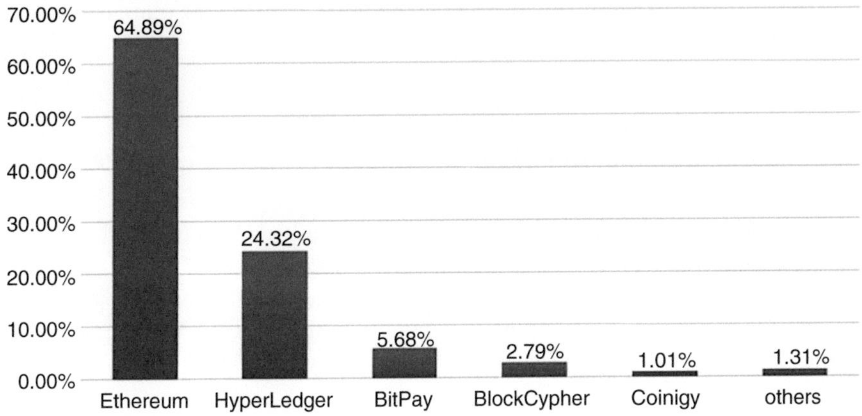

Fig. 7.1 Dominance of Ethereum in Market. Sources: https://6sense.com

7.2.4 *Border Lessness*

As it is bound to a certain geographic area and a specific fiat currency, centralized finance is finite. Cross-border capital and value movements thus frequently experience conflict and lag time. Contrarily, decentralized finance has no boundaries and is not restricted by place or fiat money, enabling transnational fundraising (Varma 2019). It is not bound to any certain geographical place and may be used by anyone anywhere in the world because it relies on a limitless number of cryptocurrencies. Furthermore, it is not controlled by any specific country's Government or central bank. Decentralized finance thus eliminates obstacles to international value transmission and enables it to be as simple as sending an email.

7.2.5 *Transparency*

The financial system becomes more transparent as a result of decentralized finance. Full transparency cannot be guaranteed by a centralized financial institution since access must be limited to safeguard the controlled ledger. In contrast, decentralized finance leverages radical transparency and decentralized consensus mechanisms to safeguard a public ledger (Singh et al. 2023). Keep documentation in a public ledger that is simple to access and examine. Decentralized finance establishes decentralized trust through the use of a public ledger, enabling trade partners to do business without the necessity for well-known associates or trustable middlemen, increasing the possible reach and size of transactions (Trump et al. 2018). Generally, Decentralized finance is developed using open-source code, so business logic may be examined by outside parties to identify biases and hidden hazards and to vouch for trading partners. can be shielded. Additionally, open-source software and transparent public ledgers support maintaining track of all earlier transactions and "getting to the bottom of every major financial crisis."

7.3 Important Business Models in Decentralized Finance

The development of new, previously unfeasible business models could be facilitated by blockchain technology. Zhang (2020) explain development of blockchain technology reduces the use of centralized finance in the present scenario. Presently, decentralized currencies are the most common model used by MNCs, Financial institutions, and individuals (Fig. 7.2).

7.3.1 Decentralized Currencies

National fiat money has been used for many years. Before a few decades ago, currencies had a central authority that managed the gold backing of the currency. These currencies were backed by precious metals like gold. However, in recent years, the gold backing of fiat currencies has disappeared (Taskinsoy 2021). The value of fiat money today is solely determined by the level of public confidence in an economy, a central bank, and a government. initial decentralized cryptocurrency, issued by decentralized technology rather than by a nation-state, is called Bitcoin. Bitcoin's supply is not controlled by any central bank so the supply of currency cannot be changed at will, in contrast to traditional currencies issued by central banks, which allows for inflation. Because of its decentralized nature, Bitcoin—often referred to as "digital gold"—is now the main store of value in the blockchain sector. Bitcoin

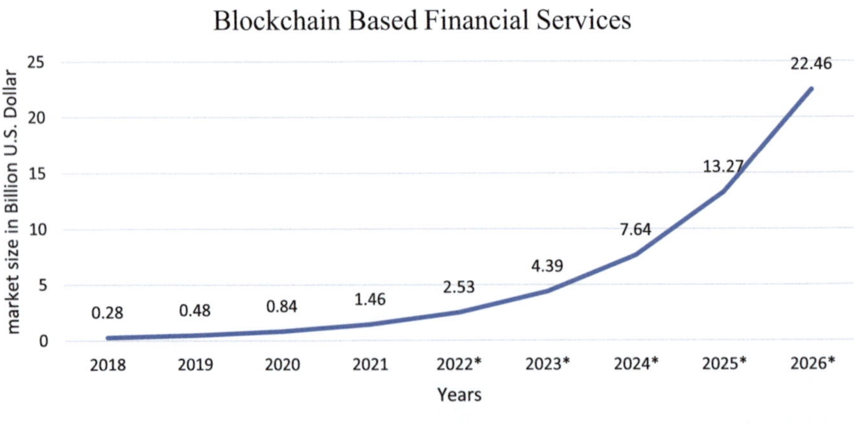

Fig. 7.2 The Emergence of Blockchain-Based Financial Services. Sources: https://www.alliedmarketresearch.com

is limitless, like gold, and may be stored and sent without the help of a centralized authority (Schar and Berentsen 2020). Other significant decentralized cryptocurrencies outside of Bitcoin include Ether, Litecoin, Monero, Dash, and Zcash.

7.3.2 Decentralized Payment Services

offline and Online commerce is made easier by centralized payment gateways like UPI, PayPal, Visa, SWIFT, and others, but their costs are often quite high, especially for cross-border payments. will cost Even if the Internet has significantly decreased the cost of many services, the cost of payment services is still expensive (Chen and Bellavitis 2020). Decentralized payment systems like Bitcoin Lightning Network and Libra aim to make cheap, immediate, international transfers possible and address the issues with conventional payment methods. The Bitcoin Lightning Network, for instance, provides immediate, irreversible, and secure payment services at reasonable costs (Mohanty et al. 2022). Merchants can dramatically save costs and increase profitability by using minimal transaction fees. Furthermore, previously underserved businesses now have affordable payment solutions at their disposal. This opens up possibilities for new business models, such as micropayments, which were previously unfeasible. Interestingly, the centralized square platform has plans to incorporate the Bitcoin Lightning Network into its payment services in the near future, offering rapid and cost-effective transactions. Additionally, Ripple, as a prominent player in this domain, collaborates with various financial institutions, including MoneyGram, to enhance the efficiency of international transfers.

7.3.3 Decentralized Funding

The fundraising process in traditional venture capital frequently involves a lot of conflicts. The funding landscape is evolving due to blockchain technology. Initial Coin Offerings are the main type of decentralized funding (ICO). A startup uses an initial coin offering (ICO) to raise money for early-stage development by creating a token tailored to the project on a public blockchain and selling it to prospective investors (Chen 2018). ICOs have arisen as a cutting-edge funding method for early-stage venture companies in recent years, allowing business owners and innovators to raise millions of dollars from international investors. A potentially effective method for projects to raise money and produce network effects is ICO. Projects are a new type of crowdfunding that may raise money from investors all across the world because of blockchain technology, smart contracts, and the openness of open-source code. Decentralized finance lowers funding friction, makes it easier to access capital, and promotes innovation and entrepreneurship by relying on the decentralized trust created by the blockchain (Allen et al. 2022). Additionally, projects can now attract stakeholders through ICOs to help drive the development of new ecosystems. When the tokens are useful only for the project's platform or product, ICOs are frequently highly valued. These tokens, also known as utility tokens, can be used as the main form of trade or to pay for particular services. Security tokens representing claims to direct ownership or cash flow may be issued by some projects. The Initial Exchange Offering (IEO), a novel variation, has lately appeared. IEOs, as opposed to ICOs, rely on cryptocurrency exchanges to verify the legitimacy of proposed projects and link top-notch companies with potential investors (Van Hijfte 2020). Cryptocurrency exchanges frequently conduct due diligence on possible projects at IEOs, offer in-depth information on projects that show promise, and lend their reputation to high-quality initiatives.

7.3.4 Decentralized Contracts

Markets, corporations, and individuals all depend on contracts since they make cooperation and transaction possible (Javaid et al. 2022b). Contracts can present challenges and incur significant expenses due to the various costs associated with their negotiation, drafting, enforcement, and renegotiation. Financial contracts, in particular, are susceptible to negative selection and moral hazard, which can lead to issues, increase transaction costs, and hinder trade opportunities.

(Momtaz 2021) In the past, financial intermediaries were frequently used by trading parties to increase confidence and lower transaction costs. Peer-to-peer financial contracts have become more prevalent in recent years as a result of smart contracts, which have started to replace financial middlemen.

"A program that automatically starts when particular criteria (such as rules of a protocol) are met" is what a smart contract is. Because of their transparency, immutability, automation, and programmability, smart contracts promise to make contracts simpler and less expensive. According to Cong and He, "Once programmable decentralized consensus is attained, contingencies become easier to converge because of their erratic, automated nature. increase efficiency, lower costs, promote innovation, and promote commerce. Smart contracts, for instance, are used by decentralized platforms like Dharma, Maker DAO, and Compound to allow decentralized borrowing and, lending which lowers costs, friction and delays in complicated operations.

7.4 Limitations of Decentralized Finance

Decentralized finance has yet to reach its full potential due to issues like volatility, regulatory uncertainty, fraud, and usability. Initially, cryptocurrencies, which are unstable and prevent stability and adoption, are frequently used as the foundation for decentralized finance (Popescu 2020). Stablecoins, whose value is frequently linked to fiat currency, have, however, found a solution to this issue. Second, regulatory uncertainty and scrutiny can stifle innovation and entrepreneurship by discouraging decentralized financing (Ferreira 2021). As an illustration, regulatory scrutiny of Facebook's ventures into cryptocurrencies and decentralized finance has been rigorous, leading several corporate partners to withdraw their support. Third, untested financial innovations and the rise of fraud are potential threats to decentralized finance (Chen and Bellavitis 2019). In order to prevent bad actors from interfering with decentralized finance, we must promote a robust environment that encourages responsible innovation.

Fourth, as opposed to market pull, decentralized finance frequently takes the technological push route. "Too many projects started with technology, tried to figure out how to get money from it, and worked from there," the Financial Times stated. 4. Friendship. Decentralized finance needs to be made more user-friendly and provide its customers with genuine value if it is to succeed in the mass market. We're all for innovation. There are many difficulties, but decentralized finance and advancements in blockchain technology can solve many of them (Hines 2020). However, there are probably some fundamental constraints that are more challenging to get around and might set a cap on what blockchain-based decentralized banking can accomplish (Ra et al. 2023) These restrictions frequently concentrate on the nature and traits of decentralized trusts and platforms. To start, it might be expensive to establish distributed trust on a decentralized platform. Decentralized networks often share information openly with all participants, validate data using distributed consensus mechanisms, and store redundant copies of information across peers to establish distributed trust utilizing Blockchain technology (Niranjanamurthy et al. 2019). How much money it costs to gather, process and store information might significantly rise when distributed trust is attained. Decentralized trust thus frequently has a high price, which can restrict its use.

Second, while decentralized platforms and decentralized trust are built on transparency, Excessive accessibility could be big a problem for privacy. The purpose of keeping and showing transaction records on open blockchains is to create decentralized trust, but these platforms can be used to threaten user privacy. Some public blockchains (like Monero and Zcash) keep a public record of all transactions while using advanced cryptography to ensure user privacy by concealing the user's identity and transaction information (Feng et al. 2019). Although this strategy can improve user privacy, it diminishes transparency and raises computing expenses, which in turn raises the cost of information processing.

Third, while smart contracts' immutability and public ledgers' increased openness and trust are positive developments, they can also result in rigidity and a lack of flexibility. Such rigidity and inflexibility are frequently inherited by decentralized finance built on blockchain technology and smart contracts, which can impede experimentation, learning, and discovery (Macrinici et al. 2018). Decentralized consensus can be used to improve smart contracts and decentralized platforms, but it can be challenging to win over key players and accomplish significant upgrades (Pal 2023).

Fourth, decentralized finance might not have adequate oversight It could be hard to determine who should be held accountable for suspected fraud in the decentralized financial environment when there is little to no engagement from a centralized authority (Beck et al. 2018). There is occasionally no one central party to turn to in challenging and controversial circumstances. There is no way for the central party to stop the transaction, fix the issue, and resume normal operations in the event of difficulty. Decentralized finance may encounter significant difficulties in the absence of adequate accountability.

Fifth, distributed reliance on information that can be collected and checked objectively is more likely to be achieved through decentralized platforms. Inputs to a decentralized trust system may not be possible for many elements of business and life since they cannot be objectively codified or made publicly available on a blockchain (Xu et al. 2018). Decentralized distributed trust systems might not make full use of all information available as a result, which would restrict their effectiveness and potential utility.

Last but not least, decentralized financial processes frequently rely more on coding rules than on human judgment. It can be advantageous to rely on coding rules since it lowers subjectivity, uncertainty, and agency costs. It may not be able to benefit from human tacit knowledge or subjective judgment, which could seriously limit the risks of decentralized finance (Roszkowska 2021). If these issues are not solved, blockchain-based decentralized finance may not be as useful as it may be.

7.5 Conclusion

Blockchain technology has the potential to lower costs, expand the scope of transaction types, facilitate peer-to-peer transactions, and introduce a novel framework for decentralized business models (Weking et al. 2020). As a result of this paradigm shift, decentralized finance has emerged as a means of utilizing blockchain technology to create a new kind of finance that is more decentralized, creative, interoperable, borderless, and transparent (Murray et al. 2023). Despite several challenges that still need to be tackled, entrepreneurs and innovators are actively exploring decentralized business models that were previously not viable without the existence of blockchain technology. These innovators are engaged in experimentation to overcome these obstacles.

References

Ahluwalia S, Mahto RV, Guerrero M (2020) Blockchain technology and startup financing: a transaction cost economics perspective. Technol Forecast Soc Change 151:119854

Allen F, Gu X, Jagtiani J (2022) Fintech, cryptocurrencies, and CBDC: financial structural transformation in China. J Int Money Financ 124:102625

Amershi S, Begel A, Bird C, DeLine R, Gall H, Kamar E, et al. (2019, May) Software engineering for machine learning: a case study. In 2019 IEEE/ACM 41st international conference on software engineering: software engineering in practice (ICSE-SEIP). IEEE, pp 291–300

Athanere S, Thakur R (2022) Blockchain based hierarchical semi-decentralized approach using IPFS for secure and efficient data sharing. J King Saud Univ Comput Inf Sci 34(4):1523–1534
Avgouleas E, Kiayias A (2019) The promise of blockchain technology for global securities and derivatives markets: the new financial ecosystem and the 'holy grail' of systemic risk containment. Eur Bus Organ Law Rev 20:81–110
Awasthi V, Kansra P (2023) Blockchain in the healthcare industry: process and applications. In: Advanced machine learning algorithms for complex financial applications. IGI Global, pp 81–93
Beck R, Müller-Bloch C, King JL (2018) Governance in the blockchain economy: A framework and research agenda. J Assoc Inf Syst 19(10):1
Braun B, Gabor D (2020) Central banking, shadow banking, and infrastructural power 1. In: The Routledge international handbook of financialization. Routledge, pp. 241–252
Chen Y (2018) Blockchain tokens and the potential democratization of entrepreneurship and innovation. Bus Horiz 61(4):567–575
Chen Y, Bellavitis C (2019) Decentralized finance: blockchain technology and the quest for an open financial system. Stevens Institute of Technology School of Business Research Paper
Chen Y, Bellavitis C (2020) Blockchain disruption and decentralized finance: the rise of decentralized business models. J Bus Ventur Insights 13:e00151
Feng Q, He D, Zeadally S, Khan MK, Kumar N (2019) A survey on privacy protection in blockchain system. J Netw Comput Appl 126:45–58
Feng S, Zhang R, Li G (2022) Environmental decentralization, digital finance and green technology innovation. Struct Chang Econ Dyn 61:70–83
Ferreira A (2021) The curious case of stablecoins-balancing risks and rewards? J Int Econ Law 24(4):755–778
Hines B (2020) Digital finance: security tokens and unlocking the real potential of blockchain. Wiley
Hsieh YY, Vergne JP (2023) The future of the web? The coordination and early-stage growth of decentralized platforms. Strateg Manag J 44(3):829–857
Javaid M, Haleem A, Singh RP, Khan S, Suman R (2021) Blockchain technology applications for Industry 4.0: a literature-based review. Blockchain Res Appl 2(4):100027
Javaid M, Haleem A, Singh RP, Suman R, Khan S (2022a) A review of blockchain technology applications for financial services. BenchCouncil transactions on benchmarks, standards and evaluations, 100073
Javaid M, Khan S, Haleem A, Rab S (2022b) Adoption of modern technologies for implementing industry 4.0: an integrated MCDM approach. Benchmarking: An International Journal
Jøsang A, Ismail R, Boyd C (2007) A survey of trust and reputation systems for online service provision. Decis Support Syst 43(2):618–644
Kumar K, Sharma H, Khan W, Kumar R (2022) Factors influencing adoption of green banking practices: evidence from commercial banks in India. J Asia Entrepr Sustain 18(1):41–57
Lahkani MJ, Wang S, Urbański M, Egorova M (2020) Sustainable B2B E-commerce and blockchain-based supply chain finance. Sustainability 12(10):3968
Macrinici D, Cartofeanu C, Gao S (2018) Smart contract applications within blockchain technology: a systematic mapping study. Telemat Inform 35(8):2337–2354
Metcalfe W (2020) Ethereum, smart contracts, DApps. Blockchain and Crypt Currency, 77
Mhlanga D (2020) Industry 4.0 in finance: the impact of artificial intelligence (ai) on digital financial inclusion. Int J Financ Stud 8(3):45
Mohanty D, Anand D, Aljahdali HM, Villar SG (2022) Blockchain interoperability: towards a sustainable payment system. Sustainability 14(2):913
Momtaz PP (2021) Entrepreneurial finance and moral hazard: evidence from token offerings. J Bus Ventur 36(5):106001
Murray A, Kim D, Combs J (2023) The promise of a decentralized internet: What is Web3 and how can firms prepare? Bus Horiz 66(2):191–202

Niranjanamurthy M, Nithya BN, Jagannatha SJCC (2019) Analysis of Blockchain technology: pros, cons and SWOT. Cluster Comput 22:14743–14757

Pal T (2023) The exploratory study of machine learning on applications, challenges, and uses in the financial sector. In: Advanced machine learning algorithms for complex financial applications. IGI Global, pp 156–165

Panigrahi RR, Dash M, Shaikh ZH, Irfan M (2023) Review of machine learning techniques in the supply chain management of Indian industry: a future research agenda. In: Advanced machine learning algorithms for complex financial applications. IGI Global, pp 199–219

Popescu AD (2020) Decentralized finance (defi)–the lego of finance. Soc Sci Educ Res Rev 7(1):321–349

Ra PK, Kumar S, Singh V (2023) Data modeling in finance challenges. In: Advanced machine learning algorithms for complex financial applications. IGI Global, pp 183–198

Rajan RG, Zingales L (2001) Financial systems, industrial structure, and growth. Oxf Rev Econ Policy 17(4):467–482

Roszkowska P (2021) Fintech in financial reporting and audit for fraud prevention and safeguarding equity investments. J Account Organ Change 17(2):164–196

Rusmita SA, An-Nafis MSA, Ramadhani I, Irfan M (2023) The effect of good corporate governance on financial distress in companies listed in sharia stock index Indonesia: machine learning approach. In: Advanced machine learning algorithms for complex financial applications. IGI Global, pp 220–251

Schar F, Berentsen A (2020) Bitcoin, Blockchain, and cryptoassets: a comprehensive introduction. MIT Press

Sharma HP, Chaturvedi A (2021, October) Adoption of smart technologies: an Indian perspective. In: 2021 5th international conference on information systems and computer networks (ISCON). IEEE, pp 1–4

Singh R, Tarkar P (2022, May) Future of work: How Artificial Intelligence will change the dynamics of work culture and influence employees work satisfaction post-covid-19. In: Proceedings of international conference on communication and artificial intelligence: ICCAI 2021. Springer Nature Singapore, Singapore, pp 239–260

Singh R, Khan S, Dsilva J, Centobelli P (2023) Blockchain integrated IoT for food supply chain: a grey based Delphi-DEMATEL approach. Appl Sci 13(2):1079

Taskinsoy J (2021) Bitcoin: a new digital gold standard in the 21st century?. Available at SSRN 3941857

Trump BD, Wells E, Trump J, Linkov I (2018) Cryptocurrency: governance for what was meant to be ungovernable. Environ Syst Decis 38:426–430

Van Hijfte S (2020) Blockchain and regulation: the legal world. Decoding blockchain for business: understand the tech and prepare for the blockchain future, pp 89–113

Varma JR (2019) Blockchain in finance. Vikalpa 44(1):1–11

Wang G, Wang Q, Chen S (2023) Exploring blockchains interoperability: a systematic survey. ACM Comput Surv

Weking J, Mandalenakis M, Hein A, Hermes S, Böhm M, Krcmar H (2020) The impact of blockchain technology on business models–a taxonomy and archetypal patterns. Electron Mark 30:285–305

Xu Q, Jin C, Rasid MFBM, Veeravalli B, Aung KMM (2018) Blockchain-based decentralized content trust for docker images. Multim Tools Appl 77:18223–18248

Zhang D (2020) The innovation research of contract farming financing mode under the block chain technology. J Clean Prod 270:122194

Chapter 8
Return Provisions Stipulated Investor Holding Periode In Islamic Banking's Share (Artificial Intelligent VS Panel Approach)

Sylva Alif Rusmita, Puji Sucia Sukmaningrum, Fadillah Mansor, and Mohammad Irfan

8.1 Introduction

Investment is an important activity to support the economy of a country. Through investment, it will make it easier for the industry to carry out company operations and expansion, so that there is an increase in company income which can later contribute to the growth of state revenue. Nowadays, the investment instruments are becoming more diverse. People are able to choose whether to invest directly or indirectly. One of the famous investment instruments is stocks, because the return from stocks is higher than others. During the Pandemic of COVID-19, many investors are starting to get restless and tend to follow and continue to observe market movements (Nugroho et al. 2021). Likewise, banking stocks are in the spotlight of investors, because during the pandemic many banks experienced default problems that made bank profits somewhat disrupted. Based on 2020 data, the change in Sharia bank profits in oil-producing countries has changed slightly so IsDB (Islamic Development Bank) took steps to provide loans to all banks in OIC countries of $ 2.3 billion to overcome the pandemic problem (Hidayat et al. 2020). This is due to a fairly massive decrease in transactions in Islamic banking. This is due to a fairly massive decrease in transactions in Islamic banking. As a result, S&P Global Market Intelligence (2023) forecast the growth of Islamic banks, especially in OIC country

S. A. Rusmita · P. S. Sukmaningrum (✉)
Faculty of Economics and Business, Airlangga University, Surabaya, Indonesia
e-mail: sylvalifr@feb.unair.ac.id; puji.sucia@feb.unair.ac.id

F. Mansor
Academy of Islamic Studies, Universiti Malaya, Kuala Lumpur, Selangor, Malaysia
e-mail: fadillah@um.edu.my

M. Irfan
Business School, NSB Academy, Bangalore, Karnataka, India

M. Irfan et al. (eds.), *Applications of Block Chain technology and Artificial Intelligence*, Financial Mathematics and Fintech,
https://doi.org/10.1007/978-3-031-47324-1_8

will rise up in 10% in 2024, which several OIC countries member such as Saudia Arabia, Qatar, Kuwait, and UAE become the majority players among others OIC members (Nagraj 2023).

The changes of profit conditions in the banking world, cannot be escaped from the investor point of view. Investors who have the freedom to choose any type of stock will anticipate loss in investment especially in the banking sector that have negative issues in Covid-19 conditions, so they may decide to sell the shares. According to Berger et al. (2019), Campbell et al. (2019), Chung and Wei (2005), Garnia et al. (2015), Ødegaard (2017), and Rashata (2022) the investor behavior could be reflected in the holding period of stock. The period of holding shares represents the risks that they confront. If an investor forecasts that the shares will be profitable, the investor will tend to hold his shares for a long period of time for diversification, with the expectation that the stock's selling price will be higher in the future. On the other side, if the company loses money and does not perform well, investors will instantly sell the shares they purchased, especially if the selling price of the shares is expected to fall; this is done to reduce the risk of loss on investment.

Based on the problems that exist in the OIC banking industry, investors must be aware of the risks that exist in unstable economic conditions. One of the impacts of this condition is that investors can accelerate to sell their Islamic banking shares and invest their money in other sectors or instruments. On the other hands another research also mention that holding period also reflect the liquidity, according to Berger et al. (2019) and Rashata (2022) stated that while stocks have low performance tend to hold longer due to difficulties in selling, in some condition where economics condition worse investor keep selling shares even though investors suffer losses in order to prevent bigger losses. Some companies, banking and insurance are invested in the capital market to maintain liquidity, therefore if the investor has a short holding period, it means that the shares can be used as liquid assets and can be easily liquidated if the company experiences financial problems.

Holding period of investors is influenced by many things. Based on previous research, there are two main factors, namely the condition of a particular company or industry and macroeconomic conditions. Previous study serves to analyze and enrich the discussion of research, and distinguish it from the research that is being done. This study included ten international journals of previous research related to the concept of Holding Period. Previous research from Campbell et al. (2019) on stock holding found that nonprofessional analysts responding to news, article publications and released holdings during the publication of articles can influence the investor holding decision in a short term. Schaub (2019) found out that the holding period between NASDAQ stocks and American Depository Receipts (ADRs) has differences whereas the NASDAQ tends to be easier to release or hold its period lower. Chung and Wei (2005) states, in China a stock that has a high spread will tend to be maintained by an investor (holding longer) with a long period of time. The other research from Hickman et al. (2001) which took place in the USA suggests that for high risk of underperformance investments tend to have a short holding time despite large yields. According to Ødegaard (2017) research in Norway, holding period is strongly influenced by the characteristics of the owner of the stock, wealthy families tend to hold the stock for a little longer, than individual investors hold the shares the longest. Cao et al. (2018) said that investment managers tend to hold less liquid stocks but as the stock price increases it gets better but remains bad in times of crisis in the USA. Other results from Garnia et al. (2015) show that stock returns play a big role in Indonesia investors' decisions to hold or divest shares they own. On the other hand, the case in Pakistan shows the fact that when economic conditions are poor investors tend to make loss-aversions or circumstances where investors think small profits are better than having to hold their funds longer even if the investment has a good chance in the future (Rashata 2022).

Based on the results of previous studies, the use of the research sample only focuses on certain countries and can only see the general picture because it uses an index consisting of many company sectors with different characteristics. More specific research, especially banking stocks, needs to be looked at further because this sector is useful as a driver of the economic sector and the capital market. Banking stocks are always known as liquid stocks and are in demand by many investors, so fluctuations in banking industry stocks need to be investigated further to find out what factors influence investors in holding banking stocks. The importance of research on holding periods, especially in OIC countries, is crucial because some OIC countries are developing countries whose investment in industry has high growth with the growth of tourism, factories, and other sectors. Therefore, many investors invest in their capital markets, especially the banking industry, which is known for maintaining liquidity. The majority of investors from OIC countries also pay attention to investment in sharia or in Islamic products such as Islamic banking because it is free from elements of uncertainty, gambling, and usury which can cause moral hazard for humans (Alrifai 2015). Therefore, researching Islamic banking stocks in the OIC is important considering that the banking sector is the main economic sector driver and from an Islamic point of view, banking stocks are also noticed by Muslim investors who are interested in the OIC country.

In this study, it presents something new, among others previous study; first is current research added cases of pandemic; before and during Covid-19 conditions that view discussed in previous study, especially in the country of OIC, which can be ascertained by many Islamic banks that go public. Second, there are inconsistencies in previous research on determinant factors that cause investors to hold shares (Chung and Wei 2005; Garnia et al. 2015; Ødegaard 2017; Rashata 2022) as well as the covid-19 cases variable that may make investors reluctant to hold banking shares because of the declining possibility of return due to the number of uncollectible receivables (Manohar and Raju 2021; Salisu et al. 2021). Third, provide robustness tests with artificial intelligence (machine learning) which have the ability to describe the most significant variable based on the historical calculation (Rusmita et al. 2023).

Given that Islamic banking stocks are one of the investment instruments considered by investors, this research needs to be done, especially when investors face uncertain macroeconomic conditions during the pandemic. This study will focus on the impact of bid-ask spreads as a repression of stock liquidity and return, as well as macroeconomic impacts measured by the exchange rate of the country's home currency against the dollar, spot gold prices on international markets, and Islamic stock period holdings before and during Covid-19 cases. The study's findings can be utilized to inform Islamic banking's corporate action strategies in the market by providing an overview of the behavior of Muslim investors in banking equities during normal times and pandemics.

8.2 Literature Review

8.2.1 Theoretical Islamic Investment

In the principle of Islamic investment, investment is an activity in instruments that are in accordance with Islamic views, namely without maisyir, gharar and riba. Therefore, investments in companies containing *ribawi* are prohibited, such as investments in conventional banks that have interest income. In addition, how to invest in the Islamic view depends on the purpose or intention of investing. In Islam, investment transactions that involve speculation and *gharar* should be avoided, such as trading without having enough purpose and insight (Alrifai 2015). In principle, when investing Islam recommends that investors really understand the trading mechanism and also the securities asset to be purchased so that the goal of the investment can be achieved without violating sharia rules (Alrifai 2015).

8.2.2 Holding Period and Liquidity

Considering that knowing the stock liquidity asset is very important for investors and companies, analyzing factors that affect the holding period is important in this study. Banking sector, banking stocks in OIC countries also play a large role in the economy of their countries because the capital market is one of their tools to maintain their liquidity, as well as showing how investors view banking stocks in OIC countries. Investors also sometimes consider other instruments to minimize risk, especially risk that is related to the market. Batten et al. (2015) explained that gold is one of the good diversification assets in the long run. Other research by Baur and McDermott (2010) suggests gold can act as a safe haven when things go badly in the market. Although the condition of gold is not always able to minimize the risk (Manohar and Raju 2021). In addition to gold, one other macroeconomic factor is that currency exchange rates are also considered by investors. The exchange rate can give an idea of the economic situation of a Country (Melvin and Prins 2015). Sgammini and Muzindutsi (2020) stated that when the exchange rate weakens, it can result in a worsening level of company liquidity. In the long run, it can have an impact on stock prices that also decline and investors choose to withdraw their investments because of liquidity risks, this is what results in a short holding period. In addition, macroeconomic factors such as covid-19 can also be considered by investors. Elton et al. (2014) explains that in bad circumstances or situations blind investors act to save their assets because it can cause losses. However, other factors such as economic capabilities, investment goals, and environmental influences are also considered by investor behavior (Ødegaard 2017).

8.2.3 Previous Study

An investor's decision to maintain, sell, or buy a stock can be influenced by economic factors derived from the company's internal affairs. Stoll (1989) explains factors such as bid-ask spreads that reflect transaction costs can be one of the causes of investors making decisions against the shares they own, the higher the bid-ask spread value means indicating expensive stock transaction costs and can have an impact on investors' period holding decisions. Another factor is the return of the stock, which is the profit that the investor gets from the difference in the current price with the previous period. The results of previous research by Garnia et al. (2015) explained that high stock returns can influence faster period holding decisions because investors prefer clear profits to be obtained and avoid the risk of losses in the future. One of the causes of the long holding period proposed by the old cause of the absence of holding periods has been raised by Garnia et al. (2015) which states that the holding period is inversely proportional to the return, if the return is high then the period holding tends to be faster because investors want to get capital gains. Meanwhile, other previous research has stated that holding periods are closely related to liquidity, return, risk and investor attention (Lyle and Wang 2015; Rusmita and Azaria 2021; Schaub 2019; Schaub and Simmons 2022).

Meanwhile, in investing activities there is an effective transaction fee arrangement and there is a careful analysis that can produce a large return, Bid-ask spread is one of the reflections of liquidity and transaction cost. Bid-ask spread is the difference between bid (buy) and ask (sell). Bid-ask spreads are a function of the three components of costs derived from inventory holding costs, processing costs (order processing costs), and adverse information costs (Stoll 1989). Research conducted by Chung and Wei (2005) states that a stock that has a high spread results in an increasingly long holding period.

Macroeconomic circumstances have an influence on investor behavior related to holding periods (H.-K. Chen and Lien 2017). Especially in other investment instruments that can replace stocks that are identical to high risk. To measure the macroeconomic state of a country. De Oliveira (2014) explained that the exchange rate can be a benchmark, where when the currency value of a country strengthens, it is an indication of a good macroeconomic situation in a country that has an impact on the high holding period. In addition, the macroeconomic situation can also be reflected in the world gold price which is an investment instrument as a means of hedging (Ciner et al. 2013). Research by Ali et al. (2020) explains that gold instruments can be an investment option when the economy is unstable, where when the stock market experiences shocks due to economic conditions, investors will turn to other investment instruments such as gold because it has a lower level of risk. This can have an impact on the low holding of the stock period because investors prefer to allocate their funds to gold instruments that are considered to have a lower risk than stocks, such as in the case of economic recession due to the Covid-19 pandemic (Salisu et al. 2021).

On the other hand, some macroeconomic variables can also know the length of time investors hold stocks (Dewanti et al. 2022). Chen and Lien (2017), said that macroeconomic risks can affect a company's return and also of course the stock market, so the holding period of the stock will change along with investors' thoughts on market conditions. Renault (2017) supports that investors have insights based on the information they receive, one of which is macroeconomics, whereas this macroeconomics will be the basis for investors to conduct Fundamental analysis. Another thing that concerns investors is how to minimize investment risk in order to obtain optimal returns. One way that can be done by investing in other instruments, such as gold (Arouri et al. 2015; Batten et al. 2015; Manohar and Raju 2021; Salisu et al. 2021).

In terms of investment, investors certainly think very much about the right instruments in investing, especially during a pandemic. Gold is a commodity that is in demand by investors, besides that gold is also used as a financial standard and is also used as foreign exchange reserves, and in various Muslim countries is used as a means of transaction. Gold is also one of the hedging or safe haven products for investors during times of crisis, therefore this gold price position can have an effect on investor stock holdings (Manohar and Raju 2021; Salisu et al. 2021). In previous research during the 1998 crisis in Indonesia, investors quibbled over stock investments and more gold. However, at normal concessions, investors will pay more attention to the possibility of earning a profit. During the crisis conditions of the Covid-19 pandemic Manohar and Raju (2021) and Salisu et al. (2021) found chaos in the stock market, where investors are trying to minimize risk either through diversification of assets or hedging. This indicates that Covid-19 has a negative impact on the stock market.

8.2.4 Hypothesis

Hypothesis is an answer to a temporary problem that is conjecture from a study (Gujarati and Porter 2009). Based on a previous study, the hypothesis was developed, so the current study analyzed the influence of bid-ask spreads, stock returns and macroeconomic factors on the holding period of Islamic banks in OIC member countries listed on stock exchanges. Based on the above description, the hypothesis in this study is:

H1: Bid-ask spread has a partial effect on the holding period of Islamic Banks in OIC member countries.

H2: Stock returns partially affect the holding period of Islamic Banks in OIC member countries.

H3: Exchange rates partially affect the holding period of Islamic Banks in OIC member countries.
H4: Gold prices partially affect the holding period of Islamic Banks in OIC member countries.
H5: Covid-19 conditions partially affect the holding period of Islamic Banks in OIC member countries.

8.3 Methodology

This study will test the bid-ask of stock spreads, stock returns, and trading volume against the holding period of shares at Islamic commercial banks in OIC member countries. The analysis used in this study is a regression with data panel and machine learning analysis by combining the classic statistics tools Eviews and R. Selection of panel data methods is used because it can identify variable mixers independent of their dependents by combining time series and cross section data types (Gujarati and Porter 2009). Research with econometric methods shows that the focus in basic and simple prediction based on research question and the panel data method with a fixed effect approach has a weakness of generalizing the research model to several latent variables (Bai 2003; Bai and Ng 2002), where this method is a factorization with the basic method in machine learning (Athey and Imbens 2019). In addition, the econometric method also has normalization assumptions on the variables to be identified.

While machine learning is used to determine the most accurate variables of the selected ML models, by comparing the lowest error values, both in traditional statistical models and ML models (Athey and Imbens 2019; Chen 2021). So, to achieve a more valid result the current studies applied ML as a robustness test. The ML method can amplify the results of the panel data to see the most influential variables, with various models, but override the direction. So that the ML method is able to strengthen the selection of variables that most appropriately affect endogenous variables, while the direction of the variables remains using the results of panel data.

The data taken in this study used access from the internet and the official website of each Islamic banking in OIC member countries, including Bangladesh, Indonesia, Qatar, Kuwait, Saudi Arabia, Egypt, and the United Arab Emirates from 2018 until 2021. The data research starts from 2018 because Islamic bank stock in Indonesia was established in 2018, as part of OIC countries. Technical purposive sampling is used to determine sharia banks that can be used as samples with the following criteria (Table 8.1):

The data in this study used daily data from May 2018 to November 2021, in addition, to prevent inequality between banks, the bank chosen is a bank that has the same market capitalization. Based on the screening of samples from 17 banks, with 5 independent variables (Bid-Ask Spread, Return, Exchange Rate, Gold Spot Price, and Covid-19 Cases) and holding periods as dependent variables. Further explanations of the research variables are described in Table 8.2:

Table 8.1 Sample screening process

Explanation	Number of Bank
Islamic Banks in OIC countries	55
Banks that own comparable stocks and market have equality in market capital	20
Data availability from May 2018-November 2021	17

Table 8.2 Variable definition

Variable	Definition
Holding Period	Data holding was calculated by the author based on the formula below: The length of time an investor spends holding the shares he owns. $Holding\ Period = \frac{Outstanding\ share}{Stock\ Trade Volume}$ Atkins and Dyl (1997) Outstanding share and stock trade volume data was collected form investing.com
Bid-Ask Spread	Bid-ask spread was calculated by the author with the formula: The difference between the highest asking price and the lowest bid price. $Bid\ Ask\ Spread = \frac{ask - bid}{(ask + bid)/2}$ Atkins and Dyl (1997) data was collected form investing.com
Return	Return is the profit obtained by investors through the difference between the price of a stock when buying and selling. $Return = \frac{Stock\ Price_t - Stock\ Price_{t-1}}{Stock\ Price_{t-1}}$ Atkins and Dyl (1997)
Exchange Rate	The exchange rate is a calculation of the conversion of the OIC Country's currency to the USD, which is the international currency standard. The conversion rate used is the free-floating rate obtained through the website https://www.exchangerates.org.uk/
Gold Price	The calculation of the price of gold is obtained through the value of gold in spot transactions denominated in USD with troy ounces. Data related to gold prices is accessed through https://id.investing.com/
Before and after Covid-19 Cases	Covid-19 cases are expressed in the dummy variable, where 0 indicates no covid-19 case from 2018 until 2019 and 1 indicates a case of covid-19 from 2020 until 2021. Data obtained through https://covid19.who.int/

8.3.1 *Panel Data Regression Analysis*

The panel data regression model is a regression analysis technique used to look at the same time period on multiple research samples. The panel regression model is considered to provide better efficiency and more accurate predictions to see the dynamics of behavior changes (Gujarati and Porter 2009). Previous research from Stereńczak (2022) stated that panel data methods can effectively identify the independent variables and minimize bias due to error in variables. Using EViews 9 software. The equations of the multiple regression model are as follows:

$$Y_{it} = \beta_0 + \beta_1 X1_{it} + \beta_2 X2_{it} + \beta_3 X3_{it} + \beta_4 X4_{it} + \beta_5 X5_{it} + \varepsilon_{it} \quad (8.1)$$

Explanation:

Y_{it} = Holding period
β_0 = Coefficient
$\beta_1 X1_{it}$ = Bid-ask spread
$\beta_2 X2_{it}$ = Return
$\beta_3 X3_{it}$ = Exchange rate
$\beta_4 X4_{it}$ = Gold Price
$\beta_5 X5_{it}$ = before and after Covid-19 cases (dummy variables)
ε_{it} = Statistic noise

8.3.2 *Partial Test*

The t test can be used to show how far an independent variable has individually influenced in describing a dependent variable (Gujarati and Porter 2009). In the statistical test t, the value of t count will be compared with t table. If t calculated > t table or probability, means the significance level less than 0.05, H0 is rejected and H1 is accepted, which indicate the independent variables affect the dependent variable, and vice versa.

8.3.3 *Coefficient Determination Test (R^2)*

The coefficient of determination or adjusted R^2 expresses the force of influence of independent variables together on dependent variables. The value of the Adjusted R^2 determination coefficient is between 0 (zero) and 1 (one). Adjusted R2 values that are minimal or close to 0 (zero) indicate that independent variables' capacity to describe dependent variables is relatively limited. If the Adjusted R2 score is large or close to 1 (one), it indicates that the independent variable almost completely explains all of the information needed to predict the dependent variable. The following formula can be used to compute adjusted R2 (Gujarati and Porter 2009).

8.3.4 *Machine Learning Data Analysis Techniques*

Before conducting a hypothesis, test using machine learning methods, it is important to do random splitting of data on subtests to be tested, this is useful to ensure the machine learning method memorizes the values in the data and can make predictions. In addition to data separation, scaling/scale data can also be done which is considered more accurate when compared to splitting. Standard scaling ensures that machine learning evaluates all variables and reports results in Gaussian z-score form. After splitting or scaling tests, it can be known the beta value of each research variable and also identify the level of significance (Athey and Imbens 2019; Chen 2021).

In machine learning models with algorithms are used to look for models that have a better level of accuracy to the understanding of data in various forms of scale. The selection of suitable machine learning models on data cannot be determined, so there is a need for the implementation of various and repeated methods. To deal with the gap between the two traditional statistical models and machine learning algorithms, the two can be combined by rendering the existing panel data in statistical software (Eviews, STATA, or SPSS) then importing data on the R application and further processing (Athey and Imbens 2019; Chen 2021).

8.3.5 Machine Learning Model Selection

According to Panigrahi et al. (2023) the advancement of technology has enhanced IT's significance in shaping the future's path. Current industries have a sense of place in the market and importance. The global digital uprising has an impact on supply, assembly, and procurement. Currently, machine learning is a prominent topic among academics and business people. Machine learning has a wide variety of models, such as Decision Trees and Forest, Bias Variance Tradeoff, Ensemble and boosting methods, Support vector machines and neural networks. Ml model selection can be accomplished by analyzing data and comparing training values and test scores in each model. The model with the highest R2 value and the lowest RMSE value is the most accurate. A high R2 suggests that the model is the most powerful or accurate, with little data bias. While the RMSE value represents the error rate, the lower the value, the less errors occur in the model (Athey and Imbens 2019; Chen 2021).

8.4 Results and Analysis

The result from the descriptive statistic table informs several variables, namely holding period (Y), bid-ask spread (X1_BA), return (X2_R), the exchange rate (X3_E), gold price (X4_G), and covid-19 cases (X5_C) (Table 8.3). The exchange rate variable uses a natural logarithm; therefore, some values show negative results, and the covid-19 cases use a dummy variable. The minimum result of Y and X1_BA is 0; this indicates that some companies are experiencing suspense on the stock exchange so that no transactions occur. The data results also show that the company's internal factors, namely stock returns, show the highest value in all aspects, meaning that returns are independent variables with a high level of volatility compared to other exogenous variables. Especially in standard deviation, a high return

Table 8.3 Descriptive statistic

	Y	X1_BA	X2_R	X3_E	X4_G	X5_C
Mean	1262.459	0.021469	133.7184	−2.269848	7.347823	0.483917
Median	196.2261	0.017094	0.000000	−1.322006	7.358774	0.000000
Maximum	308,159.8	0.258993	44,483.00	1.203603	7.632009	1.000000
Minimum	0.000000	0.000000	−0.193200	−9.623435	7.068308	0.000000
Std. Dev.	7843.492	0.017340	2411.108	2.122740	0.163242	0.499758

Source: Data processed

Table 8.4 Chow test result

Test summary	Chi-Sq. Statistic	Chi-Sq. d.f.	Prob.
Cross-section F	71.897102	(15,14,653)	0.0000
Cross-section Chi-square	1042.106623	15	0.0000

Source: Data processed

Table 8.5 Hausman test

Effects Test	Statistic	d.f.	Prob.
Cross-section random	49.191651	5	0.0000

Source: Data processed

indicates a higher risk. On the other hand, the bid-ask spread variable shows a contrary result, although it is also a part of internal factors. External factors such as gold price and covid-19 cases denote stable value; however, the exchange rate is volatile.

8.4.1 Data Panel Model Test

The panel data regression model is divided into three, namely PLS (Pooled Least Square), FEM (Fixed Effect Model) and REM (Random Effect Model). To determine the regression analysis model to be used will be done Hausman Test and Chow Test. The first test performed is the Chow test to determine the model used between PLS and FEM, following the results (Table 8.4):

Hypothesis Test:

H0: The best-used model of PLS
H1: The best used FEM model
Significance level: 5%
Critical area: H0 is rejected if the p-value is less than α Used significance level (α) of 5% and p-value of the bottom line obtained from the results at the output above at p-value <2.2e-6. The output shows that the p-value (0.000) < α (0.05) and H0 is rejected, so the best model used is FEM (Fixed Effect Model).

The next test performed is The Hausman Test to determine the model used between REM and FEM, following the results (Table 8.5):

Hypothesis Test:

H0: The best-used model of REM
H1: The best used FEM model
Significance level: 5%
Critical area: H0 is rejected if the p-value is less than α Used significance level (α) by 5% and the p-value of the bottom line obtained from the results at the above output at p-value = 0.001241. The output shows that the p-value (0.000) < α (0.05) and H0 is rejected, so the best model used is FEM (Fixed Effect Model).

8.4.2 Partial Test

Hypothesis Test: H0: No influence of Independent variables on Holding Variable Period H1: There is an influence of Independent variables on Holding Period Significance level: 5% Area of criticism: H0 rejected if the p-value is less than α Used significance level (α) of 5%. Here is a table that describes the results of the partial test (Table 8.6):

$$Y_{it} = 13512.14 - 38081.33R_{it} + 1.501381BA_{it} - 239.1613E_{it} - 1683.471G_{it} + 401.1668.8C_{it} + \varepsilon_{it} \quad (8.2)$$

Thus, the p-value for the Bid-ask spread variable, Stock Return, Exchange Rate, and Gold Price < α (0.05) and H0 were rejected, so the conclusion that all independent variables had a significant effect on variable Y (Holding Period > α).

Table 8.6 Partial test

Variable	Coefficient	Std-error	T-value	P-value	Explanation
Bid-ask spread (BA)	−38081.33	3432.058	−11.09577	0.000	Significant Negative
Return (R)	1.501381	0.022908	65.54009	0.000	Significant Positive
Exchange rate (E)	−239.1613	54.87051	−4.358650	0.000	Significant Negative
Gold price (G)	−1683.471	737.7177	−2.281999	0.0225	Significant Negative
Before and After Covid-19 cases (C)	401.1668	241.7388	1.659505	0.0970	Non-Significant

Source: Data processed

8.4.3 *Simultaneous Test*

Furthermore, the simultaneous test was carried out. This test consists of 2 types, the first is the F test to determine the effect of all variables independently simultaneously on the dependent variable. Another test is the R2 test or the coefficient of determination which is used to determine the proportion of this research model that can affect the y variable and predict the percentage size related to other variables outside the research model that can affect the y variable. The following are the results of the simultaneous test (Table 8.7):

The results showed that the p-value 0.000, indicating simultaneously the variable Bid-ask spread, Stock Return, Exchange Rate, Gold Price, and Covid-19 Case affected the holding period (Table 8.8).

The R-squared results show a value of 0.293779 or 29%, which can be interpreted that 29% of the holding period was influenced by the independent variables used in the study. The remaining 71% can be influenced by other variables outside the study model.

8.4.4 *Machine Learning*

The results of the machine learning method are displayed in Table 8.9:

Table 8.7 F test

P-Value 0.00000
Source: Data processed

Table 8.8 Coefficient Determination Test

R-Squared 0.293779
Source: Data processed

Table 8.9 Machine learning test result

	Training		Test	
Model	R^2	RMSE	R^2	RMSE
Linear	0.24373	0.85684	0.19400	0.93688
Decision Tree	0.40396	0.76068	0.32942	0.85457
Random Forest	0.86970	0.37844	0.33287	0.85241
Extra Tree	0.80578	0.45233	0.31366	0.86647
Bagging	0.91326	0.31428	0.31426	0.86516
XGBoost	0.65384	0.65713	0.26659	0.89924
Support Vector Machine	0.24286	0.85809	0.19263	0.93847
Neural Network	0.25092	0.86563	0.20308	0.94018

Source: Data processed

From Table 8.9 shows linear model or classical statistic is not the best model to see how much influence each independent variable has on the holding period, this can be seen from the largest R2 value and the smallest RMSE, both in training data and test data (Athey and Imbens 2019; Chen 2021). The results showed that the Random Forest model is the best method of making predictions. This can be seen in the highest R2 test score, which indicates the highest accuracy level and the lowest RMSE compared to other models, which shows minimal errors on the model. The equations of the random forest model can be written as:

$$\hat{\mu}_{rf}(x) = \sum_{i=1}^{n} \alpha_i(x) Y_{i,} \sum_{i=1}^{n} \alpha_i(x) = 1, \alpha_i(x) \geq 0, \tag{8.3}$$

Explanation:

$\hat{\mu}_{rf}(x)$: Random forest model estimator
$\alpha_i(x)Y_{i,}$: Weighted training model to predict the x

Equation (8.3) explains the process of the estimated random forest model by measuring the weighted training model (forest model) from the original Eq. (8.1) (panel model). Whereas the condition of must equal to 1 means the result must not have significant changes to other variables to ensure the training result will predict the real test correctly (Athey and Imbens 2019). The Random Forest model, which capture in Fig. 8.1, is able to make additional predictions to see which variables are the most influential. (Only for decision tree, random forest, extra Tree, and XGBoost) using the mean decrease gini criterion, where the greater the mean decrease gini value of a variable, the greater the effect (Chen 2021) (Fig. 8.1).

From the Random Forest graph above (Fig. 8.1), the return variable has the highest mean decrease value so that the variable is the most influential variable, both for decision tree, XGBoost, and Extra Tree models. Similarly, the variables of covid-19 cases do not show significant influence. The results showed consistency on all ML models with a mean decrease gini approach. This result is also in line with the panel data regression model in the T test contained in Table 8.6.

8.4.5 Discussion

Based on the results of the t test, it can be said that the holding period of investors in OIC countries depends on bid and ask spreads, stock returns, and macroeconomic conditions, which are represented by exchange rates, gold prices, and covid-19 cases. An interesting finding in this study, is that the Covid-19 cases that occur in various countries does not necessarily change the view of investors to hold in Islamic banking stocks. So far, when the Covid-19 stock market in various countries has decreased, so far, the decision to release the shares is due to the decline in investor confidence to invest their funds during this crisis, but in fact it does not apply to Islamic banking stocks. Based on Markowitz theory investors are concerned to

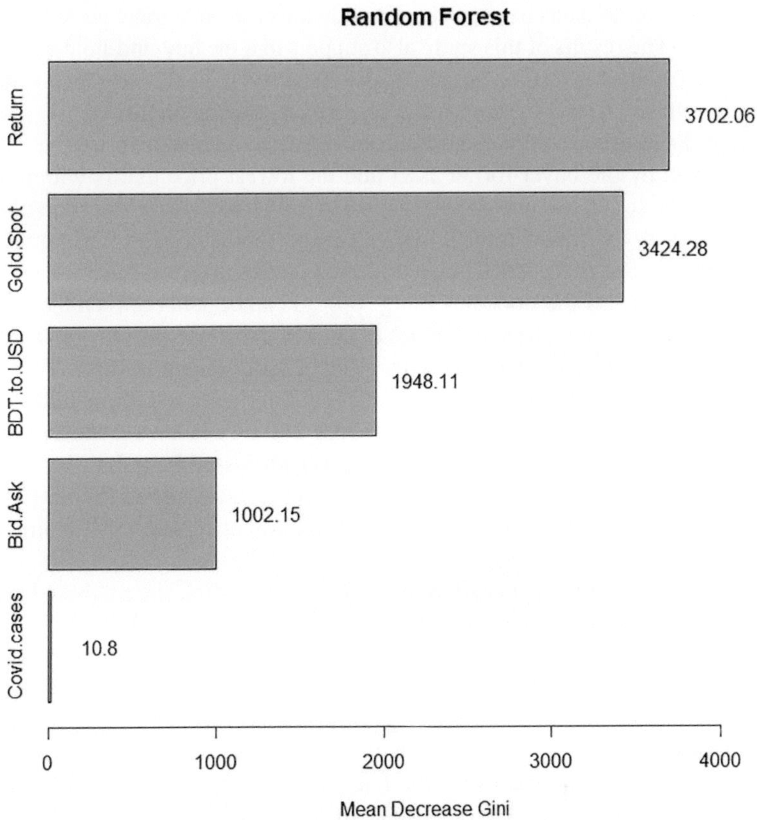

Source: Data processed

Fig. 8.1 Random forest result. Source: Data processed

invest based on the performance of the company. Some of the previous study also said that investors focus more on the macroeconomics condition rather than Covid-19, so even though OIC counties have pandemic conditions or not as long as the government has exit strategies to manage the macroeconomics condition investors will hold the shares (Fallahgoul 2021).

The decision of investors in holding the stock is more due to the profit from the return they get. Based on the results of the t test, it appears that the return will increase the holding of shares, in the sense that the more the return that is disbursed, the investor will still hold sharia banking shares, rather than releasing them and choosing other stocks that will not necessarily get the same return. According to Ødegaard (2017) the length of the holding period can be influenced by the character of investors. Furthermore, Ramasamy and Pachiyappan (2019) argues that the decision of holding period is strongly influenced by the type of company or the risks that are ready to be borne. Other research by Poretti and Das (2020) also explains that companies with good performance will result in high returns and investors prefer to maintain their investments because they already know the historical performance of the company. The results of this study also support that the buy-and-hold strategy is more widely used by investors, because it is considered to be able to generate higher long-term returns (Amédée-Manesme et al. 2016; Nowak et al. 2015).

Next, bid and ask spread is the difference between the highest price that is willing to be paid by the buyer for an asset and the lowest price that is willing to be accepted by the seller. Bid and accepted spreads can be used for a reflection of stock liquidity, in theory the smaller the liquidity, the shorter the holding. The liquidity of the stock or the ease of the stock being traded can make investors can look for a gap in obtaining profits, if the stock has a large spread can be said between the demand and the offer is too wide, the seller hopes to get a higher price than the buyer offers, while the buyer assesses at a low price therefore trading cannot be done. This result is in line with research by Al Tamimi et al. (2021), Merinkas and Prasad (2003), and Rashata (2022) who support the theory of loss-aversion behavior which assumes that investors prefer the smallest opportunity to gain losses from their investments (not maintaining the shares that investor own despite having good liquidity) rather than choosing opportunities that offer more investment returns (still maintaining liquid stocks).

However, the results of this study contradict the existing theory, based on this study, the high spread between the asking price and the price offered causes the holding period to be shorter, which means that Islamic banking shares are not liquid or large transaction costs, but holdings held short Such events can occur due to the company's internal performance that is considered less good and the occurrence of market shock so as to make the spread more and more difficult. Investors are reluctant to hold Islamic bank shares for too long and there is a rapid sale of shares. In terms of investor behavior, this result is in line with the research of Daniel and Moskowitz (2016), da Costa Jr and Goulart (2013), and Lehenkari and Perttunen (2004) which explains the tendency of investors to conduct disposition securities, namely immediately selling stocks that have high spreads and maintaining stocks with low spreads.

Furthermore, regarding the condition of the dispute of the currencies of OIC countries with dollars, currency exchange is extremely related to investment. The results showed negative results, the lower the value of the country's currency of origin with the dollar (the higher the exchange rate) then the shorter the holding of Islamic banking shares. Rising dollar prices make investing with dollars more expensive for investors, therefore investors do not switch from the stocks they already own and still continue to make transactions on other stocks to diversify portfolio and asset rescue. In terms of macroeconomic, if the value of the domestic currency weakens, it can cause debt in the form of dollars to increase, and make the performance of companies that have debt in the form of dollars will decrease and indicate poor levels of corporate liquidity (Huang and Stoll 2001; Melvin and Prins 2015), the decline in performance can lead to a decline in the company's stock price because investors are minded to divers the portfolio, looking for other stocks that perform better. Therefore, investors' perception of foreign exchange rates causes them to make a sale transaction for non-performing stocks and buy for stocks that provide benefits to investors to improve the performance of their investment portfolio (Sgammini and Muzindutsi 2020; Zolfaghari and Sahabi 2021) this asset rescue activity makes the holding time shorter.

The negative gold price means that the high world gold price can cause investors to hold stocks faster. Of course, when the price of gold rises then what happens investors still choose to hold on stocks, because if the price of gold rises then the possibility of getting less profit except for investors who already have gold and buy when the price is low, this result is also similar to Arouri et al. (2015), Batten et al. (2015), and Grassa (2015) who argue that gold can be used as investment diversification only in the long term. In terms of economic stability, countries that have gold reserves or gold mines get a favorable position when gold prices rise, their foreign exchange reserves also increase, given that the average OIC country has enough gold reserves, it will encourage the climate of investment in OIC countries to increase and will certainly make stock holdings faster because investors take advantage of the favorable investment climate in the country. Therefore, investors continue to make transactions on stocks and make holdings shorter in the hopes of obtaining benefits from a good investment climate and minimizing investment risk (Salisu et al. 2021).

8.5 Conclusion and Recommendation

This study aims to find out the factors that can affect the holding period in the case of Islamic banking stocks in OIC Countries. There are 2 results of analysis, namely regression of panel data using the FEM (Fixed Effect Model) model and Machine Learning with the Random Forest model. Both research models showed uniform results, where variable returns on stocks had a significantly positive effect on the holding period. The result shows that investor behavior concern is about profit, when the profit of stock is increasing the holding period are longer, it indicates that investor have choose investment that have high return to increase income. Contrary with variable bid-ask spreads, exchange rates, and gold prices partially show significant negative results. Investor cut of the holding to increase income by investing in other asset, such as gold or as action to risk reduction. Another variable, before and during covid-19 cases have no significant effect on the holding period. Simultaneously all variables in this study affect the holding period.

The implication of the results of this study can be useful for investors and investment managers in taking policies to maximize the performance of investment portfolios by considering macroeconomic, bid-ask spread, and stock return factors either before the pandemic or during the pandemic. As for Islamic banking, the results of this study can provide the facts related to the behavior of Islamic investors so that it can be used as a reference to make policies related to corporate actions, such as expanding the business or making new innovative products to keeping the long holding period for the investors, due to it is essential to maintain their stock return which is related to choosing best action which can generate profit significantly. However, this research is limited to the case of Islamic banking in OIC Countries. For further research, can use research samples in other sectors and countries or add other macroeconomic variables.

References

Al Tamimi HAH, Duqi A, Kanas A, Zervopoulos PD (2021) Directional distance function DEA estimators for evaluating efficiency gains from possible mergers and acquisitions. J Oper Res Soc 0(0):1–18. https://doi.org/10.1080/01605682.2021.1907243

Ali M, Alam N, Rizvi SAR (2020) Coronavirus (COVID-19) – an epidemic or pandemic for financial markets. J Behav Exp Financ 27. https://doi.org/10.1016/J.JBEF.2020.100341

Alrifai T (2015) Islamic finance and the new financial system an ethical approach to preventing future financial crises (Wiley Fina). Wiley

Amédée-Manesme CO, Barthélémy F, Prigent JL (2016) Real estate investment: market volatility and optimal holding period under risk aversion. Econ Model 58:543–555. https://doi.org/10.1016/J.ECONMOD.2015.10.033

Arouri MEH, Lahiani A, Nguyen DK (2015) World gold prices and stock returns in China: insights for hedging and diversification strategies. Econ Model 44:273–282. https://doi.org/10.1016/j.econmod.2014.10.030

Athey S, Imbens GW (2019) Machine learning methods that economists should know about. Annu Rev Econ 11:685–725. https://doi.org/10.1146/ANNUREV-ECONOMICS-080217-053433

Atkins AB, Dyl EA (1997) Transaction costs and holding period for common stocks. J Financ 3(1)

Bai J (2003) Inferential theory for factor models of large dimensions. Econometrica 71(1):135–171. https://doi.org/10.1111/1468-0262.00392

Bai J, Ng S (2002) Determining the number of factors in approximate factor models. Econometrica 70(1):191–221. https://doi.org/10.1111/1468-0262.00273

Batten JA, Szilagyi PG, Wagner NF (2015) Should emerging market investors buy commodities? Appl Econ 47(39):4228–4246. https://doi.org/10.1080/00036846.2015.1026586

Baur DG, McDermott TK (2010) Is gold a safe haven? International evidence. J Bank Financ 34(8):1886–1898. https://doi.org/10.1016/J.JBANKFIN.2009.12.008

Berger AN, Boubakri N, Guedhami O, Li X (2019) Liquidity creation performance and financial stability consequences of Islamic banking: evidence from a multinational study. J Financ Stab 44. https://doi.org/10.2139/ssrn.3278032

Campbell JL, DeAngelis MD, Moon JR (2019) Skin in the game: Personal stock holdings and investors' response to stock analysis on social media. Rev Acc Stud 24(3):731–779

Cao C, Liang B, Lo AW, Petrasek L (2018) Hedge fund holdings and stock market efficiency. Rev Asset Pricing Stud 8(1):77–116

Chen JM (2021) An introduction to machine learning for panel data. Int Adv Econ Res 27. https://doi.org/10.2139/SSRN.3717879

Chen H-K, Lien C-T (2017) Market reaction to macroeconomic news: the role of investor sentiment. Asia Pac J Financ Stud 46(6):853–875. https://doi.org/10.1111/ajfs.12198

Chung S, Wei P (2005) The relationship between bid-ask spreads and holding period: the case of Chinese A and B Shares. Glob Financ J 15:239–249

Ciner C, Gurdgiev C, Lucey BM, Ciner C, Gurdgiev C, Lucey B (2013) Hedges and safe havens: an examination of stocks, bonds, gold, oil and exchange rates. Int Rev Financ Anal 29(C):202–211. https://doi.org/10.1016/J.IRFA.2012.12.001

da Costa N Jr, Goulart M (2013) The disposition effect and investor experience. J Bank Financ 37(5):1669–1675. https://doi.org/10.1016/j.jbankfin.2012.12.007

Daniel K, Moskowitz TJ (2016) Momentum crashes. J Financ Econ 122(2):221–247. https://doi.org/10.1016/j.jfineco.2015.12.002

De Oliveira CA (2014) Investment and exchange rate uncertainty under different regimes. Estud Econ 44(3):553–577

Dewanti LA, Rusmita SA, Samad KA (2022) Sensitivity Islamic stock return in Asia: the effect of exchange rate volatility. Jurnal Ekonomi Dan Bisnis Islam (J Islam Econ Bus) 8(2):302–317. https://doi.org/10.20473/jebis.v8i2.39018

Elton EJ, Gruber MJ, Brown SJ, Goetzmann WN (2014) Modern portfolio theory and investment analysis, 9th edn. Wiley

Fallahgoul H (2021) Inside the mind of investors during the COVID-19 pandemic: evidence from the StockTwits data. J Financ Data Sci 3(2):134–148. https://doi.org/10.3905/JFDS.2021.1.058

Garnia E, Primiana I, Sudarsono R, Masyita D (2015) On the relationships among expected return, volume, holding period, and bid-ask spread in Indonesia stock market. J Comput Theor Nanosci 21(4). https://doi.org/10.1166/asl.2015.5904

Grassa R (2015) Shariah supervisory systems in Islamic finance institutions across the OIC member countries. J Financial Regul Compliance 23(2):135–160. https://doi.org/10.1108/jfrc-02-2014-0011

Gujarati DN, Porter DC (2009) Basic econometrics, 5th edn. McGraw-Hill, New York

Hickman K, Hunter H, Byrd J, Beck J, Terpening W (2001) Life cycle investing, holding periods, and risk. J Portf Manag 27(2):101–111

Hidayat SE, Farooq MO, Nasution A, Sari CA (2020) COVID-19 and its impacts on the Islamic financial industry in the OIC countries. In: The impact of COVID-19 outbreak on the Islamic finance in the OIC countries

Huang R, Stoll H (2001) Tick size, bid-ask spread, and market structure. J Financ Quant Anal 36(4):503–522. https://doi.org/10.2307/2676222

Lehenkari M, Perttunen J (2004) Holding on to the losers: Finnish evidence. J Behav Financ 5(2):116–126. https://doi.org/10.1207/s15427579jpfm0502_5

Lyle MR, Wang CCY (2015) The cross section of expected holding period returns and their dynamics: a present value approach. J Financ Econ 116(3):505–525. https://doi.org/10.1016/j.jfineco.2015.03.001

Manohar JM, Raju GA (2021) Does gold retain its hedge and safe haven role for energy sector indices during covid-19 pandemic? A cross-quantilogram approach. Int J Energy Econ Policy 11(1):233–240. https://doi.org/10.32479/ijeep.10294

Melvin M, Prins J (2015) Equity hedging and exchange rates at the London 4 p.m. fix. J Financ Markets 22:50–72. https://doi.org/10.1016/j.finmar.2014.11.001

Merinkas A, Prasad D (2003) Factors influencing Greek investor behavior on the Athens stock exchange. J Bus 66(1):1–20

Nagraj A (2023) Global Islamic finance industry to grow 10% in 2023-2024 despite economic slowdown. The National News. https://www.thenationalnews.com/business/markets/2023/05/02/global-islamic-finance-industry-to-grow-10-in-2023-2024-despite-economic-slowdown/

Nowak AZ, Winkler-Drews T, Shachmurove Y (2015) The risk of holding periods across international stock exchanges. Int J Bus 20(2):91–110

Nugroho A, Wasiaturrahma W, Anggia P (2021) Jakarta Islamic index: covid-19 pandemic and potential financial distress. Jurnal Ekonomi Dan Bisnis Islam (J Islam Econ Bus) 7(2):210. https://doi.org/10.20473/jebis.v7i2.26147

Ødegaard BA (2017) How long do equity owners hang on to their stocks? SSRN Electron J. https://doi.org/Ødegaard, Bernt Arne, how long do equity owners hang on to their stocks? (Mar 15, 2017). Available at SSRN: https://ssrn.com/abstract=2946915 or https://doi.org/10.2139/ssrn.2946915

Panigrahi RR, Dash M, Shaikh ZH, Irfan M (2023) Review of machine learning techniques in the supply chain management of Indian industry: a future research agenda. In: Advanced machine learning algorithms for complex financial applications. IGI Global, pp 199–219

Poretti C, Das P (2020) What explains differing holding periods across hotel investments? A hazard rate framework. Int J Hosp Manag 89:102564. https://doi.org/10.1016/J.IJHM.2020.102564

Ramasamy RR, Pachiyappan S (2019) Holding period for positive return from Indian mutual funds. Invest Manag Financ Innov 16(1):346–364

Rashata H (2022) Behavior in the Pakistan financial market during the COVID-19 pandemic. SSRN Electron J. https://doi.org/Rashata, Hamid, Investors' Behavior in the Pakistan Financial Market during the COVID-19 Pandemic (January 4, 2022). Available at SSRN: https://ssrn.com/abstract=4013498 or https://doi.org/10.2139/ssrn.4013498

Renault T (2017) Intraday online investor sentiment and return patterns in the US stock market. J Bank Financ 84:25–40

Rusmita SA, Azaria A (2021) Holding period of sharia and conventional bank's stock in Indonesian capital market. Int J Islam Econ 3(1):71–81

Rusmita SA, An-Nafis MSA, Ramadhani I, Irfan M (2023) The effect of good corporate governance on financial distress in companies listed in sharia stock index Indonesia. In: Irfan M, Elhoseny M, Kassim S (eds) *Advanced machine learning algorithms for complex financial applications*. IGI Global, pp 220–251. https://doi.org/10.4018/978-1-6684-4483-2.ch014

Salisu AA, Raheem ID, Vo XV (2021) Assessing the safe haven property of the gold market during COVID-19 pandemic. Int Rev Financ Anal 74. https://doi.org/10.1016/J.IRFA.2021.101666

Schaub M (2019) Holding period returns from NASDAQ traded ADRs. Appl Econ Lett 26(20):1689–1693. https://doi.org/10.1080/13504851.2019.1591593

Schaub M, Simmons G (2022) A note on the industry returns of NASDAQ-listed ADRs. Appl Econ Lett 29(4):360–365. https://doi.org/10.1080/13504851.2020.1869152

Sgammini M, Muzindutsi PF (2020) Effect of exchange rate movements on the performance of investment portfolios in South Africa. Int J Econ Financ Stud 12(2):469–486. https://doi.org/10.34109/ijefs.202012214

Stereńczak S (2022) Illiquidity and stock returns: the moderating role of investors' holding period in Central and Eastern European markets. Int J Emerg Mark, Vol. ahead-of-print No. ahead-of-print. https://doi.org/10.1108/IJOEM-01-2022-0125

Stoll HR (1989) Inferring the components of the bid-ask spread: theory and empirical tests. J Financ 44(1):115–134. https://doi.org/10.2307/2328278

Zolfaghari, M., & Sahabi, B. (2021). The impact of oil price and exchange rate on momentum strategy profits in stock market: evidence from oil-rich developing countries. In Review of managerial science 15(7). Springer Berlin. https://doi.org/10.1007/s11846-020-00413-0

Chapter 9
Transforming of the Financial Landscape from 4.0 to 5.0: Exploring the Integration of Blockchain, and Artificial Intelligence

Baranidharan Subburayan, Amirdha Vasani Sankarkumar, Rohit Singh, and Hellena Mohamedy Mushi

9.1 Introduction

Technological advancements are driving a significant transformation in the financial sector. Blockchain, artificial intelligence (AI), and financial technology (fintech) have all come together to become a potent force reshaping the financial sector. The purpose of this study is to investigate the potential advantages, drawbacks, and implications of incorporating blockchain, AI, and fintech into the financial sector. Technology advancements are driving a significant transformation in the financial sector. The integration of blockchain, artificial intelligence (AI), and financial technology (fintech) has emerged as a game-changer among the notable technological advancements, revolutionizing various aspects of global financial systems. The integration of blockchain, artificial intelligence, and fintech in the financial sector and its implications are the focus of this study, which aims to conduct a systematic literature review.

Due to its potential to disrupt conventional financial systems, blockchain technology has received a significant amount of attention over the past 10 years.

B. Subburayan (✉)
School of Business and Management, Christ University, Bangalore, Karnataka, India

A. V. Sankarkumar
Department of Corporate Secretaryship and Accounting Finance, SRMIST, Chennai, Tamil Nadu, India

R. Singh
The Assam Royal Global University, Guwahati, Assam, India
e-mail: rsingh@rgu.ac

H. M. Mushi
Mzumbe University, Mzumbe, Tanzania
e-mail: hmohamed@mzumbe.ac.tz

M. Irfan et al. (eds.), *Applications of Block Chain technology and Artificial Intelligence*, Financial Mathematics and Fintech,
https://doi.org/10.1007/978-3-031-47324-1_9

Blockchain is fundamentally a decentralized, distributed ledger technology that makes it possible for transactions to be secure and transparent without the need for middlemen. Blockchain is an ideal solution for increasing trust and security in financial transactions due to its immutability and tamper-proof nature. In addition, the addition of smart contracts to blockchain platforms has provided new opportunities for simplifying and automating intricate financial procedures. Blockchain technology, which was first introduced as the technology that underpins cryptocurrencies like Bitcoin, has gained recognition for its capacity to disrupt conventional financial systems. Blockchain is fundamentally a decentralized, immutable ledger that enables transparent and secure transactions without the need for middlemen. The integrity and privacy of financial data are guaranteed by its distributed nature and cryptographic security, in addition to facilitating cost-effective and efficient transactions. Blockchain technology has the potential to streamline processes, increase trust, and reduce friction in the financial sector.

In contrast, artificial intelligence has made significant progress in recent years, making it possible for machines to learn, reason, and carry out activities that previously required human intelligence. Machine learning, natural language processing, and computer vision are all examples of AI technologies. AI has demonstrated its value in the areas of risk assessment, fraud detection, algorithmic trading, customer service, and personalized financial recommendations in the financial industry. AI's potential impact on the financial sector is amplified by its integration with blockchain and fintech.

The innovative use of technology to provide financial services, or fintech, has been altering the financial landscape. Mobile payments, peer-to-peer lending, crowdfunding, robo-advisory services, and digital wallets are all examples of fintech solutions. Customers' experiences have been improved, operational efficiency has increased, and access to financial services has become more accessible to everyone. Fintech's integration of AI and blockchain has the potential to transform conventional financial procedures and open up new opportunities.

This investigation aims to accomplish two things. First and foremost, it aims to provide a comprehensive comprehension of the financial sector's integration of blockchain, AI, and fintech. Analyzing the potential advantages and drawbacks of integration, as well as determining the most important drivers and trends, are all part of this. Second, the research aims to investigate the effects of this integration on banking operations, investment strategies, regulatory frameworks, and customer experience in the financial industry.

A systematic review approach is used to accomplish these goals, with a focus on secondary data gathered from reliable sources like academic databases (like Google Scholar, J-gate, Ebsco, ProQuest, and ScienceDirect) and industry reports. Scholarly articles, research papers, whitepapers, and reports from 2015 to the present are included in the review. In order to guarantee the findings' reliability and validity, the inclusion criteria give preference to publications that have been reviewed by peers and authoritative sources.

Using a thematic approach, the selected articles' key insights, findings, and conclusions were extracted from the data analysis. A comprehensive synthesis of the literature is made possible by the identification of common themes and patterns. The application of AI in risk assessment and fraud detection, the role of fintech in reshaping traditional financial services, and the integration of these technologies to create innovative financial solutions are some of the topics examined in the analysis.

By providing a comprehensive overview of the integration of blockchain, AI, and fintech in the financial sector, the findings of this study are anticipated to add to the existing body of knowledge. The study aims to inform policymakers, financial institutions, researchers, and practitioners about the potential opportunities and risks associated with these technologies by examining the benefits, challenges, and implications of this integration. To fully realize the transformative potential of blockchain, AI, and fintech in the financial sector, the study may also highlight areas requiring additional research and development. In conclusion, by enhancing innovation, efficiency, and security, the integration of blockchain, artificial intelligence, and fintech has the potential to transform the financial industry. By conducting a systematic literature review, this study aims to improve our understanding of this integration. The findings and discussions that follow will provide valuable insights into the current state, trends, difficulties, and repercussions of incorporating blockchain, AI, and fintech into the financial sector.

9.2 Objectives of This Study

- To identify and analyze the key challenges and concerns associated with the integration of blockchain, artificial intelligence (AI), and fintech in the financial sector.

This objective aims to systematically review the literature and extract the major challenges and concerns that arise from the convergence of blockchain, AI, and fintech. By thoroughly analyzing the findings, discussions, and conclusions of the included articles, the study seeks to provide a comprehensive understanding of the potential obstacles and risks associated with these technologies. This objective will enable stakeholders, policymakers, and industry players to gain insights into the critical issues that need to be addressed to ensure a successful and sustainable integration of these technologies in the financial domain.

- To explore potential strategies and solutions for maximizing the benefits and mitigating the risks associated with the integration of blockchain, AI, and fintech.

Building upon the identified challenges and concerns, this objective aims to propose strategies and solutions to harness the potential of blockchain, AI, and fintech while minimizing the associated risks. By examining the literature, the study seeks to identify existing best practices, regulatory frameworks, and technological advancements that can be leveraged to address the identified challenges. This objective will help guide policymakers, researchers, and industry practitioners in formulating effective strategies, policies, and guidelines for the integration of these technologies, fostering innovation, and promoting responsible adoption in the financial sector.

9.3 Statement of the Problem

The integration of blockchain, artificial intelligence (AI), and financial technology (fintech) has garnered significant attention in recent years. This comprehensive literature review aims to delve deeper into the problem statement surrounding the intersection of these technologies in the financial sector. By examining the findings and discussions from 80 scholarly articles, we seek to identify key challenges, potential risks, and areas of concern associated with this integration. Understanding the problem statement will help pave the way for developing effective strategies and solutions to maximize the benefits and mitigate the risks associated with these technologies.

Challenges and Concerns:

1. Regulatory Frameworks: One of the primary challenges identified in the literature is the development of appropriate regulatory frameworks to govern the integration of blockchain, AI, and fintech. As these technologies disrupt traditional financial systems, regulators face the task of striking a balance between promoting innovation and ensuring consumer protection, data privacy, and cybersecurity.
2. Data Privacy and Security: The convergence of blockchain, AI, and fintech raises concerns about data privacy and security. While blockchain technology provides inherent security through its decentralized and immutable nature, the potential vulnerabilities in AI algorithms and fintech platforms pose risks to sensitive financial data. Protecting customer information and ensuring secure transactions emerge as critical areas of focus.
3. Ethical Considerations: The utilization of AI algorithms and automated decision-making processes in fintech applications demands ethical considerations. The transparency, fairness, and accountability of AI systems need to be addressed to prevent discriminatory practices, bias, and unethical behavior. Establishing ethical guidelines and frameworks becomes crucial in maintaining trust and integrity in the financial industry.
4. Risk Management: The integration of blockchain, AI, and fintech introduces new dimensions of risk in the financial sector. Managing risks associated with technological failures, cyber threats, algorithmic biases, and systemic vulnerabilities pose challenges that demand robust risk management strategies. Ensuring the reliability, accuracy, and integrity of AI models and blockchain platforms becomes paramount.
5. Talent and Skill Gap: The rapid pace of technological advancements requires a skilled workforce capable of developing, implementing, and managing these technologies effectively. However, there is a significant talent gap in the finance

industry in terms of understanding blockchain, AI, and fintech. Bridging this gap and fostering a culture of continuous learning and skill development is crucial to fully leverage the potential of these technologies.

6. Interoperability and Standardization: Achieving interoperability and standardization across different blockchain platforms and fintech applications remains a challenge. The lack of uniform standards and protocols hinders seamless integration, data exchange, and collaboration between stakeholders. Addressing interoperability issues will enable greater synergy and efficiency in utilizing these technologies.
7. Adoption and User Acceptance: Despite the numerous advantages offered by the integration of blockchain, AI, and fintech, user adoption and acceptance present a hurdle. Resistance to change, concerns about security and privacy, and the need for educating users about the benefits and risks of these technologies hinder their widespread adoption. Building trust, fostering user confidence, and enhancing financial literacy are essential for encouraging broader acceptance.

The deeper analysis of the problem statement surrounding the integration of blockchain, artificial intelligence, and fintech highlights a range of challenges and concerns. Regulatory frameworks, data privacy and security, ethical considerations, risk management, talent and skill gap, interoperability and standardization, and user adoption emerge as critical areas that require attention and proactive measures. Addressing these challenges will pave the way for the successful integration of these technologies, enabling the financial industry to unlock their full potential while ensuring transparency, security, and ethical practices.

Based on the literature review conducted in this study, the reviews can be categorized into the following categories:

1. Blockchain Technology:

   (a) Applications of blockchain in payments and remittances
   (b) Blockchain-based smart contracts and decentralized applications
   (c) Blockchain for trade finance and supply chain management
   (d) Identity management and authentication using blockchain
   (e) Blockchain in regulatory compliance and anti-money laundering

2. Artificial Intelligence:

   (a) Credit scoring and risk assessment using AI algorithms
   (b) Fraud detection and prevention with AI techniques
   (c) Natural language processing and sentiment analysis in customer service
   (d) AI-powered investment and wealth management solutions
   (e) Robo-advisory and personalized financial recommendations

3. Fintech Solutions:

   (a) Peer-to-peer lending platforms and crowdfunding
   (b) Digital payment systems and mobile banking applications
   (c) Robust financial data analytics and insights

   (d) Open banking and APIs for seamless financial integration
   (e) Fintech regulation and policy implications

4. Integration of Blockchain and AI:

   (a) AI-driven analytics on blockchain data for insights and predictions
   (b) Smart contracts with AI-enabled automation and decision-making
   (c) AI-powered identity verification and KYC processes on blockchain
   (d) AI algorithms for consensus mechanisms and scalability in blockchain
   (e) Challenges and opportunities of combining blockchain and AI

In the past few years, there has been a notable upsurge in research focused on exploring the practical applications of artificial intelligence (AI) and deep learning (DL) in the financial sector. This research has primarily emphasized areas such as compliance, prediction, detection, and automation. The goal of this literature review is to present an overview of recent studies that highlight the pivotal role of AI and DL in financial institutions, cryptocurrency, and other related domains. The financial industry has experienced a profound transformation due to the rapid advancements in AI, DL, and other related technologies. These advancements have brought forth unprecedented opportunities and challenges. This comprehensive literature review aims to provide an in-depth understanding of recent studies that delve into the transformative influence of AI and emerging technologies in finance, cryptocurrency, banking, and associated areas. The application of AI and emerging technologies in cryptocurrency markets and financial systems has captured significant attention from both researchers and practitioners. This literature review examines a collection of recent studies that explore the practical application and impact of AI in domains such as cryptocurrency, cybersecurity, fintech, and financial risk management. The integration of artificial intelligence (AI) and machine learning (ML) techniques in the field of finance has attracted considerable interest from researchers and practitioners alike. This literature review examines a selection of recent studies that investigate the role of AI and ML in finance, covering various topics such as money laundering detection, portfolio management, algorithmic trading, and the risks associated with fintech.

9.4 Fintech Solutions

A novel ensemble learning approach for cryptocurrency price prediction was proposed by Rather (2023). The study showed that AI and DL techniques can be used in the cryptocurrency industry and that ensemble models can improve prediction accuracy. Using node embeddings and artificial neural networks, Peter and Styppa (2023) proposed a method for predicting millionaires from Ethereum transaction histories. Their research demonstrated that AI and DL methods can be used to predict wealth accumulation and extract meaningful insights from cryptocurrency data. Rajeswary and Thirumaran (2023) introduced a far reaching review of mechanized site phishing identification strategies, with an emphasis on simulated intelligence and human conduct points of view. The authors highlighted the significance of AI and DL in combating cyber threats in financial contexts by discussing their roles in identifying and mitigating phishing attacks.

Wagan and others (2023) proposed enhancing the quality of experience (QoE) with an AI-enabled game-based learning framework. In order to personalize learning experiences and encourage effective knowledge acquisition and engagement in educational settings, their novel and secure framework made use of AI techniques. Tamblyn and co. (2023) presented the fintech-kMC framework, which designed and tested financial platform machine learning systems through agent-based simulations. The authors emphasized the significance of simulation-based strategies for the comprehension of intricate financial dynamics and the creation of robust machine learning models. Chen and others (2023) introduced a complicated framework for real-time news impact prediction on financial markets that is driven by event processing (CEP). The significance of CEP in processing and analyzing large amounts of news data, which enables timely and accurate predictions of market trends and impacts, was emphasized by the authors. Dziedzic (2023) investigated the relationship between AI and smart contracts in the banking and insurance industries. The author looked into how these technologies could be combined to make contractual processes more automated, transparent, and efficient in the digital revolution.

Choithani and others In the context of cryptocurrencies, a comprehensive study on the intersection of AI, cybersecurity, and the banking system was carried out by (2022). The authors looked into the difficulties and opportunities of protecting cryptocurrency transactions and the role that artificial intelligence plays in spotting and preventing cyberattacks. Bellini and Vassalli (2022) examined the initial strategies of central banks and the current state of FinTech. The creators examined the effect of FinTech on conventional financial frameworks, investigating the reception of man-made intelligence and other arising advances in upgrading monetary administrations, computerized installments, and administrative systems. Bhatt and co. (2022) deciphered the fintech, digitalization, and financial services trinity using bibliometric analysis and thematic literature reviews. Key themes, trends, and future directions in the convergence of these domains were highlighted in the study's integrated literature analysis. Guarino and others (2022) looked at how well automated, adaptive, and autonomous traders performed in cryptocurrency financial bubbles. The authors evaluated the efficacy of various trading strategies for navigating volatile cryptocurrency markets using neural computing and evolutionary algorithms.

Dempere and co. (2022) looked into methods of selecting data for training machine learning models that could predict Bitcoin prices. In order to improve the accuracy and dependability of Bitcoin price predictions, the authors compared depth-based and width-based methods for selecting relevant data features. Fritz-Morgenthal et al. 2022) looked at how AI and financial risk management can be explained, trusted, and responsible together. The authors emphasized the significance of transparency, accountability, and ethical considerations in AI-based financial decision-making when discussing the challenges and opportunities of incorporating AI techniques into risk management practices. A review was conducted by Khan and Jhapate (2022) on the use of machine learning in the investigation of suspected money laundering. The authors looked at a variety of machine learning (ML) methods that are used to find and stop money laundering in financial transactions. This showed how AI-based methods could help improve efforts to stop money laundering. Mishra and co. (2022) conducted a survey of data-driven online portfolio management strategies. The authors looked at how data analytics, machine learning, and AI can be used to improve investment decision-making by optimizing portfolio construction, asset allocation, and risk management.

Jović and Nikolić (2022) investigated the emergence of new fintech-related risks. The authors highlighted the need for proper risk management and regulation in the fintech industry as they discussed the potential negative aspects of fintech. They looked at the threats to financial stability, cybersecurity, and data privacy. Through digital financial inclusion, Mhlanga (2022) investigated the transition from an informal financial money market to a formal financial system. The author looked into the role that digital financial services like mobile banking and digital payments play in increasing financial inclusion and decreasing poverty, and he or she also talked about how AI and machine learning can help make this transition easier. Singh and others (2022) suggested a Bitcoin algorithmic trading model based on LSTM. The authors developed an automated trading strategy and used long-short-term memory (LSTM) neural networks to predict Bitcoin price changes, demonstrating the potential of AI and machine learning in cryptocurrency trading. Steiner-Otoo and Jahankhani (2022) looked into how safe smartphones are from financial sector Man-in-the-Middle (MiTM) attacks. The authors highlighted the significance of security measures in the context of financial transactions as they investigated the flaws and potential defenses that could shield smartphones from such attacks.

A paper on augmented financial intelligence was presented by Mihov et al. (2022). The authors highlighted the potential of augmented intelligence to improve financial decision-making and risk management by discussing the integration of AI and ML techniques with financial intelligence systems. The connection between cyber risk, risk-based thinking, and fintech was discussed by Vučinić and Luburić (2022). The need for a risk-based strategy to mitigate potential vulnerabilities and threats in the financial sector was emphasized by the authors as they investigated the difficulties and implications of fintech adoption in relation to cyber risk management. In a post-COVID-19 world, Madhav and Tyagi (2022) investigated the unresolved issues, obstacles, and prospects of future technologies. The authors looked into how emerging technologies like AI and machine learning could affect various industries, including finance. They also talked about how important it is to deal with the problems that come with these technologies and take advantage of the opportunities they offer. Morgan (2022) evaluated the dangers of green digital finance and suggested measures to mitigate them. The author looked at how sustainability, digital technologies, and green finance are intertwined. They talked about the potential benefits and risks and suggested ways to reduce financial and environmental risks. A bibliometric analysis was carried out by Goodell et al. (2021) to identify foundations, themes, and research clusters in the field of AI and ML in finance. In order to comprehend the key areas of focus, research trends, and interdisciplinary collaborations in the field of AI and ML in finance, the authors analyzed a substantial body of literature.

A machine learning algorithm for predicting cryptocurrency prices was proposed by Awotunde et al. (2021). The authors looked into how ML methods could be used to predict the price of cryptocurrencies, highlighting the potential of AI-based strategies in the cryptocurrency market. Kaal (2021) talked about how hedge funds and financial technology (fintech) intersect. The author looked at how fintech is changing the hedge fund industry and talked about the potential benefits and challenges of using new technology in this industry. Cumming et al. (2021) provided a comprehensive examination of hedge funds in "The Oxford Handbook of Hedge Funds." Their work shed light on various aspects of hedge fund operations, strategies, and the broader financial landscape. Hendershott & Zhang (2021) provided an overview of FinTech research frontiers. Digital currencies, crowdfunding, robo-advisors, and the impact of FinTech on financial markets and institutions were among the topics and emerging trends discussed by the authors in the FinTech research. The economics of cryptocurrencies were the subject of an investigation by Agarwal et al. (2020), with an emphasis on digital currency, blockchain, and artificial intelligence. The authors looked at market dynamics, pricing, and regulatory considerations as they discussed the impact of these technologies on the financial ecosystem. A survey of fintech research and policy discussions was presented by Allen et al. (2020). Digital banking, peer-to-peer lending, robo-advisors, and regulatory frameworks were among the topics covered by the authors' analysis of the academic literature and policy debates surrounding fintech.

Allen (2021) investigated the Middle East and North Africa (MENA) region's globalization of fintech and finance. Taking into account aspects like financial inclusion, innovation, and regulatory frameworks, the author investigated the development of fintech and its repercussions in the MENA nations. Ransomware-related data breaches were the subject of a comprehensive literature review by Reshmi (2021). The author examined the detection, prevention, and mitigation of ransomware attacks as well as their impact on information security. The rise of PropTech (Property Technology) and its impact on the real estate industry were the subject of Starr et al.'s (2021) investigation. The authors looked at how PropTech changed property investment and finance and how these new industrial technologies could be used in real estate. Farooq and Chawla (2021) looked at how data science and AI are used in finance. The authors talked about how risk management, trading strategies, credit scoring, fraud detection, and the use of data analytics, machine learning, and AI techniques in finance are used. A comparative study on financial technology regulation in Malaysia and Indonesia was carried out by Diniyya et al. (2021). The authors looked at the regulatory frameworks in both countries and found that their approaches to regulating FinTech were similar and different.

In the context of FinTech, Buckley et al. (2019) talked about the new risks that come with digital financial transformation and the rise of technology-related risks (techrisks). The authors emphasized the significance of addressing these risks and establishing appropriate regulatory frameworks to guarantee the financial system's stability and integrity. Jagtiani and John (2018) looked at how FinTech affected customers and how regulators responded. The study found that while innovations in FinTech offered a number of benefits, including increased access to financial services and cost savings, they also posed risks related to cybersecurity, consumer protection, and data privacy. In order to address these difficulties, the authors emphasized the necessity of appropriate regulations. A transcript of a discussion about making Bitcoin legal was provided by Anderson (2018). The legal aspects of cryptocurrencies, including regulatory frameworks and difficulties, were the subject of the discussion. The conclusion emphasized the significance of achieving a balance between innovation and regulation in order to effectively capitalize on cryptocurrencies' potential. Ravi and Kamaruddin (2017) looked into the advantages and drawbacks of applying big data analytics to smart financial services. Big data analytics has the potential to enhance financial sector decision-making, risk management, and customer experience, according to the study. To fully reap the benefits of big data analytics, the conclusion emphasized the need for talent development, data governance, and privacy protection.

9.5 Artificial Intelligence

Singh (2023) emphasized how AI and DL could help financial institutions in the United Kingdom meet regulatory requirements and focused on the compliance difficulties they face. The author provided insight into how these technologies could simplify compliance procedures and emphasized the significance of taking into account the burden of regulations. Bhat and co. (2023) addressed the financial sector's significance of FinTech enablers and the upcoming Internet of Things (IoT). The authors emphasized the significance of these technologies in shaping the future of finance and the transformative potential of AI and DL in various FinTech use cases. Gąsiorkiewicz & Monkiewicz (2023) explored the digital revolution's impact on banking, insurance, and capital markets. Their findings highlighted the transformative effects of technology on traditional financial institutions. Jafar and others (2023) provided a thorough evaluation of AI's role in capital markets. A valuable resource for practitioners and researchers in the field, their book provided a systematic examination of AI and machine learning (ML) techniques utilized in the context of cryptocurrency and finance. Nerurkar (2023) investigated the possibility of using DL to spot fraudulent Bitcoin transactions. The study showed that DL models are good at spotting suspicious patterns and making cryptocurrency transactions more secure. The technologies behind crypto-based decentralized finance (DeFi) were investigated by Kaur and KrishnaKumar (2023). The authors shed light on the technological foundations of this new field by discussing the role that AI and DL play in facilitating the creation of secure business models within the context of DeFi.

In addition to the idea of the InfoSphere and the upcoming metaverse, Masera (2023) gave an overview of Web 1.0, 2.0, and 3.0. The author talked about the changing landscape of digital transformation, showing how these ideas are connected and how they could affect changes in the economy, society, politics, and finances. Nguyen and co. (2023) looked at how big data, artificial intelligence, and machine learning (ML) work together in the financial technology (FinTech) industry. Their research emphasized how these technologies could complement one another and make it possible for revolutionary applications in a variety of financial fields. Balakrishna and others (2023) looked at how ML and AI are used in Industry 4.0 and the banking industry. In the context of the fourth industrial revolution, the authors emphasized the significance of these technologies in facilitating automation, optimization, and decision-making processes, thereby enhancing operational efficiency and customer experiences. Huynh-The and co. (2023) looked at how AI fits into the metaverse and looked into how AI techniques could improve virtual reality experiences. The authors presented opportunities for the integration of AI and DL in virtual financial environments by discussing various AI applications in the metaverse.

Amirzadeh and co. (2022) gave a comprehensive report on how AI is being used in the cryptocurrency markets. The authors looked into various AI strategies used in portfolio management, price prediction, risk assessment, and cryptocurrency trading. The results of the survey shed light on the current state of AI adoption in the ever-evolving cryptocurrency landscape. Gill and others (2022) gave an overview of emerging AI trends and future directions for computing of the next generation. The authors highlighted the transformative potential of AI technologies in shaping the future of computing by discussing the potential applications of AI in various fields, including finance. Aziz and others (2022) looked into how machine learning could be used in finance by using topic modeling. To shed light on the changing landscape and emerging trends in the field, the authors analyzed a large corpus of research articles to identify key research topics and subfields at the intersection of machine learning and finance. O'Hare and co. (2022) focused on Bitcoin and stablecoins in the context of ML-enabled metaverse telecollaboration to discuss the role of trust and money in the digital society. The authors discussed the potential effects on digital transactions and trust mechanisms of the intersection of cryptocurrency, machine learning, and the emerging metaverse concept.

Ahmad and co. (2022) looked at how energy systems could benefit from AI and industry 4.0. The authors highlighted the potential of AI-driven solutions in maximizing energy efficiency and sustainability by discussing the integration of AI techniques with energy systems, such as machine learning and deep learning. In the context of fintech, Dang (2021) investigated how technology trends shaped finance. The author looked at how AI, blockchain, and digital payments, among other technological advancements, are affecting the financial sector. In the banking, investment, and microfinance sectors, Ashta and Herrmann (2021) provided an overview of the opportunities and dangers associated with FinTech and artificial intelligence. The authors looked into the difficulties and ethical questions that arose when using AI to improve financial services. Using evolutionary computing and machine learning methods, Roddy (2018) investigated the application of generative music systems to ambient data monitoring. The study showed that these technologies could be combined to make interactive and adaptive music systems. The use of EC and ML methods in the creation of novel music systems was emphasized in the conclusion. Malhotra (2018a, b) talked about how AI, machine learning, and deep learning applications use generative adversarial networks (GANs) to manage model risk. Model risk management in these cutting-edge technologies was the primary focus of the investigation. The significance of efficient risk management frameworks for AI and ML models was emphasized in the conclusion.

9.6 Integration of Blockchain and AI

Jain and co. (2023) examined the most recent developments in FinTech, focusing on the financial sector's transformation. The authors talked about how new technologies like AI, DL, blockchain, and data analytics will affect risk management, payment systems, and digital banking. Biju and others (2023) examined the financial sector's AI, DL, and ML research landscape through bibliometric analysis. Their research provided a comprehensive overview of the existing literature and provided insights into the research trends, domains, and methodologies utilized in this field. Oyedele and others (2023) looked into how well boosted trees and DL worked to predict cryptocurrency closing prices. Their research shed light on the potential of DL models to identify intricate patterns in cryptocurrency data and provided investors and traders with useful insights. Myrzashova and others (2023) carried out a comprehensive examination of the convergence of federated learning and blockchain in healthcare. The authors talked about the problems and opportunities that come with integrating these technologies. They also talked about the possibility of using secure and private collaborative learning in healthcare applications.

Chae and Olson (2022) gave experiences into the advancements and uses of Industry 4.0, drawing on network examination. The authors looked at the intricate web of technologies that make up Industry 4.0 and highlighted the connections and potential synergies between new technologies like artificial intelligence (AI), the Internet of Things (IoT), and big data analytics. A systematic literature review of recent developments and future trends at the intersection of fintech and Industry 4.0 was carried out by Ferraro et al. (2022). The authors looked at how fintech and cutting-edge technologies like AI, machine learning, blockchain, and the Internet of Things can work together to benefit the financial sector. Aysan and others (2022) conducted a global empirical analysis of Islamic banks' use of fintech strategies. The authors looked at how AI, blockchain, and other technologies were used to improve services and meet the specific needs of Islamic banking by combining Islamic finance principles with fintech innovations.

Gayathri and co. (2022) utilized bibliographic analysis to comprehend the finance area of AI and ML. The authors looked at a lot of different research articles to figure out the main themes and trends in how AI and machine learning are used in the financial sector. This helped them figure out the main directions for research and how things have changed. Venkataraman and Yerchuru's (2021) outlook on the development of financial technology The authors talked about the changing landscape of fintech, including things like digital payments, blockchain, AI, and data analytics, as well as trends, problems, and opportunities in these areas. A study on blockchain and artificial intelligence-based traceable and intelligent supply chains was presented by Karadgi et al. (2021). The authors looked into how blockchain and AI could be used to improve supply chain management, focusing on increasing efficiency, transparency, and traceability. In-depth analysis was used in Anton's (2022) discussion of the integration of ML and blockchain technologies. The author looked at how blockchain and machine learning could work together to improve data security, transparency, and trust in financial systems.

Tyagi et al. (2020) examined the synergies, perspectives, difficulties, and research directions associated with combining blockchain technology and artificial intelligence. The authors looked into how healthcare, finance, and supply chain management could be improved by combining these technologies. Inbaraj and Chaitanya (2020) talked about how blockchain and machine learning work together. The authors looked into how systems and applications could be made more efficient and secure by combining blockchain technology with machine learning methods. Saigal (2020) looked into how blockchain and artificial intelligence could work together. The author talked about the possible advantages and applications of combining AI and blockchain, such as secure data sharing, privacy-enhancing methods, and decentralized AI models. Malhotra (2018a, b) talked about the importance of a national cybersecurity system and the implementation of a single digital ID for The Bahamas citizens. In order to safeguard citizens' information in e-government initiatives, the study emphasized the necessity of robust data privacy and security measures. The potential advantages of a unified digital ID system for simplified services and enhanced cybersecurity were emphasized in the conclusion.

9.7 Blockchain Technology

Serrano (2022) focused on using smart contracts and the blockchain to verify and validate data marketplaces. The author looked at the potential advantages and disadvantages of implementing blockchain-based verification mechanisms and talked about how blockchain technology could be used to increase trust and transparency in data marketplaces. Using ensemble learning and feature engineering, Jatoth et al. (2021) proposed an improved classification method for blockchain transactions. The authors focused on making transaction classification in blockchain networks more accurate and effective. In the field of augmented reality and virtual reality (AR/VR), Bhattacharya et al. (2021) investigated the possibility of combining blockchain technology with 6G technology. Utilizing the synergy between blockchain and 6G technologies to improve augmented and virtual reality (AR/VR) experiences was the topic of discussion by the authors. EthReview, an Ethereum-based product review system intended to reduce rating frauds, was proposed by Zulfiqar et al. (2021). A transparent and tamper-proof platform for genuine product reviews was developed by the authors using blockchain technology.

A smart healthcare system that makes use of blockchain technology was the topic of discussion by Shukla et al. (2020). The authors looked into how blockchain could be used to manage medical records in a secure way, make it easier to share data, and make healthcare services more efficient and private. Business process modeling and blockchain technology were investigated by Ramachandran (2019) for use in financial cloud systems. The study came to the conclusion that integrating blockchain into financial cloud platforms could have a positive effect on financial processes by enhancing security, transparency, and efficiency. IDMoB, a blockchain-based marketplace for IoT data, was proposed by Ozyilmaz et al. (2018). In the context of the Internet of Things (IoT), the study demonstrated the advantages of using blockchain for secure and transparent data exchange. The authors came to the conclusion that in IoT ecosystems, trusted and efficient transactions could be made through blockchain-based data marketplaces. In a critical examination of algorithmic thinking and its ramifications, Hurlburt (2018) In order to guarantee the responsible and ethical use of algorithms, the study emphasized the significance of comprehending the limitations and biases inherent in algorithmic systems. The conclusion required critical thinking and the evaluation of various domain-specific algorithmic systems. Vadlamani & Sk (2017) explored the opportunities and challenges presented by big data analytics in smart financial services. Their research underscored the potential benefits of harnessing big data while acknowledging the complexities involved.

Traceable supply chains, cutting-edge FinTech research, the integration of AI and FinTech in banking and microfinance, Ethereum-based product review systems, blockchain-powered smart healthcare systems, the fusion of blockchain and AI, financial technology regulation, the combination of blockchain and machine learning technologies, the convergence of AI and blockchain, and the risks associated with digital financial transformation are just some of the topics covered in these studies. These reviewed articles show, taken as a whole, how much the financial sector is affected by AI and new technologies. Compliance, prediction, security, education, decision-making, and customer experiences are just a few of the areas of finance where these technologies have the potential to change the game. The research sheds light on the opportunities and challenges posed by AI and new technologies in the financial sector and paves the way for future advancements, providing valuable insights. In general, the findings of this literature review point to the growing significance of AI and DL in the financial industry, particularly in compliance, prediction, security, and automation. The reviewed studies provide valuable insights and set the stage for future advancements in the field by demonstrating the potential benefits of integrating these technologies into financial institutions, cryptocurrency markets, and related domains. The literature review sheds light on the current state of research by highlighting the potential applications, obstacles, and future directions for using AI to improve decision-making processes and transform financial systems. It discusses security, risk management, data marketplaces, augmented intelligence, green finance, industry trends, cryptocurrency price prediction, and other aspects of AI, ML, and fintech that intersect with the financial sector. It also discusses the globalization of finance and fintech, the future of fintech, the economics of cryptocurrencies, fintech research and policy, blockchain transactions, data security breaches, PropTech, data science and AI in finance, the use of 6G and blockchain together, and the impact of technology trends on finance.

Additionally, the study will investigate the effects of fintech, AI, and blockchain on the banking, investment, insurance, and regulatory frameworks of the financial industry. Decentralized finance (DeFi), smart contracts, digital currencies, robo-advisory services, fraud detection, and regulatory compliance will all be examined in depth.

Policymakers, researchers, industry practitioners, and financial sector stakeholders will benefit from this study's insights. In order to encourage the responsible integration of these technologies, policymakers can acquire a comprehensive understanding of the implications and potential regulatory interventions. The synthesised findings will be beneficial to researchers because they can direct future investigations and highlight research gaps. The study's findings can be used by practitioners in the industry to create strategic plans, innovate their offerings, and improve the security and efficiency of financial services.

The financial sector's incorporation of fintech, artificial intelligence, and blockchain is a revolutionary development with significant repercussions. The objectives of this systematic review are to provide a comprehensive analysis of the existing literature, focusing on the benefits, drawbacks, and implications of integrating these technologies. This study will help clarify the advantages, disadvantages, and factors to take into account when using blockchain, AI, and fintech in the financial sector by combining and analyzing a wide range of scholarly articles.

9.8 Methodology

1. Research Design: This study employs a systematic review methodology to analyze secondary data gathered from various sources. It focuses on articles related to the integration of blockchain, artificial intelligence (AI), and fintech in the financial sector.
2. Data Sources: The study collects articles from reputable academic databases, including Google Scholar, J-gate, Ebsco, ProQuest, and Sciencedirect. These databases offer a wide range of scholarly literature covering the period from 2015 to 2023. By utilizing multiple sources, the study ensures a comprehensive review of the relevant literature.
3. Search Strategy: A systematic search strategy is employed to identify relevant articles. Keywords and search terms related to blockchain, AI, fintech, finance, and their intersections are used. Boolean operators such as "AND" and "OR" are utilized to refine the search and increase its specificity. The search strategy is designed to retrieve articles that discuss the integration of blockchain, AI, and fintech in the financial sector.
4. Inclusion and Exclusion Criteria: The study establishes specific inclusion and exclusion criteria to ensure the selection of relevant articles. Included articles

must focus on the integration of blockchain, AI, and fintech in the financial sector. Only peer-reviewed articles and scholarly publications are considered. Non-English articles and articles that do not address the research topic are excluded.

5. Screening and Selection: The retrieved articles undergo a two-step screening process. In the first step, titles and abstracts are reviewed to determine their relevance to the research topic. In the second step, full-text articles that pass the initial screening are assessed to ensure they meet the inclusion criteria. Any discrepancies in the screening process are resolved through discussions and consensus among the researchers.
6. Data Extraction: Relevant data from the selected articles are extracted and organized into a systematic review framework. Key information, such as author(s), publication year, research objectives, methodology, findings, and conclusions, is extracted from each article. This process ensures a standardized approach to data extraction and allows for easy comparison and analysis.
7. Data Analysis: The extracted data are analyzed using a thematic analysis approach. Similar themes and patterns across the articles are identified and categorized. The analysis focuses on the challenges, concerns, strategies, and solutions related to the integration of blockchain, AI, and fintech in the financial sector. The findings are synthesized to provide a comprehensive understanding of the research topic.
8. Synthesis and Discussion: The synthesized findings are discussed in light of the research objectives. The challenges and concerns identified are examined, and potential strategies and solutions are explored. The discussion provides insights into the opportunities and risks associated with the integration of these technologies and offers recommendations for policymakers, researchers, and industry practitioners.
9. Limitations: The study acknowledges certain limitations, such as the reliance on secondary data and the potential for publication bias. The inclusion of only peer-reviewed articles may limit the diversity of perspectives. Additionally, the study focuses on the selected time frame of 2015 to 2023, which may exclude relevant older or more recent publications. However, efforts are made to mitigate these limitations through a comprehensive search strategy and rigorous data analysis.
10. Ethical Considerations: As this study is based on secondary data, ethical considerations mainly revolve around proper citation and referencing of the included articles to give credit to the original authors. Plagiarism is strictly avoided, and appropriate referencing guidelines are followed to ensure academic integrity.

Overall, this study employs a systematic review methodology, relying on secondary data from various scholarly sources, to provide a comprehensive analysis of the integration of blockchain, AI, and fintech in the financial sector.

9.9 Findings and Discussion

The rapid advancement of technology has significantly transformed the financial industry, leading to the emergence of innovative solutions such as blockchain, artificial intelligence (AI), and financial technology (fintech). In this literature review, we delve into a comprehensive discussion of 80 articles that explore the applications, challenges, and potential implications of integrating blockchain, AI, and fintech. By examining these scholarly works, we aim to gain insights into the current landscape and future directions of these technologies in the financial sector.

1. The integration of blockchain and AI in financial cloud systems (Ramachandran 2019) offers enhanced security, transparency, and efficiency. By leveraging blockchain's distributed ledger technology and AI's analytical capabilities, financial processes can be streamlined and made more secure.
2. The impact of fintech on consumers and regulatory responses (Jagtiani and John 2018) highlights the benefits of increased access to financial services and cost savings. However, it also emphasizes the need for robust regulations to address potential risks related to data privacy, cybersecurity, and consumer protection.
3. The legal aspects of Bitcoin and cryptocurrencies (Anderson 2018) emphasize the importance of striking a balance between regulation and innovation. While regulatory frameworks are necessary to mitigate risks, overly restrictive measures could hinder the potential benefits of cryptocurrencies.
4. The proposed IoT data marketplace on the blockchain (Ozyilmaz et al. 2018) showcases the potential of blockchain for secure and transparent data exchange in the context of the Internet of Things (IoT). Such marketplaces can foster trusted transactions and unlock the full potential of IoT applications.
5. The implementation of a single digital ID for citizens of The Bahamas (Malhotra 2018a, b) underlines the importance of a national cybersecurity system to ensure data privacy and security. A unified digital ID system can streamline services and enhance cybersecurity measures, safeguarding citizens' information.
6. Critical examination of algorithmic thinking (Hurlburt 2018) emphasizes the need for understanding biases and limitations inherent in algorithmic systems. It calls for responsible and ethical use of algorithms by critically evaluating their impact in various domains.
7. The exploration of generative music systems with ambient data monitoring (Roddy 2018) demonstrates the potential of combining evolutionary computing and machine learning techniques. This integration can lead to the development of interactive and adaptive music systems, showcasing the value of these technologies in the creative domain.
8. The use of generative adversarial networks (GANs) for model risk management in AI and ML (Malhotra 2018a, b) highlights the challenges associated with model risk management in advanced technologies. It underscores the impor-

tance of implementing effective risk management frameworks to ensure the reliability and integrity of AI and ML models.

9. Big data analytics in smart financial services (Ravi and Kamaruddin 2017) presents opportunities for improved decision-making, risk management, and customer experience in the financial sector. However, it also calls for attention to data governance, privacy protection, and talent development to fully harness the benefits of big data analytics.
10. The report on the future of finance, risk, and quant in the cyber era (Malhotra 2015) emphasizes the need for proactive risk management strategies in the face of emerging challenges. It highlights the importance of effectively managing model risks to navigate the evolving landscape of the finance industry.

The reviewed literature provides a comprehensive understanding of the applications and implications of integrating blockchain, AI, and fintech in the financial sector. The findings underscore the potential benefits of these technologies, including enhanced security, transparency, efficiency, and improved customer experiences. However,

9.10 Social Implications

The integration of blockchain, artificial intelligence, and fintech in the financial sector has significant social implications. Firstly, it has the prospective to democratize contact to financial services, especially in underserved or unbanked populations. Through the use of innovative fintech solutions, individuals can access banking services, make payments, and manage their finances more conveniently and affordably. This can contribute to financial inclusion and economic empowerment.

Moreover, the increased transparency and security provided by blockchain technology can help build trust in financial transactions and mitigate fraud and corruption. This can have a positive impact on the overall integrity of financial systems, reducing the risk of financial crimes and improving the trust between financial institutions and their customers.

However, there are also social challenges that need to be addressed. The widespread adoption of blockchain, AI, and fintech may lead to job displacements and require workforce reskilling. As manual tasks are automated and AI systems take over decision-making processes, there may be a need for individuals to acquire new skills to adapt to the changing job market. Ensuring that the benefits of these technologies are distributed equitably and that individuals are adequately prepared for the digital transformation is crucial.

9.11 Practical Implications

The integration of blockchain, artificial intelligence, and fintech presents several practical implications for financial institutions, regulators, and policymakers. Financial institutions need to adapt their business models and operations to leverage the potential of these technologies. They may need to invest in infrastructure, talent, and partnerships to implement blockchain-based systems, AI algorithms, and fintech solutions effectively. Collaboration with technology providers and startups can be crucial for successful integration.

Regulators and policymakers also play a crucial role in shaping the regulatory frameworks and policies surrounding these technologies. They need to strike a balance between nurturing innovation and ensuring consumer fortification, data privacy, and systemic stability. Regulations should be agile and adaptable to keep pace with the rapid advancements in these areas, while also mitigating potential risks such as money laundering, cybersecurity threats, and market manipulation.

Furthermore, for practical implementation, interoperability and standardization across different blockchain platforms and AI systems are essential. Collaboration among stakeholders, industry consortia, and standard-setting organizations can facilitate interoperability, data sharing, and the seamless integration of these technologies.

9.12 Research Implications

This study has implications for future research in the field of blockchain, artificial intelligence, and fintech integration in the financial sector. Firstly, further research is needed to explore the specific applications and use cases of these technologies across different areas of finance, such as banking, insurance, investment, and payments. Understanding the potential benefits, challenges, and best practices in each domain can guide practitioners and policymakers in their decision-making processes.

Additionally, research should focus on addressing the ethical, legal, and societal implications of integrating these technologies. As AI systems make autonomous decisions and blockchain stores immutable records, ethical considerations, algorithmic biases, and privacy concerns need to be carefully examined. The societal impact of these technologies, including their effects on income distribution, wealth inequality, and social inclusion, requires rigorous analysis and debate.

Furthermore, research should investigate the scalability, performance, and security aspects of blockchain and AI systems. Advancements in consensus algorithms, privacy-enhancing technologies, and AI explainability can contribute to the widespread adoption of these technologies in the financial sector.

The integration of blockchain, AI, and fintech in the financial industry has enormous societal, practical, and research consequences. Understanding and addressing these implications can pave the way for responsible and sustainable integration, fostering innovation, financial inclusion, and trust in the digital economy.

9.13 Conclusion

In conclusion, this study conducted a systematic review of literature on the integration of blockchain, artificial intelligence, and fintech in the financial sector. The findings of the review shed light on the advancements, challenges, and potential implications of these technologies for various stakeholders.

Fintech, blockchain, and AI together have the power to revolutionise the financial industry by increasing efficiency, transparency, and security. Blockchain technology enables safe and open financial transactions by providing decentralised and immutable ledger capabilities. Artificial intelligence, on the other hand, brings automation, data analytics, and decision-making capabilities to financial processes. Fintech solutions leverage these technologies to provide innovative financial services, improve customer experiences, and drive financial inclusion.

The review identified several key findings. Firstly, blockchain technology is being applied in areas such as payments, trade finance, identity management, and smart contracts. Its ability to ensure trust, reduce intermediaries, and streamline processes has the potential to revolutionize financial systems. Artificial intelligence is being utilized for credit scoring, fraud detection, customer service, and investment management, among other applications. The use of AI algorithms can enhance decision-making, automate tasks, and provide personalized financial services. Fintech solutions, integrating blockchain and AI, are offering new ways of accessing financial services, such as peer-to-peer lending, robo-advisory, and digital payments.

While the integration of these technologies brings numerous opportunities, it also poses challenges. Technical challenges include scalability, interoperability, and security of blockchain networks. The complexity and explainability of AI algorithms raise ethical and regulatory concerns. Data privacy, protection, and ownership are critical issues in utilizing these technologies. Additionally, regulatory frameworks need to evolve to keep pace with the rapid advancements in fintech and ensure consumer protection, market integrity, and systemic stability.

The implications of this study are significant. The findings can guide financial institutions in adopting and implementing blockchain, artificial intelligence, and fintech solutions to improve their operations, enhance customer experiences, and drive innovation. Regulators and policymakers can use the insights to develop agile and adaptive regulatory frameworks that foster innovation while mitigating risks. In order to address the issues and support the appropriate integration of these technologies, the report also emphasises the necessity of multidisciplinary cooperation among technology suppliers, financial institutions, regulators, and academics.

Moreover, the study identifies social implications, such as financial inclusion, job displacements, and the need for reskilling in the digital era. It emphasizes the importance of ensuring equitable access to these technologies and preparing individuals for the changing job market. The study also highlights practical implications, such as the need for infrastructure investments, partnerships, and interoperability across different platforms. Lastly, it identifies research implications, including the exploration of specific applications, ethical considerations, scalability, and security aspects of these technologies.

In conclusion, combining fintech, artificial intelligence, and blockchain has the potential to transform the financial industry. This research sheds light on the advancements, difficulties, and repercussions of these technologies. By addressing the identified implications and conducting further research, stakeholders can harness the benefits of integration while addressing the associated challenges, thereby fostering innovation, financial inclusion, and trust in the digital economy.

References

Agarwal JD, Agarwal M, Agarwal A, Agarwal Y (2020) Economics of cryptocurrencies: artificial intelligence, blockchain, and digital currency. Information for efficient decision making, In, pp 331–430. https://doi.org/10.1142/9789811220470_0013

Ahmad T, Zhu H, Zhang D, Tariq R, Bassam A, Ullah F et al (2022) Energetics systems and artificial intelligence: applications of industry 4.0. Energy Rep 8:334–361. https://doi.org/10.1016/j.egyr.2021.11.256

Allen F (2021, September) Globalization of finance and fintech in the MENA region. Economic Research Forum (ERF)

Allen F, Gu X, Jagtiani J (2020) A survey of fintech research and policy discussion. doi:https://doi.org/10.21799/frbp.wp.2020.21

Amirzadeh R, Nazari A, Thiruvady D (2022) Applying artificial intelligence in cryptocurrency markets: a survey. Algorithms 15(11):428. https://doi.org/10.3390/a15110428

Anderson R (2018) Making bitcoin legal (transcript of discussion). In: Lecture notes in computer science. Security protocols XXVI, pp 254–265. doi:https://doi.org/10.1007/978-3-030-03251-7_30

Anton D (2022) Integration of blockchain technologies and machine learning with deep analysis. Int J Innov Technol Soc Sci 4(36). https://doi.org/10.31435/rsglobal_ijitss/30122022/7918

Ashta A, Herrmann H (2021) Artificial intelligence and fintech: an overview of opportunities and risks for banking, investments, and microfinance. Strateg Chang 30(3):211–222. https://doi.org/10.1002/jsc.2404

Awotunde JB, Ogundokun RO, Jimoh RG, Misra S, Aro TO (2021) Machine learning algorithm for cryptocurrencies price prediction. In: Studies in computational intelligence. Artificial intelligence for cyber security: methods, issues and possible horizons or opportunities, pp 421–447. doi:https://doi.org/10.1007/978-3-030-72236-4_17

Aysan AF, Belatik A, Unal IM, Ettaai R (2022) Fintech strategies of Islamic banks: a global empirical analysis. FinTech 1(2):206–215. https://doi.org/10.3390/fintech1020016

Aziz S, Dowling M, Hammami H, Piepenbrink A (2022) Machine learning in finance: a topic modeling approach. Eur Financ Manag 28(3):744–770. https://doi.org/10.1111/eufm.12326

Balakrishna S, Arulkumar V, Srihari M, Rohith C (2023, January 23) Usage of machine learning and artificial intelligence in industry 4.0 and banking sector. *2023 5th international conference on smart systems and inventive technology (ICSSIT)*, 1282–1287. Presented at the 2023 5th international conference on smart systems and inventive technology (ICSSIT), Tirunelveli. doi:https://doi.org/10.1109/icssit55814.2023.10060952

Bellini F, Vassalli F (2022) On the state-of-the-art of FinTech world and the initial approach of central banks. Int J Financ Innov Bank 3(2):113. https://doi.org/10.1504/ijfib.2022.124206

Bhat JR, AlQahtani SA, Nekovee M (2023) FinTech enablers, use cases, and role of future internet of things. J King Saud Univ Comput Inf Sci 35(1):87–101. https://doi.org/10.1016/j.jksuci.2022.08.033

Bhatt A, Joshipura M, Joshipura N (2022) Decoding the trinity of Fintech, digitalization and financial services: an integrated bibliometric analysis and thematic literature review approach. Cogent Econ Financ 10(1). https://doi.org/10.1080/23322039.2022.2114160

Bhattacharya P, Saraswat D, Dave A, Acharya M, Tanwar S, Sharma G, Davidson IE (2021) Coalition of 6G and blockchain in AR/VR space: challenges and future directions. IEEE Access: Pract Innov Open Solut 9:168455–168484. https://doi.org/10.1109/access.2021.3136860

Biju AKVN, Thomas AS, Thasneem J (2023) Examining the research taxonomy of artificial intelligence, deep learning & machine learning in the financial sphere—a bibliometric analysis. Qual Quant. https://doi.org/10.1007/s11135-023-01673-0

Buckley RP, Arner DW, Zetzsche DA, Selga E (2019) The dark side of digital financial transformation: the new risks of FinTech and the rise of TechRisk. SSRN Electron J. https://doi.org/10.2139/ssrn.3478640

Chae B (kevin), Olson D (2022) Technologies and applications of Industry 4.0: insights from network analytics. Int J Prod Res 60(12): 3682–3704. doi:https://doi.org/10.1080/00207543.2021.1931524

Chen W, El Majzoub A, Al-Qudah I, Rabhi FA (2023) A CEP-driven framework for real-time news impact prediction on financial markets. SOCA. https://doi.org/10.1007/s11761-023-00358-8

Choithani T, Chowdhury A, Patel S, Patel P, Patel D, Shah M (2022) A comprehensive study of artificial intelligence and cybersecurity on bitcoin, crypto currency and banking system. Ann Data Sci. https://doi.org/10.1007/s40745-022-00433-5

Cumming D, Johan S, Wood G (2021) In: Cumming D, Johan S, Wood G (Eds) The oxford handbook of hedge funds the oxford handbook of hedge funds. doi:https://doi.org/10.1093/oxfordhb/9780198840954.001.0001

Dang HNP (2021) Technology in fintech-how trends shape finance. https://www.theseus.fi/bitstream/handle/10024/501307/THESIS-Hoang%20Nam%20Phuong%20Dang.pdf?sequence=2

Dempere JM, El-Agure ZA, Memic D (2022, May 25) Data selection to train machine learning models and forecast bitcoin prices: depth vs. Width. In 2022 8th international conference on information technology trends (ITT), pp 39–44. Presented at the 2022 8th international conference on information technology trends (ITT), Dubai. doi:https://doi.org/10.1109/itt56123.2022.9863966

Diniyya AA, Aulia M, Wahyudi R (2021) Financial technology regulation in Malaysia and Indonesia: A comparative study. Ihtifaz J Islam Econ Financ Bank 3(2):67. https://doi.org/10.12928/ijiefb.v3i2.2703

Dziedzic M (2023) 12 Smart contracts and artificial intelligence. The Digital Revolution in Banking, Insurance and Capital Markets, p 12

Farooq A, Chawla P (2021, December) Review of data science and AI in finance. In: 2021 international conference on computing sciences (ICCS), pp 216–222. Presented at the 2021 International conference on computing sciences (ICCS), Phagwara. doi:https://doi.org/10.1109/iccs54944.2021.00050

Ferraro G, Ramponi A, Scarlatti S (2022) Fintech meets Industry 4.0: a systematic literature review of recent developments and future trends. Technol Anal Strateg Manag:1–17. https://doi.org/10.1080/09537325.2022.2117025

Fritz-Morgenthal S, Hein B, Papenbrock J (2022) Financial risk management and explainable, trustworthy, responsible AI. Front Artif Intell 5:779799. https://doi.org/10.3389/frai.2022.779799

Gąsiorkiewicz L, Monkiewicz J (2023) In: Gasiorkiewicz L, Monkiewicz J (Eds.) The digital revolution in banking, insurance and capital markets the digital revolution in banking, insurance and capital markets. Retrieved from https://books.google.at/books?id=r2GmEAAAQBAJ

Gayathri S, Purushothaman G, Dhongde S, Manju SD, Aboli PN, Ranjan V, Ahmad J (2022) Bibliographic analysis to understand the field of artificial intelligence and machine learning in finance by establishing its core ideas, major topics, and related studies. NeuroQuantology 20(22):1

Gill SS, Xu M, Ottaviani C, Patros P, Bahsoon R, Shaghaghi A et al (2022) AI for next generation computing: emerging trends and future directions. Int Things 19(100514):100514. https://doi.org/10.1016/j.iot.2022.100514

Goodell JW, Kumar S, Lim WM, Pattnaik D (2021) Artificial intelligence and machine learning in finance: identifying foundations, themes, and research clusters from bibliometric analysis. J Behav Exp Financ 32(100577):100577. https://doi.org/10.1016/j.jbef.2021.100577

Guarino A, Grilli L, Santoro D, Messina F, Zaccagnino R (2022) To learn or not to learn? Evaluating autonomous, adaptive, automated traders in cryptocurrencies financial bubbles. Neural Comput Applic 34(23):20715–20756. https://doi.org/10.1007/s00521-022-07543-4

Hendershott T, Zhang X (michael), Zhao JL, Zheng Z (eric) (2021) FinTech as a game changer: overview of research frontiers. Inf Syst Res 32(1): 1–17. doi:https://doi.org/10.1287/isre.2021.0997

Hurlburt GF (2018) Thinking critically about algorithmic thinking. IT Prof 20(2):5–10. https://doi.org/10.1109/mitp.2018.021921644

Huynh-The T, Pham Q-V, Pham X-Q, Nguyen TT, Han Z, Kim D-S (2023) Artificial intelligence for the metaverse: a survey. Eng Appl Artif Intell 117(105581):105581. https://doi.org/10.1016/j.engappai.2022.105581

Inbaraj XA, Chaitanya TR (2020) Need to know about combined technologies of BlockChain and machine learning. In: Krishnan S, Balas VE, Julie EG, Robinson YH, Balaji S, Kumar R (eds) *Handbook of research on blockchain technology*, pp 417–432. https://doi.org/10.1016/b978-0-12-819816-2.00017-4

Jafar SH, Hemachandran K, El-Chaarani H, Moturi S, Gupta N (2023) In: Jafar SH, Hemachandran K, El-Chaarani H, Moturi S, Gupta N (Eds) Artificial intelligence for capital markets. Retrieved from https://books.google.at/books?id=la61EAAAQBAJ

Jagtiani J, John K (2018) Fintech: the impact on consumers and regulatory responses. J Econ Bus 100:1–6. https://doi.org/10.1016/j.jeconbus.2018.11.002

Jain R, Prajapati D, Dangi A (2023) Transforming the financial sector: A review of recent advancements in FinTech. Retrieved from https://papers.ssrn.com/abstract=4380348

Jatoth C, Jain R, Fiore U, Chatharasupalli S (2021) Improved classification of blockchain transactions using feature engineering and ensemble learning. Future Internet 14(1):16. https://doi.org/10.3390/fi14010016

Jović Ž, Nikolić I (2022) The darker side of FinTech: the emergence of new risks. Zagreb Int Rev Econ Bus 25(s1):45–63. https://doi.org/10.2478/zireb-2022-0024

Kaal, W. A. (2021). Decentralization–Why We Need Technology Infrastructure Upgrades. U of St. Thomas (Minnesota) Legal Studies Research Paper, 21–13.

Karadgi S, Kulkarni V, Doddamani S (2021) Traceable and intelligent supply chain based on blockchain and artificial Intelligence. J Phys Conf Ser 2070(1):012158. https://doi.org/10.1088/1742-6596/2070/1/012158

Kaur G, KrishnaKumar A (2023) Technologies behind crypto-based decentralized finance. In: Building secure business models through blockchain technology, pp 149–166. doi:https://doi.org/10.4018/978-1-6684-7808-0.ch009

Khan NS, Jhapate A (2022) Review on machine learning in suspecting money laundering. Res J Eng Technol Med Sci 5(3):61–65. http://www.rjetm.in/RJETM/Vol05_Issue03/Review%20on%20Machine%20Learning%20in%20Suspecting%20Money%20Laundering.pdf

Madhav AVS, Tyagi AK (2022) The world with future technologies (post-COVID-19): open issues, challenges, and the road ahead. In: Intelligent interactive multimedia systems for e-healthcare applications, pp 411–452. doi:https://doi.org/10.1007/978-981-16-6542-4_22

Malhotra Y (2015) Risk, uncertainty, and, profit for the cyber era: model risk management of cyber insurance models using quantitative finance and advanced analytics (Doctoral dissertation, State University of New York, Polytechnic Institute)

Malhotra Y (2018a) Bahamas e-Government: single digital ID for citizens of the Bahamas: toward a national Cybersecurity system to ensure data privacy and Security. Retrieved from https://papers.ssrn.com/abstract=3739258

Malhotra Y (2018b) Model risk management in AI, machine learning & deep learning (presentation slides). SSRN Electron J. https://doi.org/10.2139/ssrn.3167035

Masera R (2023) Web 1.0, 2.0, 3.0; InfoSphere; Metaverse: an overview. Monetary, financial, societal and geopolitical transformation cusps. SSRN Electron J. https://doi.org/10.2139/ssrn.4337362

Mhlanga D (2022) The transition from an informal financial money market to a formal financial system through digital financial inclusion. In: Palgrave studies in impact finance, pp 137–161. doi:https://doi.org/10.1007/978-3-031-16687-7_8

Mihov A-H, Firoozye N, Treleaven P (2022) Towards augmented financial intelligence. SSRN Electron J. https://doi.org/10.2139/ssrn.4148057

Mishra R, Haridas AC, Khunduru N, Chundru A, Mahbub S, Ramljak D (2022) Online portfolio management: a survey of data-driven approaches. In: City, society, and digital transformation, pp 357–373. doi:https://doi.org/10.1007/978-3-031-15644-1_27

Morgan PJ (2022) Assessing the risks associated with green digital finance and policies for coping with them. In: Economics, law, and institutions in Asia Pacific, pp 51–68. doi:https://doi.org/10.1007/978-981-19-2662-4_3

Myrzashova R, Alsamhi SH, Shvetsov AV, Hawbani A, Wei X (2023) Blockchain meets federated learning in healthcare: a systematic review with challenges and opportunities. IEEE Internet Things J 1–1. https://doi.org/10.1109/jiot.2023.3263598

Nerurkar P (2023) Illegal activity detection on bitcoin transaction using deep learning. Soft Comput 27(9):5503–5520. https://doi.org/10.1007/s00500-022-07779-1

Nguyen DK, Sermpinis G, Stasinakis C (2023) Big data, artificial intelligence and machine learning: a transformative symbiosis in favour of financial technology. Eur Financ Manag 29(2):517–548. https://doi.org/10.1111/eufm.12365

O'Hare JJ, Fairchild A, Ali U (2022) Money & trust in digital society – Bitcoin, nostr, stablecoins, digital objects and machine learning in B2B telepresent mixed reality. Retrieved from http://arxiv.org/abs/2207.09460

Oyedele AA, Ajayi AO, Oyedele LO, Bello SA, Jimoh KO (2023) Performance evaluation of deep learning and boosted trees for cryptocurrency closing price prediction. Expert Syst Appl 213(119233):119233. https://doi.org/10.1016/j.eswa.2022.119233

Ozyilmaz KR, Dogan M, Yurdakul A (2018, June) IDMoB: IoT data marketplace on blockchain. In: 2018 crypto valley conference on blockchain technology (CVCBT), pp 11–19. Presented at the 2018 crypto valley conference on blockchain technology (CVCBT), Zug. doi:https://doi.org/10.1109/cvcbt.2018.00007

Peter FJ, Styppa K (2023) Predicting millionaires from Ethereum transaction histories using node embeddings and artificial neural nets. Expert Syst Appl 223(119834):119834. https://doi.org/10.1016/j.eswa.2023.119834

Rajeswary C, Thirumaran M (2023, March 23) A comprehensive survey of automated website phishing detection techniques: a perspective of artificial intelligence and human behaviors. 2023 International conference on sustainable computing and data communication systems (ICSCDS), pp 420–427. Presented at the 2023 international conference on sustainable computing and data communication systems (ICSCDS), Erode. doi:https://doi.org/10.1109/icscds56580.2023.10104988

Ramachandran M (2019, May 3) Application of business process modelling and blockchain technology for financial cloud. INSTICC. Presented at the FMIB 2019 1st Intl. conference on finance, economics, and IT Business, Crete. Retrieved from http://eprints.leedsbeckett.ac.uk/id/eprint/6388/

Rather AM (2023) A new method of ensemble learning: case of cryptocurrency price prediction. Knowl Inf Syst 65(3):1179–1197. https://doi.org/10.1007/s10115-022-01796-0

Ravi, V., & Kamaruddin, S. (2017). Big data analytics enabled smart financial services: opportunities and challenges. In Big Data Analytics: 5th International Conference, BDA 2017, Hyderabad, India, December 12-15, 2017, Proceedings 5 (pp. 15–39). Springer International Publishing.

Reshmi TR (2021) Information security breaches due to ransomware attacks - a systematic literature review. Int J Inf Manag Data Insights 1(2):100013. https://doi.org/10.1016/j.jjimei.2021.100013

Roddy S (2018) Ambient data monitoring w/generative music systems using EC & ML techniques. Retrieved from http://www.tara.tcd.ie/handle/2262/98073

Saigal P (2020) Merger of artificial intelligence and blockchain. In: Blockchain technology and applications, pp 139–158. doi:https://doi.org/10.1201/9781003081487-8

Serrano W (2022) Verification and validation for data marketplaces via a blockchain and smart contracts. Blockchain Res Appl 3(4):100100. https://doi.org/10.1016/j.bcra.2022.100100

Shukla RG, Agarwal A, Shukla S (2020) Blockchain-powered smart healthcare system. In: Krishnan S, Balas VE, Julie EG, Robinson YH, Balaji S, Kumar R (eds) Handbook of research on blockchain technology, pp 245–270. doi:https://doi.org/10.1016/b978-0-12-819816-2.00010-1

Singh C (2023) Artificial intelligence and deep learning: considerations for financial institutions for compliance with the regulatory burden in the United Kingdom. J Financ Crime. https://doi.org/10.1108/jfc-01-2023-0011

Singh J, Thulasiram R, Thavaneswaran A (2022, December 4) LSTM based Algorithmic Trading model for Bitcoin. 2022 IEEE symposium series on computational intelligence (SSCI), pp 344–351. Presented at the 2022 IEEE symposium series on computational intelligence (SSCI), Singapore. doi:https://doi.org/10.1109/ssci51031.2022.10022021

Starr CW, Saginor J, Worzala E (2021) The rise of PropTech: emerging industrial technologies and their impact on real estate. J Prop Invest Financ 39(2):157–169. https://doi.org/10.1108/jpif-08-2020-0090

Steiner-Otoo D, Jahankhani H (2022) An investigation into how smartphones can be secured against MiTM attacks: financial sector. In Blockchain and other emerging technologies for digital business strategies, pp 171–215. doi:https://doi.org/10.1007/978-3-030-98225-6_7

Tamblyn I, Yu T, Benlolo I (2023) fintech-kMC: agent based simulations of financial platforms for design and testing of machine learning systems. Retrieved from http://arxiv.org/abs/2301.01807

Tyagi AK, Rekha G, Sreenath N (2020) Beyond the hype: Internet of things concepts, security and privacy concerns. In Learning and analytics in intelligent systems, pp 393–407. doi:https://doi.org/10.1007/978-3-030-24322-7_50

Vadlamani R, Sk K (2017) Big data analytics enabled smart financial services: opportunities and challenges. In Lecture notes in computer science. Big data analytics, pp 15–39. doi:https://doi.org/10.1007/978-3-319-72413-3_2

Venkataraman R, Yerchuru SK (2021) Future of financial technology—a perspective. CSI Trans ICT 9(4):207–213. https://doi.org/10.1007/s40012-021-00341-8

Vučinić M, Luburić R (2022) Fintech, risk-based thinking and cyber risk. J Cent Bank Theory Pract 11(2):27–53. https://doi.org/10.2478/jcbtp-2022-0012

Wagan AA, Khan AA, Chen Y-L, Yee PL, Yang J, Laghari AA (2023) Artificial intelligence-enabled game-based learning and Quality of experience: a novel and secure framework (B-AIQoE). Sustainability 15(6):5362. https://doi.org/10.3390/su15065362

Zulfiqar M, Tariq F, Janjua MU, Mian AN, Qayyum A, Qadir J et al (2021) EthReview: an Ethereum-based product review system for mitigating rating frauds. Comput Secur 100(102094):102094. https://doi.org/10.1016/j.cose.2020.102094

Chapter 10
The Potential Application of Blockchain in Green Finance

Kavita Singh, Komal, Yaditi Singh, and Seyedeh Shabnam Jazaeri

10.1 Introduction

Green finance mostly uses debt and equity as its financial vehicles. There are a number of distinguishing features of financial instruments, including the seniority of the holder (junior equity verses preferred stock), the structure of the flow of finance and intermediary actors (investor types and investment the vehicles), the nature of the contract, and the sources of the funds. Debt and equity-related products, and also a guarantees risk management solution, are the primary focus of this brochure. In the early stages of the project's or a business's development, equity financing, in which money is invested in the stock of an organization in exchange of a stake in the company, is commonly used. Shares of ownership in a company might be either preferred or common equity. There are two main distinctions between the various types of stocks. The preferred shareholders get paid first in the event of a corporate liquidation, ahead of unsecured creditors and bondholders. In the event of a surplus, common investors will be compensated. Second, preferred stock pay-outs are not the same as regular stock dividends and are often higher.

K. Singh (✉) · Komal
Faculty of Management, SRM University Delhi-NCR, Sonepat, Sonipat, Haryana, India
e-mail: kavita.s@srmuniversity.ac.in

Y. Singh
Zakir Husain College, University of Delhi, Delhi, India

S. S. Jazaeri
Department of Computer Engineering, North Tehran Branch, Islamic Azad University, Tehran, Iran

M. Irfan et al. (eds.), *Applications of Block Chain technology and Artificial Intelligence*, Financial Mathematics and Fintech,
https://doi.org/10.1007/978-3-031-47324-1_10

Green Finance: When money is invested in programs that help promote long-term sustainability, we call this "green finance." Investments in environmentally friendly areas including renewable energy, water purification, industrial pollution prevention, energy conservation, and carbon reduction are just a few examples. Green finance includes things like public funding for environmental policy. Green financing seeks to increase the amount of money that flows to sustainable development goals from the private, non-profit, and public sectors through banking, microcredit, insurance, & investment. An integral part of this is taking increasing responsibility for environmental and social matters, seeking out opportunities that have a positive impact on the environment and a decent rate of return.

Definition of Green Financing: Financing for ecologically positive endeavours, like the purchase of environmentally friendly products and services or the creation of green infrastructure, is known as "green finance." Green finance is growing in popularity as the dangers connected with environmentally harmful goods and services rise.

10.2 Types of Green Financing

Now that we've defined green finance, let's look at how it manifests itself in practice.

1. **Green Mortgages:** For properties that have been rated highly for environmental sustainability, or if the buyer agrees to make investments to boost the property's environmental performance, lenders can offer more favourable conditions.
2. **Green Loans:** Loans like this help finance green projects like installing solar panels on homes, switching to electric cars, and improving insulation.
3. **Green Credit Cards:** A tree is planted for every transaction made with a green credit card like the Aspirations Zero card. Customers are able to reallocate their funds toward green finance, which has a positive and long-lasting effect on the environment.
4. **Green Banks:** Green banks serve a similar purpose to traditional banks, but they instead employ public funding to encourage private investment in green energy and other sustainable projects. Research by 2020 estimates that there will be twenty green banks on the United States by then, having invested a total of $7 billion into renewable energy between 2011 and 2020.
5. **Green Bonds:** Most environmentally friendly financing comes from green bonds. Bond investments are included, with the interest going toward funding various environmentally friendly projects like those dealing with renewable energy, eco-friendly transportation, and conservation. Green money is booming. By 2023, the market for green bonds could be valued $2.36 trillion globally. It is thought to be a means of concurrently addressing the demands of capitalism and conservation. The US, China, and France are the top three countries that issue green bonds (Fleming 2020).

10.3 Advantages of Green Finance

To improve resource management over the long term, increase national competitiveness, and channel private capital in local green markets, governments in developing countries are establishing necessary infrastructure.

1. **Produces a Comparative Advantage:** Adapting to the growing threats poses by changes in the climate and other economic and environmental issues may necessitate shifting from a voluntary to a required approach to low-carbon green development. If environmental restrictions are strengthened, you can gain a competitive advantage by increasing green funding.
2. **Adds Business Value:** Businesses can boost their portfolio value by increasing (and advertising) their involvement in green financing. It gives them a leg up on the competition and attracts eco-conscious customers and investors.
3. **Enhances Economic Prospects:** Governments that back green financing help safeguard their populations from the effects of resource depletion. They do this by expanding into lucrative new markets and establishing strong foundations for existing renewable energy industries in their respective regions.

10.4 Challenges of Green Finance

Assuming we set aside the tricky problem of varying degrees of green, it isn't unexpected that a lot of the other difficulties in green financing have plagued the sector for decades.

- Provide funding for a wide range of initiatives including SMEs, infrastructure, and those with positive social benefits but low private returns.
- funding public investments in the face of competing priorities
- Deal with the many obstacles and inefficiencies that appear even in ostensibly well-functioning financial markets.

10.5 Green Finance vs. Sustainable Finance

Financial tools that promote environmental and social goals are called sustainable finance; those that support environmental goals only are called green finance. According to Bloomberg, in 2018, sustainability & green finance accounted for one-third of the tracked flows of cash in assets under management, totalling $30.7 trillion.

Lending, investing, and other forms of finance account for the vast bulk of investors' emissions. A carbon footprint in financial organizations is impacted by sponsored emissions 700 times more than operational emissions. Green investment, loan, and credit card finance can help with emission reduction goals.

10.6 How Does the Digitization of Financial Instruments Impact Finance?

The blockchain's advantages are multiplied by the digitalization of financial instruments like digital property, smart contracts, and programmed money, which creates new degrees of interoperability and programmability amongst commodities, services, property, and holdings. These digital tools will usher in a new era in the commercial & financial markets, one in which value is generated at every touchpoint.

The benefits of using digital financial instruments for firms include:

- **Authenticity and scarcity:** Data integrity is maintained, asset provenance is revealed, and a whole transactional history can be retrieved from one authoritative location thanks to digitization.
- **Programmable capabilities:** Assets can be imbued with code that handles governance, compliance, data protection, identity (Know Your Customer/Anti-Money Laundering traits), system incentives, & features that control stakeholder engagement (for voting and other rights).
- **Streamlined processes:** Increasing the number of automated processes boosts efficiency. Reduces processing timeframes, mistake and delay risk, and the number of steps & intermediaries needed to attain similar levels of trust in traditional processes while enabling real-time settlement, auditing, & reporting.
- **Economic benefits:** Automatic, more efficient procedures lead to lower infrastructure, operating, and transaction costs.
- **Market reactivity:** In addition to being issued more quickly, digital securities also allow for greater personalization than their traditional counterparts. Using investor demand as a guide, issuers can create digital financial products.

10.7 Opportunities and Solutions to Financial Services Challenges

Today's financial service providers face a wide range of challenges. Most of this can be attributed to the rapid development of technology. Although many banks have embraced the digital age, traditional banking institutions still confront several challenges.

The seven most significant obstacles facing the financial services industry today are discussed in this article.

1. **Eliminating Data Breaches:** Banks and other financial organizations are common targets for cybercriminals. They are more vulnerable because of the sensitive nature of their data. The financial sector was hit 300 times more often than any other sector. Attacks on financial institutions increased to 819 in 2018 from 69 the year before. While we won't know how many hacks have been launched until 2023, there have been numerous data breaches already in 2018. Between March 22 and 23, 2005, a hacker gained access to the capital One's consumers & small-company credit card applications. On the American continent, Capital One thinks that 80,000 banking account numbers and 140,000 social security information were hacked. It's estimated that about a million Canadians' social insurance number was exposed.
2. **Keeping Up with Regulations:** Regulations have recently increased for the financial services sector. Compliance costs consume a sizable amount of a bank's earnings. A sizable chunk of a bank's revenue goes toward regulatory compliance (Panigrahi et al., 2023)..

   In order to stay up with fast altering consumer & shareholder expectations, technological innovations, and industry rules, traditional banks must regularly analyze and improve their operations.

   KPMG predicts that in 2023, financial services companies will face ten key regulatory issues. They include:
   (i) **Geopolitical change:** Expect disruption and change if you run a business.
   (ii) **Divergent regulation:** Expect variations in state, federal, & international rules due to protectionist and regional public policy goals of the US and other countries.
   (iii) **Data protection and governance:** Keep your information safe at all costs.
   (iv) **Operational resilience:** Plan for the worst. As a result,
   (v) **Credit quality:** The lessons learned from previous credit cycles must be implemented by businesses.
   (vi) **Capital and liquidity shifts:** Firms should not weaken their risk management in anticipation of a possible drop in regulatory liquidity and capital requirements.
   (vii) **Compliance agility:** Need a flexible and simple approach to compliance
   (viii) **Financial crime:** It's fine to be creative, but not if it means leaving the door open to more financial fraud.
   (ix) **Customer trust:** Maintaining customer trust is essential for every business.
   (x) **Ethical conduct:** Act morally no matter what the situation providers of monetary services need to formulate an approach to both innovation and regulation.
3. **Exceeding Consumer Expectations:** Still, customers have high standards for banks and other financial services providers. Many customers of financial institutions would want individualized attention. Half of consumers want situation-specific tailored banking advice, citing a 2019 Accenture, a World Financial Services Consumer Study. They desire an analysis of their spending patterns and financial management recommendations. Sixty-four percent of those who take part are considering insurance discounts for good behaviour like not getting in

accidents. In addition to a digital banking experience, 50% of survey respondents desire an in-person encounter.

4. **Organizing Big Data:** Financial service companies face a double-edged sword in the form of big data. Multiple sources are generating a substantial amount of data, causing big data to grow. Existing data management infrastructures are overwhelmed by the volume and variety of this new data, which is both organized and unstructured. The numerous varieties of incoming data present one of the greatest obstacles for financial service providers. According to EMC's forecasts, there shall be 44 zettabytes of information in existence by 2023. In other words, that's 44 quadrillion megabytes.

5. **Crowdfunding:** Crowd funding refers to the process of soliciting a large number people to contribute small amounts of money to a new business venture. Crowdfunding has a potential to improve entrepreneurship by widening the pool for investors beyond the usual owners, relatives, and a venture capitalist. It does this by capitalizing on the widespread availability of social networks and crowdfunding sites to bring together investors and entrepreneurs (Belleflamme 2014).

Key Takeaways

- There are limits on who can and will contribute to a startup's funding round.
- Crowdfunding allows investors to put money into a wide variety of projects with just ten dollars.
- Crowdfunding platforms earn income by charging service fees.
- Equity-based crowdfunding initiatives in the US are regulated by the Commission on Securities and Exchanges (SEC).
- Kickstarter, Indiegogo, & GoFundMe are three of the most well-known crowdfunding sites.

10.8 Types of Crowdfunding

Donors provide financial support to each of the four kinds of crowdsourcing. A breakdown of each is as follows:

- **Donation:** People who donate to a crowdfunding effort, business, or individual expect nothing in return for their generosity. Let's say you decide to obtain new tools for your company through a crowdfunding campaign. Only if they believe they will benefit from your company's growth will they consider giving you money.
- **Debt:** Debt-based donations are a kind of crowdfunding that takes the form of peer-to-peer (P2P) lending. Donations made in the form of debt constitute a loan that, along with interest, must be repaid by a certain date.
- **Rewards:** This happens when donors are recognized for their generosity. Donors are incentivized to give more by receiving ever better benefits for their money.

Participants might get a T-shirt, a good or service, generally at a discounted price, in exchange for financial contributions to a campaign.

- **Equity:** Equity-based crowdfunding allows startups and small companies to give away a share of their firm in exchange for money, in contrast to campaigns that do not allow backers to possess a stake in the organization they are supporting. These payments are investments in the form of which the giver is given a stake in the business equal to the amount given.

10.9 How Does Crowdfunding Work?

Crowdfunding platforms are online meeting places for donors and fundraisers that is Internet is used to raise money from large number of investors by donating small contributions (Bradford 2012). Financial contributions can be pledged and collected using the crowdfunding website.

Most crowdfunding platforms will charge the campaign's organizers a fee if the campaign is successful. In return, these crowdfunding sites need to offer a reliable and user-friendly service.

Many sites only accept whole payments or none at all. In other words, if your goal is met, you get the money, and if it isn't, everyone gets their money back and nobody loses anything.

Several different kinds of crowdsourcing are explained in detail below. By far the most popular forms of crowdfunding among profits SMEs and startups are peer-to-peer, equity, and incentives crowdfunding. In order to better understand these three categories, this handbook offers objective guidance.

10.10 Crowdfunding Models

There are four primary types of crowdfunding campaigns that business owners might adopt during the planning stages (Asano 2013; HiveWire 2014) (Table 10.1).

Table 10.1 Crowdfunding models

Model	Type	Value proposition	Examples
Donation	Non-tangible feel good return; philanthropy or social sponsorship	Support a social cause or give to charity	Katipult, Weeve
Reward or Perk	Supporters receive non-financial rewards for their contributions--e.g. Clothing, product, lunch with founders; legally a donation	The greater the contribution the better the reward; pre-purchase discount on innovative products or services	Fundo, Fundrazr, Rockethub, Kickstarter, Indigogo
Lending	Specific precent return; peer-to-peer (P2P) lending, peer-to-business (P2B) lending	Borrowing or lending at attractive rates	Prosper, Lendingclub
Equity	Ownership/ Profit sharing	Company ownership; profit sharing and dividend rights; some companies offer equity for time	Crowdcube (UK),Wiseed (France), ASSOB (Australia)

Note: Adapted from "Crowdfunding s essential for SME innovation and job creation "by Asano (2013), NCFA Canada; and "Crowdfunding 101" workshop by Hivewire (2014)

10.11 Risks and Lessons of Crowdfunding

Although crowdfunding has many potential benefits and advantages, it is nevertheless not without some risks that should be avoided. Because crowdfunding is still a relatively new idea, there is a lack of in-depth data and instances about the dynamics of successful crowdfunding, general distribution and utilization of crowdfunding mechanism, and whether crowdfunding contradicts or confirms existing theories about how ventures raise capital and succeed (Mollick 2014). Concerns have been raised about the weather existing crowdfunding campaigns will actually produce the promised products in the long run. The majority of crowdfunding campaigns either succeed marginally or fail spectacularly. If a project has a high-quality demonstration and is linked to a sizable online social network, it has a better chance of being funded (Mollick 2014). Over 75% of project founders run behind schedule in getting outcomes to investors, with the length of delay being correlated to the amount of money invested. Overfunded projects are more vulnerable to delay because of the additional complexity as well as expectations associated with bigger tasks. In conclusion, there are a few key takeaways for business owners thinking about crowdfunding (Mollick 2014):

- Businesses must actively look for ways to show they are prepared to provide high-quality projects.
- A large number of the founder's and project's social network connections is crucial.
- To ensure on-time delivery of a finished product, a successful crowdfunding campaign is essential, as is the founder's readiness to quickly go forward with the endeavor they've promised backers.

10.12 Reasons that Explain Why Blockchain Can Power the Growth in Green Finance

10.12.1 Blockchain Can Power Green Finance (Lim 2021)

The Blochchain and machine learning are considered as fast growing technologies which are being used by businesses for growth. Good Corporate Governance by means of Machine Learning is used as a tool to overcome financial crisis in many companies across the globe and Sharia Stock Index Indonesia is such an example of the same. Rusmita et al. (2023) "The missing traceability element is where Blockchain can really help," Greenfield explained. "Tokenization is an excellent method for achieving this because it is a component of the traceability solution in which a token can only be generated if, first, the project exists and, second, it meets certain criteria." In-person deployments of IoT sensors and AI could improve tracking and reporting to investors by gathering data in real time (Schletz et al. 2020).

In the words of Oi Yee Choo, the chief commercial officer in Singapore's iSTOX digital security exchange, Blockchain technology and electronic securities are the game-changer for every private the market issues as they automate manual processes, enabling securities to get shared in smaller units and easily traded on an exchange.

"Even in markets where demand for green bonds is high because investors are motivated by ESG considerations, tokenization enables investors to diversify their portfolio across various bonds due to smaller subscription sizes, " in Choo's words. Tokenization has the potential for playing a larger role in markets where the demand for green debt has still room to develop since it increases the pool of financing available for new green bond issues by inviting smaller individual and institutional investors at the table.

Meaning of Blockchain: Information in a corporate network can be shared securely and openly thanks to blockchain technology, which is the sophisticated data structure. The blocks of information in a blockchain database are linked together. Due to the impossibility of erasing or changing a chain without network consensus, the data is always in chronological order. With blockchain technology, you may make an immutable ledger to track transactions like payments, purchases, and account balances. The system has safeguards built in to stop illegitimate transactions from being added and to keep the consolidated picture of all transactions accurate. Since Bitcoin's inception in 2009, numerous other cryptocurrencies, decentralized financial apps, non-fungible tokens (NFTs), and the smart contract have emerged as possible uses for the blockchain technology. The buzz surrounding Bitcoin and cryptocurrencies was intense (Narayanan et al. 2016).

Key Takeaways

- A blockchain is a distributed database that stores information in encrypted blocks, as opposed to in a linear fashion like traditional databases.
- Blockchains can store many different sorts of information, but the most common application is as a public transaction ledger.
- Because Bitcoin's blockchain is decentralized, no single entity or group of individuals has authority over it; rather, control remains with the users as a whole.
- Since Bitcoin's blockchain is decentralized, no single entity or organization exercises authority over its operations; rather, all users retain control in a collective sense. After being recorded, information in a blockchain that is distributed cannot be changed. Bitcoin transactions are public and accessible forever.

10.12.2 Five Important Blockchain Benefits

1. **Enhanced security:** Blockchain technology has the potential to profoundly transform the way sensitive and crucial data such as yours is viewed. Blockchain aids in the prevention of fraud and unlawful conduct by creating immutable, end-to-end encrypted data. When it comes to protecting sensitive information, the blockchain can be used to obfuscate identities and enforce access controls. By spreading data among multiple computers in a network rather than using a central server, security is improved.
2. **Greater transparency:** Without blockchain, every company needs its own separate database administrator. Transactions and information are kept in numerous locations, and are all the same, because blockchain uses a distributed ledger. There is total openness because all authorized participants in the network have instantaneous access to the same data. The time and date of each transaction are logged, making them permanent records. By letting members view the entire transaction history, the opportunity for fraud is greatly reduced.
3. **Instant traceability:** Blockchain technology creates a log of an asset's history, or "provenance," at each step of its journey. Proof like this is especially useful in sectors plagued by counterfeiting & fraud, or when buyers worry about the product's effect on the environment and human rights. Using blockchain, you may provide consumers with traceability details without intermediaries. Weak points in any chain of supply, like when goods are sitting on a loading dock, can be revealed by analysing traceability data.
4. **Increased efficiency and speed:** Paper-heavy processes take a long time, leave room for error by humans, and frequently call for a mediator. Blockchain technology streamlines these procedures, allowing for faster and more reliable transaction processing. The blockchain may preserve not only transaction data but also supporting documentation, making physical paper documents obsolete.

Because clearing and settlement don't have to wait while different ledgers are reconciled, they can happen more quickly.

5. **Automation:** Using "smart contracts," you can even automate transactions, which will boost your efficiency and speed things up. When a certain condition is met, the system will move on to the subsequent action in a series. With the use of smart contracts, verifying compliance with a contract's terms no longer necessitates the involvement of a human or a third party. For instance, after an insurance claim form and supporting documents have been completed, the claim can be processed and reimbursed immediately.

10.12.3 Types of Blockchain's

1. **Public Blockchain:** The public blockchain was a decentralized ledger that doesn't require special permissions to access. Anyone with access to the internet can sign up on a platform that uses blockchain technology and join the network as an authorized node. Anyone with access to the public blockchain can view the most recent records, check the legitimacy of transactions, validate the effort required to create new blocks, and mine. Mining and trading cryptocurrencies are the cornerstone uses for public blockchains. Therefore, the two most popular public blockchains are Bitcoin and Litecoin. If users adhere closely to security processes and laws, public blockchains are relatively secure. The only way it might be harmful is if people didn't follow the security rules honestly. Example: Bitcoin, Ethereum, Litecoin
2. **The private blockchain** is one that only users in a certain network may access. Private blockchains are used in settings where access to the network should be limited to a select group of users, such as a company or organization. Security, permissions, and access privileges are all set by whoever has power. As a result, private blockchains offer features similar to those of public blockchains but operate on more constrained networks. Use cases for private blockchain networks include digital identity, voting, supply chain management, and asset ownership. Examples of private blockchains are; Multichain and Hyperledger projects (Fabric, Sawtooth), Corda, etc.
3. **Consortium Blockchain:** Consortium blockchains are a semi-centralized variety wherein numerous organizations work together to manage a distributed ledger system. This is in stark contrast to what we saw in private blockchains, which are under the control of a single entity. Multiple entities can participate as nodes in this blockchain model, exchanging data and even mining. Consortium blockchains are widely used by financial institutions, governments, etc. Examples of consortium blockchain are; Energy Web Foundation, R3, etc.
4. **Hybrid Blockchain:** There are two types of blockchains: public and private. It incorporates features from both permissioned and permissionless blockchains, making it possible to build both closed and open networks. Users of this hybrid network structure can regulate who gets access to what blockchain information.

Some of the information or records stored in a blockchain can be kept secret while the rest are accessible only within a private network. The flexibility of the hybrid blockchain technology allows users to easily link the private blockchain to several public blockchains. For the most part, a hybrid blockchain's private network is where a transaction is verified. However, users can verify it by disclosing it to the public blockchain. The number of the nodes and volume of hashing in a public blockchain are both larger. The transparency and safety of the blockchain system benefits from this. Dragonchain is a good example for a hybrid blockchain.

10.12.4 Challenges Faced by the Blockchain Technology

- **Initial Costs for Setup:** Because of the fact that each company's implementation of blockchain must be unique, the upfront costs associated with doing so might be substantial. In addition, after the initial setup, specialists are needed to meet the requirements, but as we all know, there are very few resources available. Therefore, organizations pay a premium for qualified human resources.
- **Consumption of Energy:** In order to verify and process transactions using the proof-of-work algorithm, miners must expend a significant quantity of computing power, making the process extremely energy-intensive.
- **Compatibility with the current infrastructure:** Since blockchain technology is cutting edge, it might be challenging to get it in sync with legacy infrastructure. This is because it is more expensive to meet blockchain needs if older software or systems must be modified to handle the changes. Therefore, it may require a lot of resources (both monetary and human) and time.
- **Safety and Confidentiality:** It has been established that Blockchain is a safe platform due to the fact that transactions are only accessible to people on the network; but, if the vast majority of users condone fraudulent practices, this could present issues. Therefore, it is one of the issues that the blockchain community must examine to prevent data manipulation and misuse.
- **General Knowledge:** A lot of money is being put into blockchain, but it will be a while before regular people start using it. Still, there are many people who have yet to learn about Blockchain and the many benefits they might reap if they adopt this technology. To get around this, they need to make a serious effort to educate the public about blockchain technology.

In spite of these limitations and obstacles, Blockchain remains robust, and with the adoption of Blockchain increasing daily, a prosperous future based on Blockchain is anticipated.

10.12.5 Limitations of the Blockchain Technology: The Limitaions of Blockchain Technology may be Enlisted as Follows:

- **Disparity in Technical Understanding:** Despite Blockchain's rising popularity, many investors still don't understand all the technical concepts, and there is no adequate documentation to help customers acquire such a thorough understanding. As a result, investors cannot directly pose questions or have their doubts resolved.

  The majority of people have yet to learn about ICOs, and Initial Coin Offerings, that are a non-equity method of raising cash.
- **Reduced Availability of Qualified Professionals:** There is a shortage of Blockchain experts in comparison with the growing demand for this technology. A barrier to the further growth of the blockchain is the fact that there are people in markets who can supply information for the system but who are unfamiliar with its technological workings.
- **Scalability:** As every single transaction on a network must be validated by every single node, it's one of the biggest drawbacks to the blockchain network. This slows down the processing of transactions. Distributed Ledger technologies, like Hyperledger Fabric, built on blockchains, are still under development to address the scalability problems.
- **Less Privacy:** While users' identities remain concealed, their transaction histories can still be parsed to learn more about them thanks to the distributed nature of the ledger.
- **Security Concerns:** Humanity itself constitutes the blockchain. A majority of the people advocating for data tampering might lead to the deception being accepted as fact, which could be a major factor in the breakdown of the network as a whole. This necessitates vigilance on their part to safeguard the data and prevent its unauthorized usage.
- **Complexity:** As it requires a large number of mathematical calculations, blockchain technology is difficult for novices to grasp. As a somewhat complex industry, it cannot be understood immediately.
- **Increased Transaction cost:** Initially, transaction costs are nearly zero, but as the network expands, transaction costs continue to rise.
- **Human Errors:** Outdated log information and mismatched data might occur from human error while entering information into a database. It is crucial to validate data before using it to avoid any mistakes. This is why 'garbage in, trash out' is often a phrase used when discussing blockchain technology.

10.12.6 Peer-to-Peer Financing and Investing

Direct borrowing between individuals eliminates the need for a centralized lending institution to mediate repayment of debts. The availability of peer-to-peer lending platforms online has greatly expanded the potential of this method of obtaining capital.

Other names for P2P lending include "social lending" and "crowd lending." Competitors like Prosper, a Lending Club, Upstart, & Street Shares have been around longer than its 2005 inception.

ICOs: In the cryptocurrency world, an ICO, or initial coin offering, was the equivalent of a traditional IPO. An Initial Coin Offering (ICO) is a means by which a business generates capital to develop its cryptocurrency, app, or service.

The goal of an ICO is to raise capital by selling newly created cryptocurrency tokens to investors. Tokens can represent a share in a company or a project, or they can simply be used to access a service or product. A risk-taker, 31-year-old French socialist named Olivier Stern recently put a third of his life savings, or 10,000 euros, or roughly $11,000, into a cryptocurrency start-up that faces legal challenges and regulatory roadblocks but has the potential to completely disrupt the opaque world of virtual investing (Popper 2016).

10.12.7 How an Initial Coin Offering (ICO) Works

The first stage in raising capital through an ICO for a cryptocurrency project is deciding how the coin will be formed. ICOs can be structured in a number of ways, including:

- **Fixed resources and prices**: Every token sold in an ICO has a predetermined price, and the total number of tokens is limited to the amount the firm has decided to raise through the ICO.
- **Stable demand and fluctuating prices**: It's possible for an ICO to have fixed supplies of tokens and a variable financing goal, in which case the total cost per token is set by the amount of money raised.
- **Constant demand and fluctuating prices**: In some ICOs, the number of tokens available is tied to the total amount raised, while the price remains constant.

10.12.8 Exchange Platforms for Peer-to-Peer Trading

Before Bitcoin, P2P or decentralized networks already existed. They involve the exchange or sharing of data, funds, or assets over the internet between two or more parties without the involvement of a central authority. Since late April, almost $100 million worth of cryptocurrency has been gathered by a new organisation called The

DAO, which was founded utilising the Ethereum financial platform, which was modelled after Bitcoin. The DAO plans to utilise the money to support sharing economy initiatives. The DAO is being hailed as a prototype for a novel form of organisation, one that is built and managed via blockchain technology as opposed to traditional corporate frameworks (Morris 2016).

P2P exchanges tend to have less stringent verification requirements, and trading directly with another party allows users to choose their own payment method and the best available exchange rate while savouring lower transaction fees.

Users can establish trust in participants on P2P platforms through a rating system that demonstrates their repute as a guarantee of their dependability.

The role of the P2P platform is to match buyers and sellers for a small fee, but it does not retain assets, and users transfer their funds to their own wallets immediately after a transaction.

10.12.9 Carbon Credit Trading

Carbon markets function as exchanges in which carbon credits may be bought and sold. Individuals and companies can utilize carbon markets to offset their carbon footprints by acquiring carbon credits from organizations working to eliminate or reduce emissions of greenhouse gases.

Carbon credits are units of measurement for the reduction, sequestration, or avoidance of greenhouse gas emissions equal a single tonne of carbon dioxide. Credits are no longer transferable once they have been used as offsets to cut, sequester, or avoid emissions.

10.12.10 The Challenges Facing Blockchain Adoption

Before we delve into the obstacles surrounding the adoption of Blockchain, let's first define what Blockchain is. Transactions are recorded in a distributed, secure, and publicly verifiable digital ledger called blockchain. It resembles a digital version of a traditional ledger, except that instead of being stored in one central location, it is distributed across a network of computers.

10.12.11 The Potential of Blockchain

Now that we have a fundamental comprehension of Blockchain, it is essential to recognize its potential. Many sectors, such as banking, healthcare, and logistics, stand to benefit greatly from blockchain technology. Productivity, cost savings, and openness are all boosted by this method.

10.12.12 The Challenges of Adoption

However, as with any new technology, obstacles must be surmounted for widespread adoption to occur. Let's examine some of the most significant obstacles to Blockchain adoption:

1. **Regulatory Uncertainty:** Regulatory unpredictability is one of the greatest obstacles to Blockchain adoption. Governments across the globe are still attempting to determine how to regulate this new technology. This uncertainty can impede the adoption of Blockchain by businesses and consumers equally.
2. **Scalability:** Scalability is a further significant obstacle to Blockchain adoption. As it is, the transaction throughput of Blockchain networks is now quite low. This is a major issue in fields like financing, where many thousands transaction occur per second.
3. **Interoperability:** Interoperability is the capacity of various Blockchain networks to communicate with one another. Currently, there are a variety of Blockchain networks, each with its own protocols and standards. This lack of interoperability hinders the widespread adoption of Blockchain by enterprises.
4. **Security:** Despite Blockchain's reputation for security, it is not immune to attack. In recent years, there have been several high-profile breaches of Blockchain networks. This presents a difficulty for businesses contemplating the adoption of Blockchain, as they must ensure the security of their data.
5. **Education and Awareness:** Education and awareness represent a significant barrier to Blockchain adoption. Although Blockchain technology has existed for more than a decade, many people still do not fully comprehend what it is or how it operates. This lack of comprehension can result in reluctance to implement the technology.
6. **Wrapping Up:** While there are definitely obstacles to Blockchain adoption, it is essential to remember the technology's potential. As governments around the world work to regulate Blockchain and as businesses work to overcome the challenges of scalability, interoperability, and security, we can anticipate a rise in Blockchain's adoption in the future years.

References

Asano C (2013) Crowdfunding is essential for SME innovation and job creation. NFCA Canada. Retrieved from http://ncfacanada.org/crowdfunding-is-essential-for-sme-innovation-and-job-creation

Belleflamme P, Lambert T, Schwienbacher A (2014) Crowdfunding: tapping the right crowd. J Bus Ventur 29(5):585–609. https://doi.org/10.1016/j.jbusvent.2013.07.003

Bradford CS (2012) Crowdfunding and the federal securities laws. Colum Bus Law Rev 1(1):1–150

Fleming S (2020 Nov 9) What is green finance and why is it important? Retrieved from https://www.weforum.org/agenda/2020/11/what-is-green-finance/

HiveWire (2014) Crowdfunding 101. Workshop, Centre for Social Innovation

Lim M (2021) How blockchain can power the growth of green finance. Retrieved from https://forkast.news/blockchain-green-finance-bonds-china-asia/

Mollick E (2014) The dynamics of crowdfunding: an exploratory study. J Bus Ventur 29(1):1–16. https://doi.org/10.1016/j.jbusvent.2013.06.005. ISSN 0883-9026. https://www.sciencedirect.com/science/article/pii/S088390261300058X

Morris DZ (15 May 2016) Leaderless, blockchain-based venture capital fund raises $100 million, and counting. Fortune. Archived from the original on 21 May 2016. Retrieved 23 May 2016

Narayanan A, Bonneau J, Felten E, Miller A, Goldfeder S (2016) Bitcoin and cryptocurrency technologies: a comprehensive introduction. Princeton, NJ: Princeton University Press. ISBN 978-0-691-17169-2

Panigrahi RR, Dash M, Shaikh ZH, Irfan M (2023) Review of Machine Learning Techniques in the Supply Chain Management of Indian Industry: A Future Research Agenda. In: Advanced Machine Learning Algorithms for Complex Financial Applications. IGI Global, pp 199–219

Popper N (2016 May 21) A venture fund with plenty of virtual capital, but no capitalist. The New York Times. Archived from the original on 22 May 2016

Rusmita SA, An-Nafis MSA, Ramadhani I, Irfan M (2023) The effect of good corporate governance on financial distress in companies listed in sharia stock index Indonesia: machine learning approach. In: Advanced machine learning algorithms for complex financial applications. IGI Global, pp 220–251

Schletz M, Nassiry D, Lee M-K (2020) ADBI working paper series blockchain and tokenized securities: the potential for Green Finance Asian Development Bank Institute. https://www.adb.org/sites/default/files/publication/566271/adbi-wp1079.pdf

Chapter 11
Bibliometric Analysis of Publications on Artificial Intelligence and Finance in the Databases of Scopus

Mahammad Habeeb, T. V. Sriram, and Syed Muhammad Abdul Rehman Shah

11.1 Introduction

Artificial Intelligence and Finance are some of the recent fields dominating the financial sector and the technological sector respectively. Artificial Intelligence (AI) has no one definition but according to Collins et al. (2021) the most popular definition often cited is given by Russel and Norvig (2013). Russell given the term AI for systems that mimic cognitive functions usually associated with human attributes that include learning, speech, and problem-solving. However, in the field of finance, learning and problem-solving skills are valued highly. AI has transformed the financial sector through various means through the help of different methods of application such as machine learning and deep learning. The application of AI in finance has led to new opportunities and challenges for businesses, investors, and regulators. AI-powered systems have the potential to improve financial decision-making, automate tasks, and enhance customer experiences. However, the adoption of AI in finance has also raised concerns about privacy, ethics, and the potential impact on employment. Through a bibliometric analysis of the industry's structure and

M. Habeeb (✉)
SVKM's NMIMS deemed to be University, Bengaluru, Karnataka, India
e-mail: mahammad.habeeb@nmims.edu

T. V. Sriram
Symbiosis School of Economics, Symbiosis International (Deemed University), Pune, Maharashtra, India
e-mail: sriram.tv.2022@sse.ac.in

S. M. A. R. Shah
Economics & Islamic Finance, University of Engineering and Technology, Taxila, Pakistan
e-mail: a.rehman@uettaxila.edu.pk

M. Irfan et al. (eds.), *Applications of Block Chain technology and Artificial Intelligence*, Financial Mathematics and Fintech,
https://doi.org/10.1007/978-3-031-47324-1_11

developments, the study attempts to understand how artificial intelligence and finance are related.

A bibliometric analysis is thought to be a well-liked and accurate technique for perusing and analysing huge volumes of scientific data. It allows us to explore the evolutionary nuances of a particular area while providing insight on the emerging areas of that particular area (Donthu et al. 2021).

The author seeks to analyze the evolution of AI in the financial sector, due to its multifaceted nature throughout the industry. Artificial intelligence is increasingly used in the financial sector for a variety of purposes, including lending, payments, asset and wealth management, risk and regulation management, accounting, and auditing (Cao 2021). To address issues in areas ranging from banking to behavioural finance, advanced AI concepts like machine learning, natural language processing, and other focus-based applications and tools are frequently used in the context of economics research.

11.2 Objectives

The main objectives of the present bibliometric study are as follows:

1. To measure the collaboration index and collaboration coefficient of the field.
2. To analyze the pattern of authorship and citations amongst the article.

11.3 Research Methodology

The current paper is based on 284 articles sourced from Scopus. The search terms utilized were "Artificial Intelligence" and "finance". Articles without an affiliation or author name attached, were dropped from the analysis. The scope of the study spans from 1990 to 2023. The data was then downloaded and analyzed using MS-Excel. The author analyzed the data utilizing the Performance Analysis framework set down by Donthu et al. (2021). A Performance Analysis can simply be described as an analysis aimed at evaluating (groups of) scientific actors on the basis of bibliographic data (Noyons et al. 1999).

With a focus on describing information related to Publication Metrics, Citation Metrics and a combination of the both Citation and Publication Metrics.

Under the aegis of the Citation and Publication Metrics, two key indices were calculated by the author, the Collaboration Coefficient and Collaboration Index. The latter was calculated by referring (Donthu et al. 2021a) and (Donthu et al. 2021), the index given by them is:

$$CC = TCATA - 1$$

Where CC is the collaboration Cofficient, TCA represents total contributing authors in number, TA is the total no. of articles identified from Scopus and the −1 signifies the lead author of the paper. (Donthu et al. 2021) states that the The collaboration index shows how many authors a primary contributor works with to produce a piece of research. It is calculated as the proportion of articles with at least one contributing author to all articles sans 1.

The collaboration index as defined by Donthu et al. (2021) is an index that calculator the extent of collaboration of research constituents. This is calculated by the following formula:

$$\text{CI} = (\text{NCATP})\text{TP}$$

Where CI is the collaboration Index, NCA is Number of Contributing Authors, TP represents total publications. The TP in the denominator is representative of total publications in the respective research constituent.

11.4 Data Analysis

11.4.1 Publication Related Metrics

There are a total of 284 publications analyzed in the field of Artificial Intelligence and Finance that were available from Scopus. The scope available was between the years 1990 and till date 2023.

From Chart 11.1 we can see that the majority of publications cataloged by scopus are from the constituents representing Academia (239 papers) followed by the mix of 'Academia and Industry (42) with publications from industry (3) being the least.

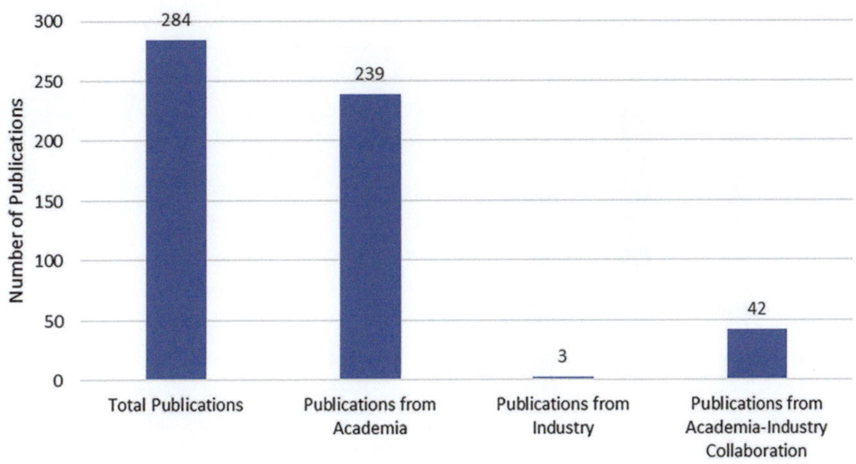

Chart 11.1 Publication related metrics. Source: Authors calculations

From the 204 publications there are a grand total of 665 authors.

From publications it's interesting to note that a majority of the papers are authored by individual researchers and almost equally by two researchers. On the tail end of the analysis we also have papers authored by a maximum of seven–eight researchers.

This can further be broken down to view the count of co-authors according to their research constituent (please refer to the Chart 11.8 in the appendix). Following the trend from the graph below, a majority of research done in the field is carried out by solo researchers in academia. In terms of collaborative research work between Academia and Industry, most work tends to lean towards publications with two or three co-authors. Research publications from Industry are an outlier in this dataset as there have been only 3 publications in total and all three conducted between three co-authors in the same year (Chart 11.2).

From Chart 11.3, we can see that the number of publications peaked in the year 2011 before initially hitting a stagnant three year period between 2014 and 2016 before a rapid rise post 2017. The Productivity per year of publications reaches a value of about 10.9, this value is calculated by taking the average of total publications and dividing it by the number of years. This can be interpreted as, the productivity of publications being 10.9 papers a year, however this number is driven by the peak in the year 2011.

The recent surge in the number of publications at first glance can be attributed to advancements in the field of artificial intelligence and better awareness of the same amongst participants in the financial sector.

Image 11.1 visualizes the most relevant recurring keywords, throughout the dataset, with finance, artificial intelligence and application gaining a higher relevance score. The relevant score is based on an index calculated by the software. It is also curious that some other common keywords that recur are China and machine learning respectively on each side.

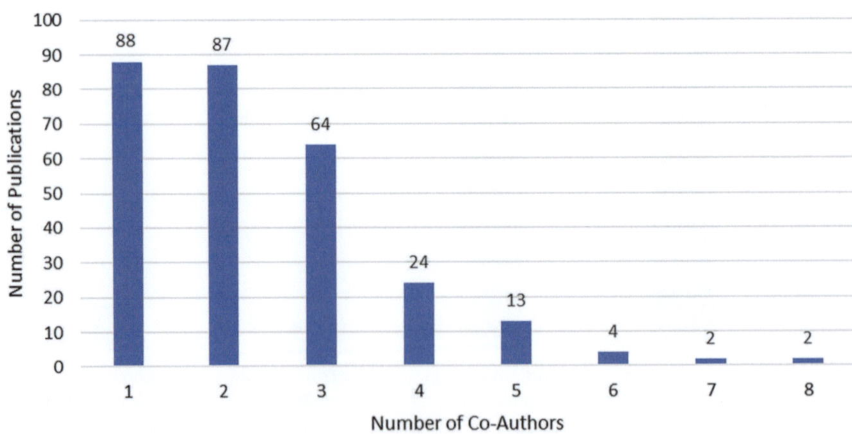

Chart 11.2 Count of number of co-authors. Source: Authors calculations

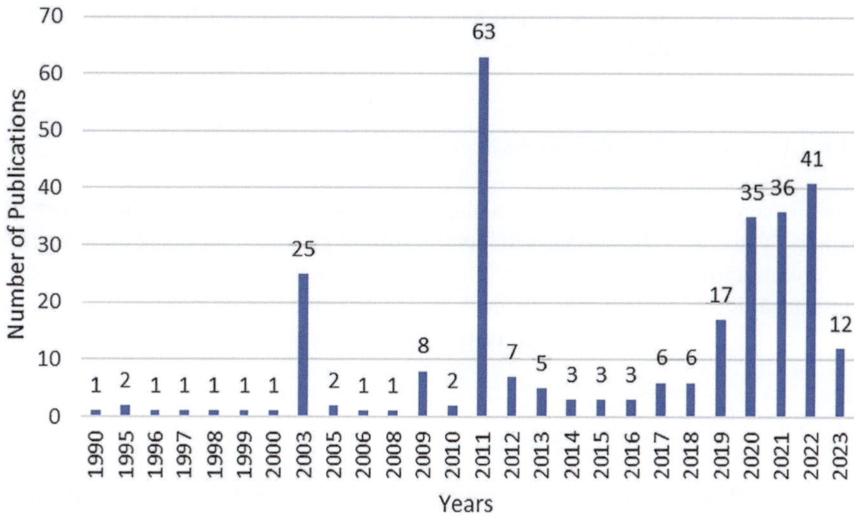

Chart 11.3 Total number of publications per active year. Source: Authors calculations

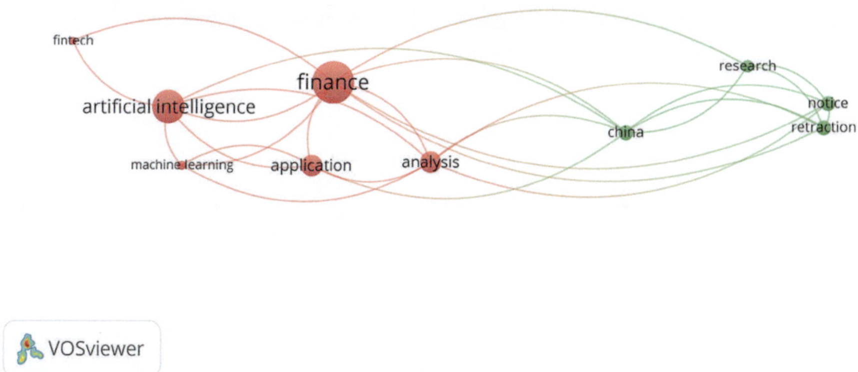

Image 11.1 Map of key terms. Source: Authors calculations with the help of VOS Viewer

11.4.2 Network Analysis of Social Structures

A network analysis of the social structures within a field are a key set of visualizations to better understand the vast amount of data one can collect from research journal databases. Three kinds of social structures are analyzed below, author collaborations, country collaborations and institutional collaborations. Only the largest clusters from each subsection have been analyzed due to the vast nature of the data and more importantly the emerging nature of the field which has not led to as many collaborations across different authors and institutions. The preferred tool was the VOS viewer, which was created by the Centre for Science and Technology Studies at Leiden Universities. A software tool for creating and visualising bibliometric networks is called the VOS viewer.

11.4.2.1 Author Collaborations

From Image 11.2 with the help of the VOS viewer, we can view the cluster of authors. The aim of the cluster is to identify further groups of papers or authors who are collaborating with each other. The largest cluster of authors centers around, Hassan M. K's contribution to the field. It is to be noted that he also has the greatest number of citations in the queried results from Scopus.

11.4.2.2 Country Collaborations

It is pretty obvious from the nation collaborations image that nearly all of the major economies in the world take part in the study of this new frontier (Image 11.3). The United States, China, United Kingdom, and India may appear to have a huge circle, indicating the scope of publications, but it is important to keep in mind that these nations also have the strongest measured link strengths (a measure of how many other countries they are connected to).

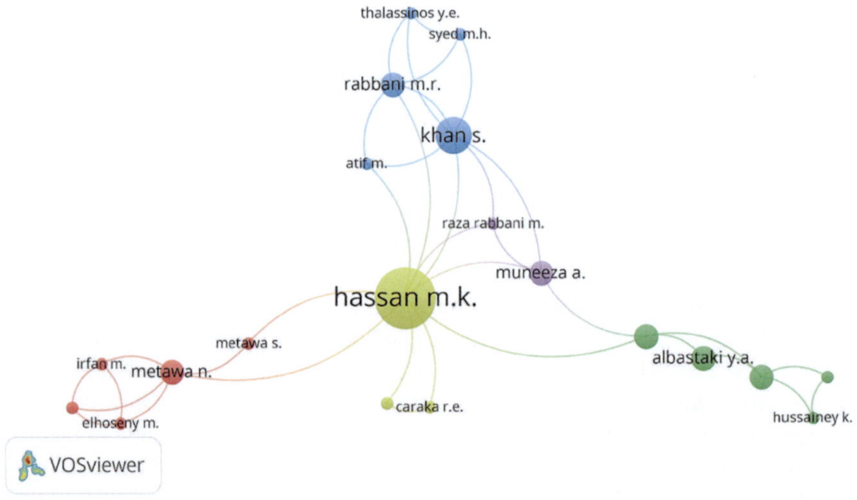

Image 11.2 Network analysis of author collaboration (largest cluster imaged below). Source: Authors calculations

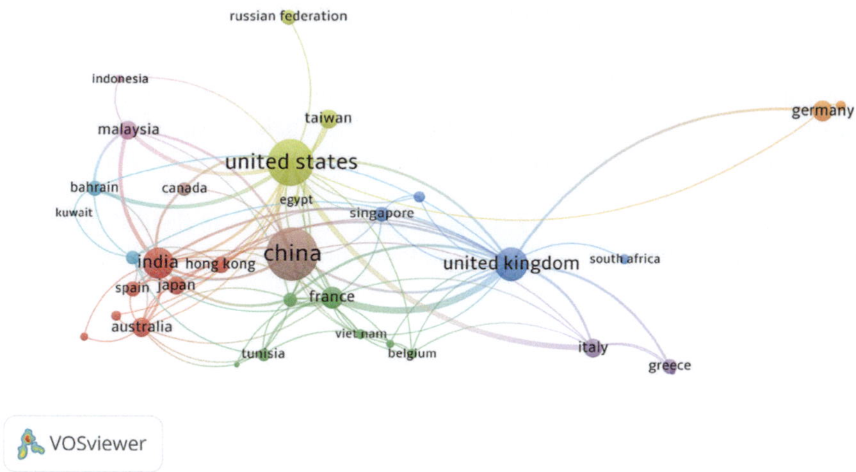

Image 11.3 Network analysis of country collaboration (largest cluster imaged below). Source: Authors calculations

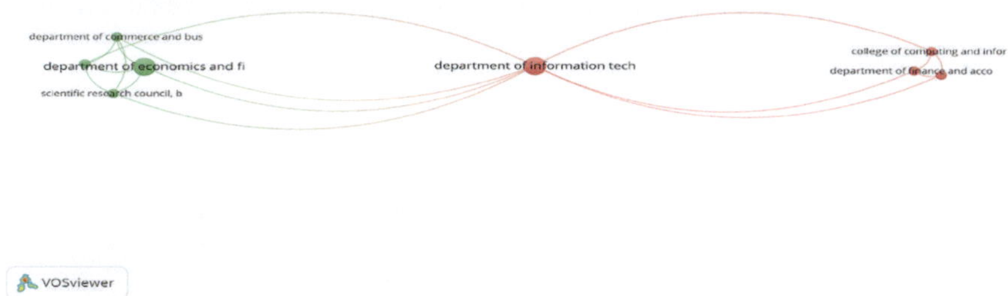

Image 11.4 Network analysis of institutional collaboration (Largest cluster imaged below). Source: Authors Calculations

11.4.2.3 Institutional Collaborations

Image 11.4, clusters together various institutions that collaborate with each other. The institutions in the left cluster are primarily from the Middle East representing Bahrain, Saudi Arabia and Kuwait. The institutions in the right cluster are from the United States, India, Bahrain and Saudi Arabia. Most of the institutions in the above clusters are departments of business studies, finance or economics of their respective universities.

From this section the key insight is rather paradoxical that countries like the U.K, U.S, China & even India emerge as the top country contributors but in terms of institutional collaborators, institutions in the middle east appear to be more to collaborate with each other.

11.4.3 Citation Related Metrics

There are a total of 1484 citations across the three research constituents. Academic papers earn the most citations overall, followed by articles produced in collaboration, and publications produced exclusively for the industry come in last. This is represented by a pie chart in Chart 11.4.

The average citation per paper as seen in Table 11.1, comes to about 57 per paper, with academia leading with 52 citations followed by collaborative publications with 4 citations on average.

11.4.4 Citation and Publication Related Metrics

11.4.4.1 Collaboration Index

By dividing the total number of authors who make up the index by the total number of publications, the index is calculated. Chart 11.5 shows the derived index graphically. Below. We can see that the index value is highest in the year 2000, followed

Chart 11.4 Total number of citations. Source: author's calculations

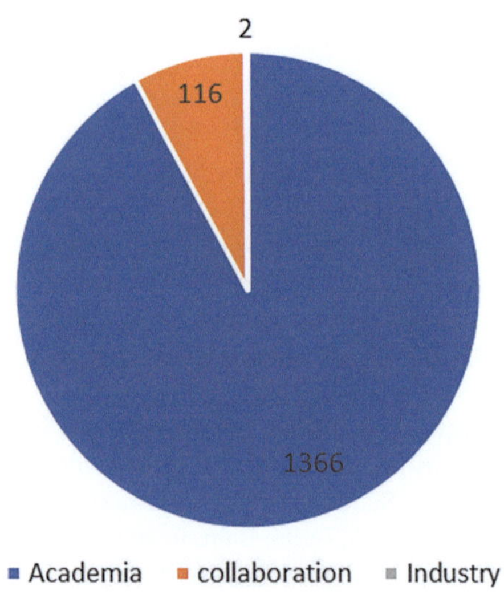

Table 11.1 Average citations per research constituent

Research Constituent	Average per Year
Academia	52.53846
collaboration	4.461538
Industry	0.076923
Grand Total	57.07692

Source: authors calculations

by a decline. However this is not the case as the year 2000 is an outlier, when we increase the Total number of authors on the right side of the chart, this becomes apparent. Let's say the goal of the collaboration index is to demonstrate that an article only has one primary author; in this case, the result shows that each author cooperated with at least six other authors to contribute to the field. However in the year 2000 there were only 1 paper published. A better time period to look at the field is between 2010 and 2022, which saw the most change in terms of number of authors, but still the index follows a declining trend of collaboration. This tallies with our earlier observation of the large number of solo authors. In 2023, every one author collaborated with 0.068 authors to publish one paper.

11.4.4.2 Collaboration Coefficient

The collaboration coefficient (CC), according to Yadav et al. (2019), is a measurement of research collaboration that may be understood as the combination of the average number of authors per paper and the percentage of multi-authored papers. One sign of the validity of scientific articles is the co-authorship of the articles.

From our analysis, represented visually below by Chart 11.6, the overall collaborative coefficient for the field is −1.355 and in 2022 it was −1.088, indicating a severe lack of collaboration, this goes hand in hand with our results from the collaboration index and publication related metrics.

11.4.4.3 Number of Cited Publications

From all of the publications, a total of 53% (150) of the papers have been cited once and 47% (134) have not been cited yet. For every cited paper there is an average of 9.89 citations. The proportion of cited to uncited is about 1:1.27, for every cited paper there are 1.2 uncited papers (Chart 11.7).

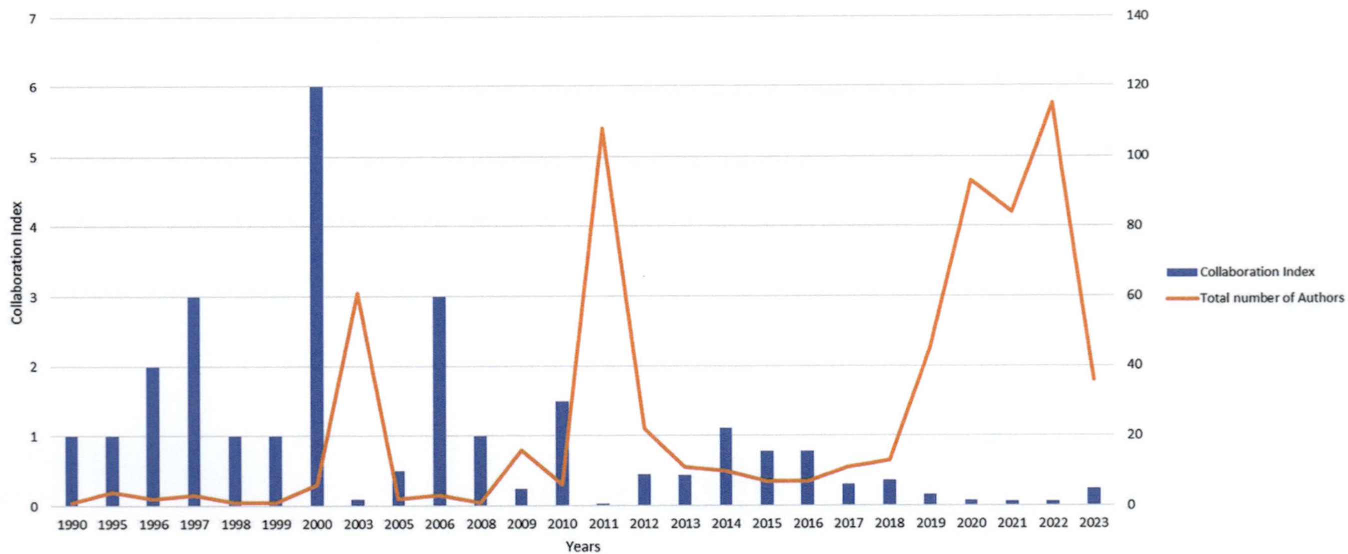

Chart 11.5 Collaboration Index per Year (LHS represents the Index and the RHS represents the number of Authors). Source: Authors Calculation

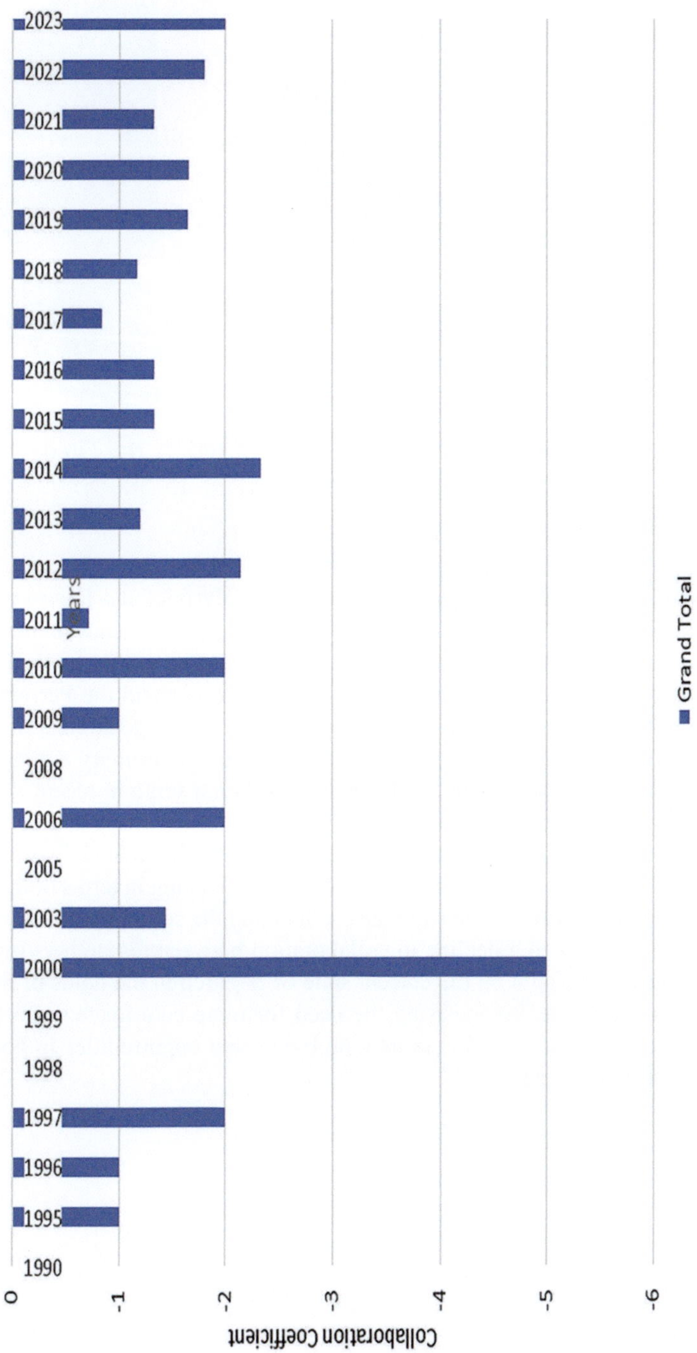

Chart 11.6 Collaboration coefficient per year. Source: Authors calculations

Chart 11.7 Number of cited publications. Source. Authors calculation

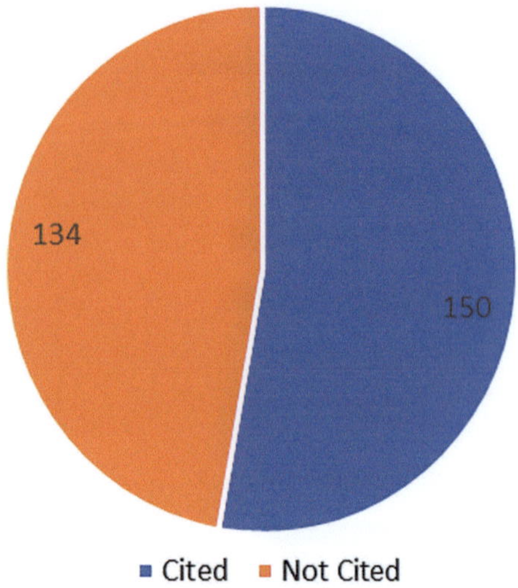

11.5 Conclusion

In conclusion, the study conducted a bibliometric analysis of the intersection of artificial intelligence and finance to determine the structure and trends in the industry along with visualizations of collaboration networks. The author analyzed 284 articles sourced from Scopus using MS-Excel, VOS Viewer and the Performance Analysis framework. The findings of the study show that the large number of publications were from academia, with solo researchers being the primary authors. The number of publications peaked in 2011. There was then a surge in recent times in the number of publications due to advancements in the field of artificial intelligence and better awareness of the same amongst participants in the financial sector. The average citation per paper was about 57, with academia leading in terms of the most number of citations received. The study also calculated the collaboration index and coefficient, which showed a decline in collaboration between researchers over the years. The study sheds light on the current state of research in the fields of finance and artificial intelligence, emphasising the need for more coordinated efforts and research in the field that can address new problems and opportunities in both the financial and technological sectors.

Appendix

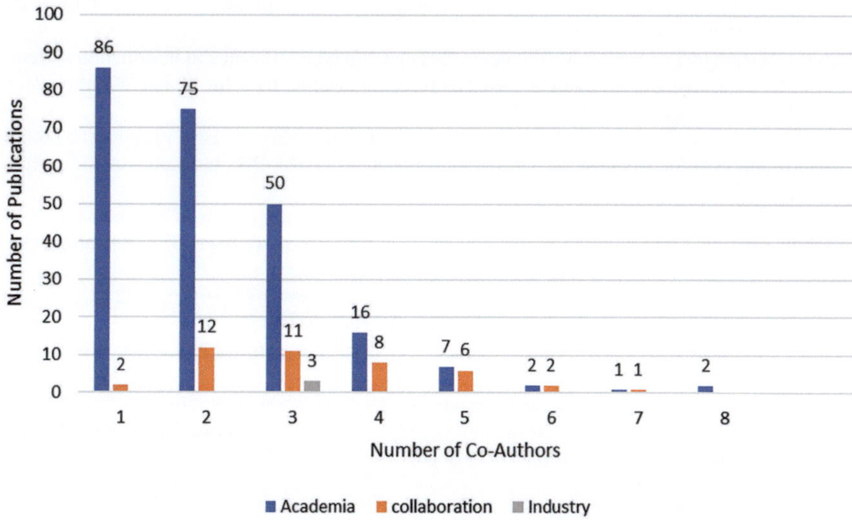

Chart 11.8 Count of co-authors according to their respective research constituent. Source: Authors calculations

References

Cao L (2021, June) AI in finance: challenges, techniques and opportunities. Arxiv. doi:https://doi.org/10.1145/nnnnnnn.nnnnnnn

Collins C, Dennehy D, Conboy K, Mikalef P (2021) Artificial intelligence in information systems research: a systematic literature review and research agenda. Int J Inf Manag 60. https://doi.org/10.1016/j.ijinfomgt.2021.102383

Donthu N, Kumar S, Mukherjee D, Pandey N, Lim WM (2021, May 14) How to conduct a bibliometric analysis: An overview and guidelines. J Bus Res 133:285–296. https://doi.org/10.1016/j.jbusres.2021.04.070

Donthu N, Reinartz W, Kumar S, Pattnaik D (2021a) A retrospective review of the first 35 years of the International Journal of Research in Marketing. Int J Res Mark 38(1):232–269. https://doi.org/10.1016/j.ijresmar.2020.10.006

Noyons E, Raan TV, Moed HF (1999, November) Integrating research performance analysis and science mapping. Scientometrics. https://doi.org/10.1007/BF02459614

Russel S, Norvig P (2013) Artificial intelligence: a modern approach, vol 256. Pearson Education Limited, London

Yadav SJ, Singh SN, Verma MK (2019, Spring) Authorship and collaboration pattern in SRELS Journal of Information Management during 2008-2017: an evaluation. RELS J Inf Manag. https://core.ac.uk/download/pdf/188141446.pdf

Chapter 12
Understanding the Need of BlockChain Technology and Artificial Intelligence, and Transformation of Financial Services: A Conceptual Framework

Manoj Kumar, Sumit Kumar, and Rubina I. Ahmed

12.1 Introduction

The world is witnessing an unprecedented wave of technological advancements reshaping industries and revolutionizing how we live and work. Two of the most transformative technologies leading this wave are Blockchain and Artificial Intelligence (AI). Individually, these technologies have already made significant strides, but when combined, they have the probability to explain a novel age of revolution. In this essay, we will explore the convergence of Blockchain technology and AI, their synergistic potential, and the implications for various sectors.

Blockchain technology, at its core, is a decentralized and immutable ledger that enables secure and transparent transactions. It operates through a link of computers that are recognized as nodes, where each node keeps a duplicate of the ledger. By removing the necessity for mediators, Blockchain ensures trust, and security, besides efficiency in transactions, making it suitable for various applications beyond cryptocurrencies.

Artificial Intelligence, on the other hand, refers to the development of intelligent machines that simulate human cognitive abilities. AI covers a range of techniques, counting machine learning, language processing, computer vision, and robotics. With the exponential growth of data and computing power, AI has become a driving force behind automation, decision-making, and predictive analysis across multiple industries.

M. Kumar (✉) · S. Kumar
Department of Economics, Guru Jambheshwar University of Science and Technology, Hisar, Haryana, India

R. I. Ahmed
Adjunct Academic, School of Business, Charles Sturt University, Bathurst, NSW, Australia
e-mail: rahmed@csu.edu.au

M. Irfan et al. (eds.), *Applications of Block Chain technology and Artificial Intelligence*, Financial Mathematics and Fintech,
https://doi.org/10.1007/978-3-031-47324-1_12

Artificial intelligence (AI) is a new concept and significant advancement innovation in technology. The main aim of this research is to explore how AI is implemented in the insurance industry. The insurance industry has become the major adopter of artificial intelligence technology. Artificial intelligence in the insurance sector to the overawed current problem for enhanced customer satisfaction. Many companies use artificial intelligence models not only insurance sector as well as the finance industry, particularly the banking sector. Artificial intelligence and its use are to help reduce functioning costs and increase the productivity of the insurance sector.

Artificial intelligence can create visions from big multifaceted data sets to enlarge decision making e.g. risk assessment.AI have positively helped businesses in transforming their business competencies and we can increase efficiency. Various study has been done theoretically to measure the connection between artificial intelligence and the insurance sector.

Artificial has transported an innovative change in the way the insurance sector was achieved a few years ago. Insurance was usually related to loads of bookkeeping, time-consuming conferences, filling complex claims processes, and waiting for months for a decision. Artificial insurance has carried out robotics that has started upgrading the trust toward insurance providers. Artificial intelligence can investigate and estimate huge amounts of data to modernize and shorten insurance processes.AI influence on insurance will reduce time and save expenses for policyholders and insurers.

While it's an industry that has established resistance to transformation for centuries, insurance is undergoing a digital revolution. Artificial intelligence helps insurers evaluate risk, identify scams, and reduced human faults in the application process. Found that insurers are enhanced and well-appointed to sell customers the plans most suitable for them. Customers' advantage of the modernized service and claim processing that artificial intelligence affords. Many insurance companies' websites include chatbots. Artificial intelligence tools can monitor customers through many queries without human intervention. They are also available twenty-four to seven.

12.1.1 The Synergy Between Blockchain and AI

Data Security and Privacy: Blockchain technology provides a decentralized and tamper-proof data storage mechanism. When integrated with AI, it enhances data security and privacy. By leveraging Blockchain's immutability, AI systems can securely store and access sensitive data, ensuring that personal information remains confidential and protected against unauthorized modifications.

Trust and Transparency: Trust is a crucial factor in AI adoption, especially in areas like finance, healthcare, and supply chain management. Blockchain's transparency and verifiability can augment AI algorithms by providing a reliable and auditable trail of data. This enables stakeholders to validate the integrity of AI models, ensuring fairness, and accountability, and mitigating the risk of bias.

Data Sharing and Collaboration: The combination of Blockchain and AI can revolutionize data sharing and collaboration. Blockchain's distributed nature allows for secure and efficient data exchange between parties, eliminating intermediaries and reducing costs. AI algorithms can analyze data from multiple sources stored on the Blockchain, leading to more accurate insights and improved decision-making across industries such as research, finance, and logistics.

This paper explores the need for AI in the recent financial sector from different dimensions therefore the improved products can be structured. First, need a review of the literature that helps us to understand the given works.

12.2 Literature Review

Mikko Riikkinen et al. (2018) examined the using artificial intelligence to construct value in the insurance industry. This study aims to reveal how insurance chatbots support customers' value formation. The approach used in this paper is three speculative viewpoints- artificial intelligence, service logic, and reverse use of customer data. Found that Chatbots denote a new type of interface through which companies can inspire customers' value formation by providing supplementary resources.

Kevin H. Kelley et al. (2018) inspect artificial intelligence's capability to enhance the insurance industry value chain by changing the connection, reinventing business stages, and increasing unseen data. Insurance companies use artificial intelligence to significantly enhance huge data analytics, and transaction data faster and combine data in new ways to determine better-endorsing risks. This research explores artificial intelligence's significant impact on workers, jobs as well as insurance industry such as its customers.

Naman Kumar et al. (2019) observed the artificial intelligence used in the insurance sector. This paper explores the opportunity and market penetration of artificial intelligence in insurance service overawed current difficulties for better customer satisfaction in the hospitality industry. The theoretical models have been used to measure the connection between artificial and it is use cases in the insurance sector.

Stephan Bredt (2019) discussed the potential impact of artificial intelligence on the financial sector. The authors have explained the theoretical importance of artificial intelligence's future aspects in the financial sector. This paper has not used any specific method. This paper highlights the enhanced and more tailor–made service, price decrease, and the development of new business models in the financial industry through the use of artificial intelligence. Artificial intelligence offers a major foundation for future technological innovation.

Sunitha. B.K. et al. (2020) investigate the change in the financial sector mainly focusing on the banking sector. This paper motive analyzes the impact of artificial intelligence in the financial sector particularly banking in India. Secondary data has been used in this research paper from different sources like articles, journals, case studies, etc. The time series that has been used is from 2014–2020 progress and artificial developments in India. The authors has been explore the different sections of banking such as trading, lending, security, credit rating, and fraud detection. This study concludes that there is a major occurrence of artificial intelligence in the case of Indian financial markets but there is still an untouched marketplace and application which give a scope of development and growth. Found that accuracy of artificial intelligence techniques is greater than the traditional statistical approaches used for dealing with monetary complications.

A. Geetha (2021) studies artificial intelligence in banking and financial services. The data has been collected from secondary sources constructed by the literature of review. Primary data was collected by customers' structured questions about artificial intelligence applications. The result found that private banks and private financial institutions are using various services for the customer's benefit and are satisfied with their services. In addition, this study suggests that customers are more committed to representing financial services and the development of innovative structures to improve artificial intelligence.

Martin Eling et al. (2021) analyzed the influence of artificial intelligence on the insurance sector. The author has used Porter's value chain and Berliner's insurability criteria. This research used a data set of 91 papers and 22 industry studies to the analysis the influence of artificial intelligence on the insurance sector. Found that cost competencies and new income streams can be realized. The insurance commercial model lifted from damage recompense to lose perdition and anticipation.

Pinky Soni (2021) explains artificial intelligence in the financial sector. This study has been constructed on secondary data and descriptive nature. The data has been collected from various reports and journals. This research explains tests and their influence with merits and demerits in the financial sectors. This study exposes how artificial intelligence makes variations in the financial field in the future with few blessings.

V. Padmanabhan and V. Princy Matil (2021) examined the influence of artificial intelligence in Indian banking industries. This paper is based on descriptive nature and relevant data has been gathered from various journals, magazines, and websites. This study gives the positive and negative impact of artificial intelligence used in the Indian banking sectors. The authors have included the positive impact like improvement in online and mobile banking, Risk assessment process, protection of personal data, security, and swift transactions. On the other hand negative impacts included High costs, bad calls, distribution of power, and irrational behavior in ecosystems.

Sharan Kumar Shetty (2022) examined the impact of artificial intelligence on the banking sector particularly private banks in the context of India. In this research data has been used from both primary and secondary. The primary data was collected from 170 customers and 30 samples from bankers. The secondary data was gathered from books, magazines, and websites. In this study, the authors used statistical techniques such as chi-square, correlation, and regression for the investigation. The correlation test results show positive relation because the age of the customer special affects human inaccuracy it's difficult to use old people compare to young ones. The reason behind artificial intelligence is a new conception.

12.3 Methodology and Descriptive Analysis

The study is based on the existing written documents on descriptive nature of the insurance sector in India. In this chapter, the author explores the various benefits and opportunities of AI used by number of companies and numbers of customers in India. The already published research contents from different source of information indicates the uses of artificial intelligence in the financial industry. The analysis of numerous papers, different source of information offline and online have reviewed and captor an innovative change of technology in the insurance industry. Fintech is a good initiative that adopted by companies for the future of the insurance sector in India. That also highlighted the already given results for improvement in the existing sector.

12.3.1 Improved Digital Acceptance

Insurance companies must adopt different digital platforms and make operational work more suitable. Life Insurance Corporation introduces LIC Mitra as a virtual associate, New India Assurance launched the BIMA Bot.

Table 12.1 shows that customers spend a lot of time on the Internet and demand personalized experience across insurance premiums, valuing, and claims. Internet penetration has increased from 45% to 75% until the 2030P years. Smartphone penetration has increased from 39% to 75%. Although as well as payment wallet penetration and online shopping penetration have increased.

12.3.2 AI in the Context of Financial Services

Artificial Intelligence (AI) has revolutionized various industries and has played a significant role in changing the face of the financial services industry. AI has made the insurance sector more effective and efficient, increasing the quality of service,

Table 12.1 Favorable online consumer trends: All value in (%)

Years	2020	2025 P	2030 P
Internet penetration	45	63	75
Smartphone penetration	39	57	72
Payment wallets penetration	14	23	38
Online shopping penetration	14	22	36

Source: RBI, Frost, & Sullivan Analysis

reducing costs, and enhancing customer experience. The Indian insurance industry is no exception, with the increasing adoption of AI transforming the way insurance companies operate. This endeavor explores AI's impact on the Indian financial services industry, with a focus on the insurance sector. It provides authentic data and facts in a tabular format, supported by references.

12.3.3 Impact of AI on the Indian Insurance Sector

The use of AI in the Indian insurance industry is transforming the way insurance companies operate and interact with their customers. The following are some of how AI is impacting the Indian insurance industry:

Enhanced customer experience: AI is enabling insurance companies to provide personalized and seamless customer experiences. By analyzing customer data, AI-powered systems can recommend products and services that meet the unique needs of individual customers. Additionally, AI-powered chatbots are being used to provide quick and efficient customer support.

Improved underwriting: AI-powered systems are helping insurance companies to improve their underwriting processes. By analyzing vast amounts of data, AI-powered systems can identify patterns and insights that help insurers to assess risk more accurately. This enables insurers to offer more competitive pricing and better coverage to their customers.

Fraud detection: AI-powered systems are being used to detect and prevent insurance fraud. By analyzing data from multiple sources, AI can identify patterns and anomalies that indicate fraudulent activity. This helps insurers to prevent losses and maintain the integrity of their products and services.

Claims processing: AI-powered systems are being used to automate and streamline the claims processing process. By analyzing data from multiple sources, AI can verify claims and calculate payouts quickly and accurately. This helps insurers to provide better customer service and reduce the time and cost associated with claims processing.

Product development: AI is enabling insurance companies to change new goods and services that encounter the changing requirements of their consumers. By analyzing customer data and market trends, AI-powered systems can identify new opportunities for innovation and growth (Table 12.2).

The use of AI in the Indian insurance industry is transforming the way insurers operate and interact with their customers. By leveraging AI-powered systems, insurers can improve their operational efficiency, develop new products and services, optimize existing products, and enhance their risk management capabilities. Additionally, AI is enabling insurers to provide better customer service, reduce costs, and increase profitability.

While there are some challenges that insurers in India may face in adopting AI, therefore lack of skilled AI experts and concerns around data privacy and safety, the benefits of AI far outweigh the challenges. As the Indian insurance industry

Table 12.2 Impact of AI on the Indian Insurance Industry

Area	Impact of AI
Customer experience	Enhanced customer experiences through personalized and seamless interactions with customers
Underwriting	Improved underwriting processes through more accurate risk assessment and better pricing and coverage options for customers
Fraud detection	Enhanced fraud detection and prevention through analysis of data from multiple sources
Claims processing	Streamlined and automated claims processing, reducing time and cost and improving customer service
Product development	Identification of new opportunities for innovation and growth through analysis of customer data and market trends

Table 12.3 Cost savings due to AI adoption

Type of Cost	Estimated cost savings
Customer Service	Up to 30%
Claims Processing	Up to 25%
Underwriting	Up to 20%

Reference: Accenture

continues to adopt AI-powered systems, it is expected to become more efficient, customer-focused, and profitable.

12.3.4 *Cost Reduction*

AI has played a significant role in reducing operational costs for insurance companies. With AI-powered chatbots, insurance companies can handle a large number of customer queries and complaints, reducing the need for human intervention. According to a report by Accenture, AI-powered chatbots can reduce customer service costs by up to 30%. Additionally, AI-powered systems can automate many of the back-end processes, including claim processing and underwriting, reducing the need for manual intervention and saving time and resources (Table 12.3).

12.3.5 *Improved Customer Experience*

With AI-powered chatbots and computer-generated assistants, insurance companies can provide 24 into 7 support to their customers, reducing response times and improving the overall customer experience. Additionally, AI can help insurance companies personalize their services by analyzing customer data and providing personalized recommendations. According to a survey by Capgemini, 70% of customers are willing to share their data with insurance companies if it results in personalized service (Table 12.4).

12.3.6 Fraud Detection

Insurance fraud is a significant problem for insurance companies, with the Insurance Regulatory and Development Authority of India (IRDAI) estimating that insurance frauds account for approximately 10% of all claims made. AI can help insurance companies detect fraud by analyzing data patterns and identifying anomalies. According to a report by PwC, AI-powered fraud detection systems can reduce insurance fraud by up to 80% (Table 12.5).

12.3.7 Product Development

AI can help insurance companies develop new products and services by analyzing customer data and identifying new trends and market opportunities. Additionally, AI-powered systems can help insurance companies optimize their existing products and services by analyzing customer feedback and usage patterns (Table 12.6).

12.3.8 Improved Underwriting

AI-powered systems can handle large volumes of data quickly and correctly, helping insurance companies make more informed underwriting decisions. Additionally, AI can help insurance companies automate the underwriting process, reducing the time and resources required (Chart 12.1, Table 12.7).

Table 12.4 Improved customer experience due to AI adoption

Parameter	Improvement
Response time	Up to 90% reduction
Personalization	Up to 70% increase
Satisfaction	Up to a 50% increase

Reference: Capgemini

Table 12.5 Reduction in insurance fraud due to AI adoption

Type of fraud	Estimated reduction
Claims Fraud	Up to 80%
Application Fraud	Up to 70%

Reference: PwC

Table 12.6 Benefits of AI in product development

Benefit	Description
New Product Development	AI can analyze customer data and identify novel trends and market opportunities, helping insurance companies develop new products and services.
Product Optimization	AI can help insurance companies improve their existing products and services by analyzing customer feedback and usage patterns, leading to improvements in customer satisfaction and retention.

Reference: Deloitte (2018)

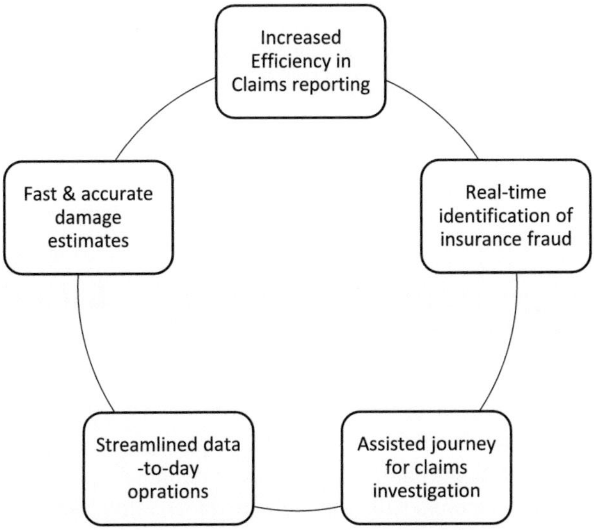

Chart 12.1 Flow chart of artificial intelligence key benefits in the insurance industry

Enhanced Claims Processing: AI-powered systems can automate the claims processing process, reducing the time and resources required. Additionally, AI can help insurance companies detect fraudulent claims and expedite legitimate claims, improving customer satisfaction (Table 12.8).

12.3.9 Improved Risk Management

AI can help insurance companies analyze data in real time and identify potential risks, enabling them to take proactive measures to mitigate those risks. Additionally, AI can help insurance companies predict future trends and events, enabling them to make informed decisions about their risk exposure (Table 12.9).

Table 12.7 Benefits of AI in underwriting

Benefit	Description
More Informed Underwriting Decisions	AI-powered systems can handle large amounts of data fast and correctly, leading to better underwriting decisions.
Automated Underwriting Process	AI can help insurance companies automate the underwriting process, reducing the time and resources required.
Improved Risk Assessment	AI-powered systems can help insurance companies assess risk more accurately, leading to reduced losses.

Reference: EY

Table 12.8 Benefits of AI in claims processing

Automated claims processing	AI-powered systems can automate the claims processing process, reducing the time and resources required.
Fraud detection	AI can help insurance companies detect fraudulent claims, reducing losses.
Expedited legitimate claims	AI can help insurance companies expedite legitimate claims, improving customer satisfaction.

Reference: PwC

12.4 Challenges and Future in the Insurance Industry to Adopt AI

The potential benefits of AI in the insurance industry are significant, while lot of challenges in connection with AI adoption. Data quality is one of them that the AI algorithms depend on high-quality data to generate valuable insights and predictions. However, many insurers still struggle with data quality issues, such as incomplete or inaccurate data. This can limit the effectiveness of AI algorithms and make it difficult to achieve the desired results.

Another challenge is the need for specialized skills and expertise. Developing and implementing AI algorithms requires specialized skills and expertise, which may not be readily available within the insurance industry. Insurers may need to invest in training and development programs to build the necessary skills and expertise in-house, or they may need to partner with external experts.

While AI has the potential to transform the insurance industry, some challenges and considerations must be taken into account when implementing the technology. One major challenge is the need for high-quality data. AI algorithms rely on large amounts of data to learn and make predictions. If the data used is incomplete, inaccurate, or biased, the results produced by AI can be flawed. Ensuring the quality and fairness of data is therefore critical to the success of AI implementation.

Another challenge is the potential for job displacement. As AI is used to automate processes such as claims processing and underwriting, there is a risk that some jobs may become redundant. It will be important for companies to consider the impact of AI on their workforce and implement strategies to reskill and redeploy affected employees. Privacy and security are also important considerations for AI

Table 12.9 Benefits of AI in risk management

Benefit	Description
Real-Time Risk Analysis	AI can help insurance companies analyze data in real time and identify potential risks, enabling them to take proactive measures to mitigate those risks.
Predictive Analytics	AI can help insurance companies predict future trends and events, enabling them to make informed decisions about their risk exposure.
Improved Risk Mitigation	AI can help insurance companies identify and mitigate risks more effectively, reducing losses.

Reference: Deloitte

implementation in the insurance industry. Insurers must ensure that customer data is handled securely and ethically and that privacy regulations such as GDPR and CCPA are followed.

Finally, there is the challenge of regulation. The insurance industry is heavily regulated, and there are concerns that regulators may not be equipped to handle the complexities of AI. It will be important for insurers to work closely with regulators to ensure that AI is implemented in a way that is compliant with regulations.

Overall, while there are challenges and considerations associated with AI implementation, the potential benefits are significant. It will be important for insurers to carefully consider these challenges and work to address them to fully realize the potential of AI in the insurance industry.

Additionally, there are concerns about data privacy and security. However, this data must be collected, stored, and processed in a secure and privacy-compliant manner to protect customer data and ensure compliance with regulatory requirements.

Finally, there has been a concern about the potential effect of AI on jobs in the insurance industry. While AI can help automate routine tasks and improve efficiency, it may also lead to job losses in some areas. The next section will explore strategies for addressing these challenges and successfully implementing AI in the insurance industry.

12.4.1 Strategies in the Insurance Industry for Successful Implementation of AI

To overcome the challenges associated with the adoption of AI in the insurance industry, insurers can consider several strategies. One strategy is to focus on improving data quality. This can involve implementing data governance frameworks to ensure data accuracy, completeness, and consistency. Insurers can also leverage advanced analytics tools to identify and correct data quality issues in real time (Irfan et al. 2023).

Another strategy is to build the necessary skills and expertise within the organization. This can involve investing in training and development programs to build in-house capabilities or partnering with external experts to fill gaps in expertise.

Data privacy and security can be addressed by implementing robust data privacy and security policies and procedures. This can involve implementing encryption and access controls, as well as conducting regular security audits to identify and address potential vulnerabilities.

Finally, insurers can address concerns around job losses by focusing on upskilling and reskilling employees. This can involve investing in training programs to help employees develop new skills and expertise, as well as offering job placement services to help employees transition to new roles within the organization. By implementing these strategies, insurers can successfully overcome the challenges associated with AI adoption and realize the potential benefits of AI in the insurance industry. In the next section, we will explore some of the ethical considerations associated with AI in the insurance industry.

12.4.2 Best Practices for Implementing AI in the Insurance Industry

To effectively implement AI, insurers must follow certain best practices. Here are some of the key best practices: Develop a clear AI strategy: Insurers must have a clear understanding of how they plan to use AI to achieve their business goals. They must also identify the specific areas where AI can add the most value.

Invest in data governance frameworks: A robust data governance framework is essential for effective AI implementation. Insurers must ensure that data is collected, stored, processed, and analyzed consistently and securely. Hire skilled AI professionals: Insurers must invest in hiring skilled professionals in areas such as data science, machine learning, and natural language processing.

Address data privacy and security concerns: Insurers must ensure that customer data is stored and processed in a secure and compliant manner. They must also ensure that they are transparent about how customer data is being used.

Upskill employees and invest in workforce development programs: Insurers must ensure that their employees have the necessary skills to work effectively with AI. They must also invest in workforce development programs to help employees transition into new roles.

By following these best practices, insurers can effectively implement AI and realize its potential benefits. However, insurers must also remain mindful of the challenges associated with AI adoption and work to address these challenges as they arise.

In the next section, we will explore some of the potential future developments in the insurance industry as a result of AI.

12.4.3 Ethical Considerations of AI

As with any technology, there are ethical considerations associated with the use of AI in the insurance industry. One key ethical consideration is the potential for bias in AI algorithms. The accurate data can be trained by the AI algorithms as unbiased. Despite the data used to train the algorithm being biased, the algorithm may perpetuate that bias in its predictions and decision-making.

There is another ethical consideration is the potential for AI to undermine customer privacy. AI algorithms rely on large amounts of data to generate insights and predictions. However, this data must be collected, stored, and processed in a secure and privacy-compliant manner to protect customer data and ensure compliance with regulatory requirements. Additionally, there are concerns about the potential for AI to perpetuate discrimination. For example, if an AI algorithm is trained on data that reflects historic patterns of discrimination, it may perpetuate that discrimination in its decision-making (Irfan et al. 2023).

To address these ethical considerations, insurers must ensure that their AI algorithms are developed and trained using unbiased data and that they are transparent in their decision-making processes. They must also ensure that they comply with all relevant data privacy and security regulations and take steps to prevent discrimination in their algorithms. In the next section, we will explore some of the potential future developments in AI in the insurance industry.

12.4.4 Future Developments of the Insurance Industry Within the AI Frame

As AI continues to evolve, it can be likely to have a significant impact on the insurance industry in the coming years. One potential future development is the use of AI-powered chatbots to provide personalized customer service. Chatbots can help insurers to provide quick and efficient customer service by answering customer queries and guiding customers through the claims process.

Another potential future development is the use of AI-powered risk assessment tools to improve underwriting accuracy. AI algorithms can analyze vast amounts of data to identify patterns and trends that may be indicative of risk. By leveraging these insights, insurers can make more informed underwriting decisions and reduce the likelihood of losses.

AI is poised to have a significant impact on the insurance industry in the years to come. Here are some of the potential future developments: Increased Personalization: With the help of AI, insurers can better understand customer needs and preferences, and offer more personalized products and services.

Improved Claims Processing: AI can be used to automate claims processing, reducing the time and cost associated with manual processes. This can improve the customer experience and lead to faster claims resolution. Enhanced Fraud Detection:

AI can help insurers detect and prevent fraud by analyzing data patterns and identifying suspicious behaviour. Better Risk Management: AI can be used to analyze vast amounts of data to identify emerging risks and develop more effective risk management strategies.

Automation of Underwriting: AI can be used to automate the underwriting process, improving efficiency and accuracy. Usage-Based Insurance: AI can help insurers better understand customer behaviour and usage patterns, allowing for more accurate and personalized pricing models. These are just a few examples of the potential future developments in the insurance industry as a result of AI. As the technology continues to evolve, we can expect to see even more exciting developments in the years to come.

AI-powered fraud detection is another potential future development in the insurance industry. AI algorithms can analyze vast amounts of data to identify potential fraudulent activities and alert insurers to potential fraud. This can help insurers to detect and prevent fraudulent activities before they result in significant losses.

Finally, the use of AI-powered drones for risk assessment and claims management is also a potential future development in the insurance industry. Drones can be used to survey the damage and assess risks in areas that are difficult to access, such as remote or hazardous locations.

In conclusion, AI is transforming the insurance industry in India by enabling insurers to make more informed decisions, improve operational efficiency, and enhance the customer experience. While there are challenges associated with AI adoption, insurers can overcome these challenges by implementing data governance frameworks, building in-house capabilities, addressing data privacy and security concerns, and upskilling employees. As AI continues to evolve, it is likely to have an even more significant impact on the insurance industry in the coming years.

12.5 Conclusion

According to a report by PwC, the use of AI in the Indian financial services industry is expected to grow at a compound annual growth rate (CAGR) of 28.8% from 2019 to 2025. This growth is expected to be driven by the increasing adoption of AI-powered systems by insurance companies to improve their operations and customer service.

However, there are some challenges that insurance companies in India may face in adopting AI. One of the primary challenges is the lack of skilled AI professionals in the industry. According to a report by EY, only 22% of the Indian financial services industry has the necessary AI skills and expertise to adopt AI-powered systems.

Additionally, there are concerns around data privacy and security that insurance companies need to address when adopting AI. Insurance companies need to ensure that the data they collect and use is secure and compliant with regulatory requirements.

Despite these challenges, the benefits of AI in the Indian insurance industry far outweigh the challenges. As the Indian insurance industry continues to adopt AI-powered systems, it is expected to become more efficient, customer-focused, and profitable.

The paper concludes that the positioning of Artificial intelligence systems in the financial industry particularly the insurance sector will continue to increase due to advances in technology and transformation toward digital financial services. The use of artificial intelligence systems can bring substantial benefits such as reduced cost, productivity increase, and better risk management. AI has likely to interrupt the insurance sector and insurance companies can use different business models to influence artificial intelligence and data-determined technologies. It highlights the artificial intelligence transforming the way organization function and interrelate with users. The paper evaluates the implementation of artificial intelligence computerization in the insurance industry and identifies emerging business models in the financial sector. It does not provide specific details about the future works. Though it suggests that insurance companies should explore and adopt artificial intelligence to stay competitive and progress their business models. So, it can be contingent that future works may involve more research on the implementation of artificial intelligence in the insurance sector and its impact on the financial industry.

References

Naman K, Jayant DS, Harshit B (2019) Artificial Intelligence in Insurance Sector, Journal of The Gujarat Research Society, 21(7):79–91.

Deloitte (2018) Insurance in the age of AI. https://www2.deloitte.com

Eling M, Nuessle D, Staubli J (2021) The impact of artificial intelligence along the insurance value chain and on the insurability of risks. Geneva Pap Risk Insur 47(2):205–241. https://doi.org/10.1057/s41288-020-00201-7

Geetha A (2021) A study on artificial intelligence (Ai) in banking and financial services. Int J Creat Res Thoughts 9(9):110–114. http://ijcrt.org/viewfull.php?&p_id=IJCRTG020019

Irfan M, Elhoseny M, Kassim S, Metawa N (2023) Advanced machine learning algorithms for complex financial applications [online]. Available at https://www.igi-global.com/book/advanced-machine-learning-algorithms-complex/290048#table-of-contents

Kelley K, Fontanetta LM, Heintzman M, Pereira N (2018) Artificial intelligence: implications for social inflation and insurance. Risk Manag Insur Rev 21(3):373–387. https://doi.org/10.1111/rmir.12111

Padmanabhan V, Metilda VP (2021) An impact of artificial intelligence in Indian banking industries. International Research Journal of Education and Technology 1(4):9–45

Riikkinen M, Saarijärvi H, Sarlin P, Lähteenmäki I (2018) Using artificial intelligence to create value in insurance. Int J Bank Mark 36(6):1145–1168. https://doi.org/10.1108/ijbm-01-2017-0015

Shetty SK, Spulbar C, Birau R, Filip RD (2022) Impact of artificial Interlligence in banking sector with reference to private banks in India. Physics AUC 32:59–75

Soni P (2021, May 1) A study on artificial intelligence in the finance sector. www.ijcrt.org. http://ijcrt.org/viewfull.php?&p_id=IJCRT2105683

Sunitha BK, Nukala Jahnavi, Khushi V Tibrewal (2020) Banking sector transformation - artificial intelligence in the modern digital era, International Journal of Advances in Engineering and Management (IJAEM) 2(9):404–411

Stephan Bredt (2019) Artificial intelligence (AI) in the financial sector- potential and public strategies, Frontiers in Artificial Intelligence, Vol.2-2019, https://doi.org/https://doi.org/10.3389/frai.2019.00016

Chapter 13
Future Trends and Opportunities in Machine Learning and Artificial Intelligence for Banking and Finance

Anand Kumar Mishra, Amit Kumar Tyagi, and Micheal Olaolu Arowolo

13.1 Introduction

13.1.1 Evolution of Machine Learning (ML) and Artificial Intelligence (AI) in Banking and Finance

The evolution of AI and ML in banking and finance has been a transformative journey, revolutionizing the industry and paving the way for new possibilities (Carlo et al.). Here is a summary of the key stages in the evolution of ML and AI in banking and finance:

- Early Adoption and Rule-Based Systems: In the early stages, banks and financial institutions started using rule-based systems to automate simple tasks and decision-making processes. These systems relied on predefined rules and logical statements to perform tasks such as credit scoring, fraud detection, and risk assessment.
- Data-Driven Approaches and Predictive Analytics: With the advent of large-scale data collection and advancements in computing power, financial institutions began leveraging data-driven approaches and predictive analytics. This enabled more accurate risk modeling, improved credit scoring, and enhanced fraud detection capabilities.

A. K. Mishra (✉)
Computer Science and Engg, NIIT University, Neemrana, Rajasthan, India

A. K. Tyagi
Department of Fashion Technology, National Institute of Fashion Technology, Hauz Khas, New Delhi, India

M. O. Arowolo
Department of Computer Science, Landmark University, Omu-Aran, Nigeria

M. Irfan et al. (eds.), *Applications of Block Chain technology and Artificial Intelligence*, Financial Mathematics and Fintech,
https://doi.org/10.1007/978-3-031-47324-1_13

- Advanced ML Techniques: As ML techniques evolved, financial institutions started adopting more sophisticated algorithms such as decision trees, random forests, support vector machines, and neural networks. These techniques enabled complex pattern recognition, nonlinear modeling, and improved accuracy in various applications, including investment analysis, algorithmic trading, and customer segmentation.
- Natural Language Processing and Chatbots: The integration of natural language processing (NLP) with ML and AI opened doors to automated customer interactions and support systems. Chatbots and virtual assistants equipped with NLP capabilities emerged, providing personalized customer service, answering queries, and facilitating routine banking tasks.
- Deep Learning and Neural Networks: The rise of deep learning, a subfield of ML, brought about significant advancements in the banking and finance industry. Deep neural networks, showed remarkable capabilities in areas like image recognition, speech processing, and text analysis. These techniques find applications in fraud detection, anti-money laundering (AML), sentiment analysis, and credit risk assessment.
- Robotic Process Automation (RPA) and Intelligent Automation: Robotic Process Automation (RPA) and intelligent automation technologies entered the banking and finance landscape, automating repetitive tasks and streamlining operational processes. RPA bots perform rule-based tasks, while intelligent automation combines ML and AI techniques to handle more complex and cognitive tasks, such as document processing, data extraction, and regulatory compliance.
- Explainable AI and Ethical Considerations: As ML and AI became more prevalent, concerns regarding bias, transparency, and accountability emerged. The focus shifted towards developing explainable AI models that provide clear explanations for decisions and ensure fairness. Ethical considerations, such as data privacy, responsible use of AI, and algorithmic transparency, gained prominence, leading to regulatory frameworks and guidelines.
- Future Directions: The future of ML and AI in banking and finance holds immense potential. Areas like explainable AI, federated learning, quantum computing, and reinforcement learning are poised to make further advancements. Additionally, the integration of ML and AI with emerging technologies like blockchain, Internet of Things (IoT), and edge computing is expected to open new avenues for secure transactions, personalized services, and improved risk management.

In summary, the evolution of ML and AI in banking and finance has transformed the industry by enabling more accurate decision-making, enhancing customer experiences, improving operational efficiency, and managing risks effectively. Continued innovation and responsible implementation will shape the future of ML and AI, ensuring a dynamic and technologically advanced banking and finance sector.

13.1.2 Importance of Identifying Future Trends and Opportunities in Banking and Finance

Identifying future trends and opportunities in banking and finance is of paramount importance for several reasons (Richard et al.):

- Strategic Planning: Understanding future trends and opportunities helps financial institutions develop robust strategic plans. By anticipating market shifts and technological advancements, banks can align their resources and investments accordingly. This proactive approach enables them to stay ahead of the competition, seize emerging opportunities, and adapt to changing customer expectations.
- Innovation and Differentiation: Identifying future trends allows financial institutions to innovate and differentiate themselves in the market. By embracing new technologies, business models, and customer-centric approaches, banks can deliver unique and tailored solutions. This fosters customer loyalty, attracts new customers, and positions the institution as an industry leader.
- Risk Management: Future trends impact risk profiles in banking and finance. By understanding these trends, institutions can proactively manage risks associated with technology, market shifts, and regulatory changes. This includes identifying potential vulnerabilities, evaluating risk exposure, and implementing appropriate risk mitigation strategies.
- Enhanced Customer Experience: Anticipating future trends helps financial institutions meet evolving customer expectations. By understanding changing preferences, needs, and behaviors, banks can tailor their products and services to deliver exceptional customer experiences. This customer-centric approach fosters customer satisfaction, loyalty, and long-term relationships.
- Technological Transformation: The banking and finance industry is undergoing significant technological transformations. Identifying future trends helps institutions harness the power of emerging technologies, such as artificial intelligence, blockchain, and data analytics. This enables them to streamline operations, improve efficiency, and unlock new revenue streams.
- Regulatory Compliance: Regulatory frameworks in banking and finance are continuously evolving. Identifying future trends helps institutions stay compliant with upcoming regulations and adapt their processes and systems accordingly. This proactive approach minimizes compliance risks, avoids penalties, and ensures adherence to industry standards.

In summary, identifying future trends and opportunities in banking and finance is crucial for strategic planning, innovation, risk management, customer experience, technological transformation, regulatory compliance, talent acquisition, and sustainable growth. By staying ahead of the curve, financial institutions can navigate industry disruptions, leverage emerging opportunities, and position themselves as leaders in an ever-evolving landscape.

13.1.3 Literature Survey and Organization of the Work

In literature (Nenad et al.) they examine the likelihood of effective implementation while also considering potential obstacles, issues, and solutions. Machine learning and artificial intelligence can support mitigation strategies for the current financial and economic issues, especially those brought on by the COVID-19 crisis. This article primarily examines credit risk management, but it also analyses the use of artificial intelligence and machine learning in other risk management domains. The final choice would be made by a human being, but technological advancements. Making wise financial decisions will be aided by such a training session in this situation. Further analyses of the effects of personality characteristics and anomalies in the financial markets on financial decision-making to advance financial knowledge and provide new avenues for its development for the benefit of both individuals and society. The study by Ranjan et al. (2020) will examine how people's brains interpret financial information and how and what decisions are made there. So that we may create a training module based on whether a student is an investor, an advisor, or a manager.

The paper provided by Fares et al. (2022) offers a comprehensive and methodical analysis of the literature on artificial intelligence (AI) use in the banking industry from 2005. The authors of this study used a systematic literature review approach to review 44 papers, and then they analyzed them thematically and in terms of substance. The results show how three important study areas—Strategy, Process, and Customer—are covered by the literature on AI and banking. Currently, artificial intelligence (AI) is thought to be the most significant and disruptive new technology for big businesses. However, outside of technology startups, the technology is mostly absent from smaller businesses and is still in an early stage in large ones. According to surveys, less than half of significant AI initiatives in large organizations are now under progress, while the ratio is rising with time. The essay "AI in organizations: current state and future opportunities" discusses the existing issues and ramifications that could result from using AI applications, as well as how to get over them and fully utilize this cutting-edge technology. The paper (Benbya et al. 2020) starts off by giving a quick overview of AI typologies and a brief history of AI. We talk about the difficulties, effects, and potential of AI today.

The study (Mogaji and Nguyen 2022) aimed to better understand managers' awareness offers empirical understanding of the possibilities, chances, and difficulties associated with applying AI to the marketing of financial services. Additionally, it specifically challenges some assumptions about AI's function in financial services, the use of chatbots in the provision of financial services, and the contribution of marketing managers to the development of AI.

Currently, terminology like digital marketing, decision-making, industry 4.0, and corporate digital transformation are associated with notions like neural networks, machine learning, or deep learning. The purpose of the study (Ruiz-Real et al. 2021) is to evaluate contemporary studies on artificial intelligence in the workplace. To do this, a bibliometric study employing the Web of Science and Scopus online databases has been implemented. This study analyses 11 clusters and the most prevalent phrases used in Artificial Intelligence research using a fractional counting method. The current analysis outlines the key trends in business AI research and suggests new areas for investigation. The essay presented by Zhang et al. (2020) examines the development of the accounting profession in light of current technological advancements and evaluates their potential effects. Examined are the inherent opportunities and challenges that these new technologies present for accounting educators and professionals, including the rising need for IT specialists with accounting experience as opposed to accounting degree graduates.

The study presented by Hentzen et al. (2022) provide an overview of contexts that have been explored and research focuses, identify gaps in the body of knowledge, and establish a thorough agenda for future research. More individualized, sophisticated, better, safer, and more recent mainstream and alternative economic-financial mechanisms, goods, models, services, systems, and applications are being made possible by AI. In addition to providing a comprehensive, multidimensional, and problem-driven research landscape of the roles, research objectives, and potential of AI in new-generation FinTech and finance (Cao 2020). The paper by Rabbani et al. (2023) offers a current overview of peer-to-peer and the report suggests a P2P lending model for small and medium-sized businesses that is based on Islamic Fintech. To revitalize industry and save hundreds of thousands of jobs, it is past time to create and promote new lending mechanisms.

The cyber threat has made it difficult for researchers to discover a remedy as the industry trend has changed to intelligent internet-based enterprises on the same side. By selecting the most pertinent research papers (126) from the four most reliable online repositories based on the four stated research topics, a systematic search is done by (Khan et al. 2022) to fill the gaps in the available literature. To preserve high security and authenticity within the financial sectors of the GCC countries, these research subjects seek to examine the current situation from a variety of angles and present fresh research directions for future investigation.

More financial gains result from the trade credit. The major goal of the study (Shah et al. 2022) is intelligent and secure revenue creation because of the influence of TC with predictive machine learning algorithms. That is, the fusion of the financial data and technology (FinTech) sectors may lead to more sales and, ultimately, higher profits. Although the insured party benefits by this situation, the insurance firms face several risks because of the drop in insurance premiums. To improve their precarious position, insurance companies are being forced to undergo several changes in their business strategies. Given this, it is crucial to address these breakthroughs to avoid falling behind these advancements. From this vantage point, the chapter (Erem Ceylan 2022) investigates AI developments and their impacts on the insurance industry, exposing use cases of both Turkish and worldwide insurance in the application of AI (Fig. 13.1).

The next section discusses the advancement in machine learning and artificial intelligence in banking and finance. This section has covered many areas various aspects such as deep learning, neural network, reinforcement learning, and others in the same area. Further, the integration of machine learning and deep learning with

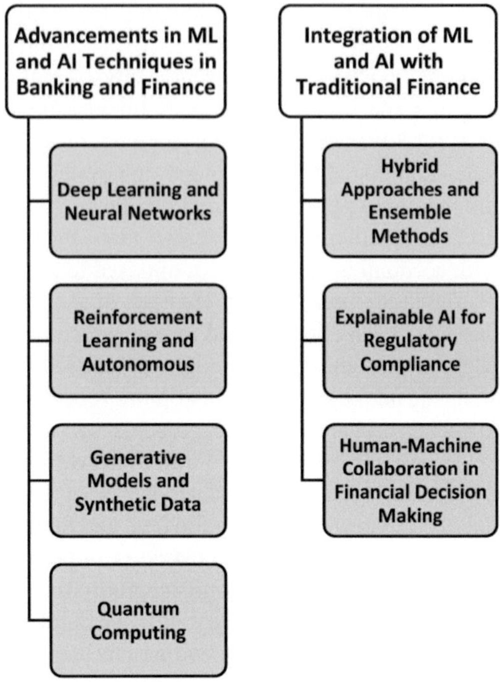

Fig. 13.1 Various aspects of machine learning and artificial intelligence in banking and finance

various traditional finances has been discussed. The next section discusses the emerging applications in banking and finance. The contribution of big data and cloud computing for banking and finance has been explored in the next section whereas the later section covers the ethical considerations and responsible artificial intelligence for banking and finance. Considering the importance of industry in the banking and finance area the next section discusses the industry adoption and transformation for banking and finance. To explore various research aspects in the area the next section provides a detailed discussion about the research and investment opportunities in banking and finance. The later section concludes the chapter.

13.2 Advancements in ML and AI Techniques in Banking and Finance

13.2.1 Deep Learning and Neural Networks in Banking and Finance

Deep learning and neural networks have emerged as powerful tools in the banking and finance industry, revolutionizing various aspects of operations, decision-making, and customer experiences (Huang et al. 2020). Here are some key applications of deep learning and neural networks in banking and finance:

- Fraud Detection: These models can analyze large volumes of transaction data, identify patterns of fraudulent activities, and flag suspicious transactions in real-time, improving the accuracy and speed of fraud detection.
- Credit Scoring and Risk Assessment: Neural networks can be employed to develop credit scoring models that predict creditworthiness and assess the risk associated with lending. By analyzing historical data on borrowers, these models can learn complex patterns and provide more accurate risk assessments, leading to better lending decisions and reduced default rates.
- Customer Relationship Management: To improve customer segmentation and personalized marketing. By understanding customer preferences and needs, financial institutions can offer targeted product recommendations, personalized offers, and tailored customer experiences, leading to increased customer satisfaction and retention.
- Sentiment Analysis: Deep learning models, such as RNNs and long short-term memory (LSTM) networks, are used in sentiment analysis to analyze textual data from social media, news articles, and customer feedback. This analysis helps financial institutions gauge public sentiment towards specific stocks, companies, or financial products, aiding in investment decision-making and risk management.
- Risk Management: Deep learning models can enhance risk management by analyzing complex data sources and identifying potential risks. For example, neural networks can be trained to analyze unstructured data such as news articles, regulatory filings, and economic indicators to assess systemic risks and make predictions about market trends.

13.2.2 Reinforcement Learning and Autonomous Agents in Banking and Finance

Reinforcement learning (RL) and autonomous agents have started to gain attention in the banking and finance industry, offering promising applications for decision-making, risk management, and trading (Sen et al. 2021). Here are some key aspects of reinforcement learning and autonomous agents in banking and finance:

- Portfolio Management: Reinforcement learning can be used to develop autonomous agents that learn optimal portfolio management strategies. These agents learn from historical market data, optimize asset allocation, and dynamically adjust investment decisions based on market conditions and predefined objectives. RL algorithms, such as Q-learning and deep Q-networks (DQNs), enable agents to adapt and improve their strategies over time.
- Algorithmic Trading: RL techniques can be applied to build autonomous trading agents that learn and optimize trading strategies. These agents learn from market data, historical price patterns, and order book dynamics to make trading decisions. RL allows agents to adapt to changing market conditions, identify profitable trading opportunities, and execute trades autonomously.
- Risk Management: Autonomous agents trained through RL can aid in risk management. These agents learn risk models and decision-making policies that optimize risk-reward trade-offs. RL can be used to model and predict market volatility, manage risk exposures, and dynamically adjust risk management strategies based on evolving market conditions.
- Pricing and Market Making: RL can be applied to develop autonomous agents for pricing financial instruments and participating in market making activities. These agents learn optimal pricing strategies by considering market trends, liquidity conditions, and other relevant factors. By continuously learning and adapting, these agents can improve pricing accuracy and liquidity provision.
- Fraud Detection and AML: RL techniques can be employed to develop autonomous agents for fraud detection and anti-money laundering (AML) activities. These agents learn from historical data, transaction patterns, and known fraud cases to identify anomalous behaviors and detect potential fraudulent activities. RL allows agents to adapt to evolving fraud patterns and improve detection accuracy over time.
- Customer Service and Chatbots: Reinforcement learning can be used to train autonomous agents, such as chatbots and virtual assistants, for customer service in banking and finance. These agents learn from customer interactions, historical data, and predefined service policies to provide personalized and efficient customer support. RL enables agents to improve their responses and service quality based on user feedback and reinforcement signals.

It is worth noting that the application of RL and autonomous agents in banking and finance requires careful consideration of risks, ethics, and regulatory frameworks. The development and deployment of these technologies should be accompanied by proper validation, monitoring, and human oversight to ensure responsible and compliant use.

Reinforcement learning and autonomous agents offer exciting possibilities in enhancing decision-making. As the field continues to advance, further research, collaboration, and experimentation will shape the practical application of RL and autonomous agents in the industry.

13.2.3 Generative Models and Synthetic Data in Banking and Finance

Generative models and synthetic data have gained attention in the banking and finance industry as valuable tools for data generation, privacy preservation, and augmenting limited datasets (Assefa et al. 2020). Here are some key aspects of generative models and synthetic data in banking and finance:

- Data Generation and Augmentation: Mimics the characteristics and statistical properties of real financial data. This synthetic data can be used to augment limited datasets, create realistic scenarios for testing models, and address data scarcity issues.
- Privacy Preservation: In the era of stringent data privacy regulations, generative models and synthetic data offer a way to generate privacy-preserving data without exposing sensitive information. By learning from real data and generating synthetic counterparts, these models can ensure privacy protection while maintaining the utility and statistical properties of the original data.
- Fraud Detection and Anomaly Detection: Synthetic data generated by generative models can be used to train and test fraud detection and anomaly detection algorithms. By creating realistic but synthetic fraud scenarios, these models can help improve the accuracy and robustness of fraud detection systems, enabling financial institutions to detect new and emerging fraud patterns.
- Stress Testing and Risk Assessment: Generative models and synthetic data can be employed to simulate various stress-testing scenarios and assess risk exposures. By generating synthetic data that captures different market conditions, economic shocks, or extreme events, financial institutions can evaluate their risk management strategies and assess potential vulnerabilities.
- Model Training and Validation: Synthetic data generated by generative models can be used to augment training datasets for building and validating ML models. By incorporating synthetic data that represents a wide range of scenarios and edge cases, financial institutions can enhance the generalization and performance of their models.
- Backtesting Trading Strategies: Synthetic data can be useful in backtesting trading strategies and evaluating their performance. By generating synthetic historical price and market data, generative models allow traders and quantitative analysts to test their trading strategies in simulated market conditions before deploying them in real-world scenarios.
- Data Sharing and Collaboration: Generative models and synthetic data provide a means of sharing and collaborating on sensitive financial data without compromising privacy and confidentiality. Financial institutions can generate synthetic versions of their datasets and share them with external partners, researchers, or regulators, facilitating collaborative efforts while protecting sensitive information.

It is important to note that the application of generative models and synthetic data in banking and finance requires careful validation and consideration of the quality, representativeness, and statistical properties of the generated data. Additionally, regulatory and compliance requirements must be taken into account to ensure the appropriate use of synthetic data.

Generative models and synthetic data offer valuable solutions for data generation, privacy preservation, and addressing data limitations in the banking and finance industry. Continued research and advancements in this field will contribute to the development of more robust and reliable synthetic data generation techniques tailored to the specific needs of the industry.

13.2.4 Quantum Computing in Banking and Finance

Aspects of banking and finance by providing computational power that surpasses classical computers (Egger et al. 2020). Here are some key areas where quantum computing can have an impact:

- Portfolio Optimization: Quantum computing can significantly enhance the efficiency of portfolio optimization by solving complex optimization problems. Quantum algorithms can explore a large number of potential asset combinations simultaneously, enabling more accurate and efficient portfolio allocation strategies.
- Risk Analysis and Modeling: Quantum computing can improve risk analysis and modeling by enabling the analysis of vast amounts of data and complex risk scenarios. Quantum algorithms can provide faster and more accurate simulations of market conditions, allowing for better risk assessment and stress testing.
- Option Pricing and Derivatives: Quantum computing can help in the accurate pricing and valuation of complex financial derivatives and options. Quantum algorithms can provide faster and more accurate pricing models, considering multiple variables and market conditions simultaneously.
- Machine Learning and Data Analysis: Quantum algorithms can handle massive datasets and perform complex data analysis tasks, enabling more accurate predictions, fraud detection, and customer segmentation.
- Cryptography and Security: Quantum computing has implications for cryptography and security in banking and finance. Quantum cryptography can provide enhanced security for financial transactions and data protection.
- Monte Carlo Simulations: Quantum computing can accelerate Monte Carlo simulations, commonly used in pricing financial instruments, risk analysis, and derivative valuation. Quantum algorithms can perform multiple simulations in parallel, reducing the computational time required for complex simulations.
- Optimization of Trading Strategies: Quantum computing can improve the optimization of trading strategies by considering multiple variables, constraints, and market conditions simultaneously. Quantum algorithms can quickly identify optimal trading strategies based on real-time market data, enhancing trading efficiency and profitability.

It is important to note that quantum computing is still in its early stages, and significant technical and practical challenges need to be overcome before widespread adoption in banking and finance. These challenges include error correction, scalability, and the development of quantum algorithms specifically tailored to financial applications. However, with ongoing research and advancements, quantum computing has the potential to revolutionize the way financial institutions operate and make decisions.

13.3 Integration of ML and AI with Traditional Finance

13.3.1 Hybrid Approaches and Ensemble Methods for Finance

Hybrid approaches and ensemble methods are widely used in finance to improve the accuracy and robustness of predictive models and decision-making processes (Faris et al. 2020). Here are some key aspects of hybrid approaches and ensemble methods in finance:

- Hybrid Models: Hybrid models combine multiple modeling techniques to leverage their individual strengths and overcome their weaknesses. For example, a hybrid model could combine statistical models with machine learning algorithms or combine quantitative models with qualitative expert opinions. By integrating different modeling approaches, hybrid models aim to achieve more accurate predictions and better capture the complexity of financial markets.
- Model Stacking: Model stacking, also known as stacked generalization, involves training multiple individual models and then combining their predictions using a meta-model. Each individual model may have different characteristics or specialize in different aspects of the data. The meta-model learns to weigh the predictions from each individual model and generates the final prediction. Model stacking can improve prediction accuracy and reduce the risk of relying on a single model.
- Ensemble Methods: Ensemble methods combine multiple models to make collective predictions. Boosting focuses on iteratively training models that correct the mistakes of previous models. Random forests build an ensemble of decision trees and aggregate their predictions. Ensemble methods can improve prediction accuracy, reduce model variance, and handle complex relationships in financial data.
- Expert Systems: Expert systems combine human expertise with machine learning algorithms to make informed decisions. They leverage the knowledge and insights of domain experts to complement the predictive capabilities of models. Expert systems can be used in areas such as credit risk assessment, investment recommendation systems, and fraud detection. By incorporating human expertise, these systems aim to enhance the accuracy and explainability of the decision-making process.

- Meta-Learning: Meta-learning involves using machine learning techniques to learn how to best combine and utilize multiple models. It learns the optimal weights or strategies for combining predictions from different models based on past performance. Meta-learning can adapt and optimize the ensemble approach dynamically, considering the specific characteristics of the data and the performance of individual models.
- Genetic Algorithms: Genetic algorithms are optimization techniques inspired by the process of natural evolution. They involve creating a population of candidate solutions, evolving them through successive generations, and selecting the best-performing solutions. In finance, genetic algorithms can be used for portfolio optimization, parameter tuning of models, or feature selection. They can explore a large search space efficiently and find near-optimal solutions.

The use of hybrid approaches and ensemble methods in finance helps address the limitations of individual models, improve prediction accuracy, and enhance decision-making processes. However, it is important to carefully design, validate, and interpret these approaches, considering factors such as model compatibility, data quality, and the specific objectives of the problem at hand.

13.3.2 Explainable AI for Regulatory Compliance for Finance

Explainable AI (XAI) plays a crucial role in regulatory compliance for the finance industry, where transparency, accountability, and interpretability are of utmost importance (Kuiper et al. 2022). Here are some key aspects of using XAI for regulatory compliance in finance:

- Regulatory Interpretability: Financial institutions are subject to various regulations and compliance requirements imposed by regulatory bodies. XAI techniques can help interpret these regulations and translate them into understandable rules or decision criteria for AI systems. This ensures that AI models and algorithms operate within the boundaries set by regulatory frameworks.
- Model Explainability: XAI techniques provide explanations for the decisions made by AI models, allowing stakeholders to understand the factors influencing those decisions. This is particularly important in finance, where regulatory compliance often requires justification and documentation of decision-making processes. By providing transparent explanations, XAI helps build trust in AI systems and enables regulators, auditors, and compliance officers to assess the fairness, ethics, and compliance of the models.
- Bias and Fairness Assessment: XAI can help identify and mitigate biases in AI models that may lead to unfair outcomes. XAI techniques enable the examination of the underlying features and variables that contribute to decision-making, helping detect and address biases related to race, gender, or other protected attributes. By providing insights into model behavior, XAI assists in ensuring fair

and unbiased outcomes, aligning with regulatory requirements and ethical standards.

- Auditing and Governance: XAI supports the auditing and governance processes in financial institutions by providing transparency and traceability of AI systems. XAI techniques allow for the monitoring and documentation of model behavior, inputs, and outputs, enabling compliance officers and auditors to assess the compliance of AI systems with regulatory standards. This facilitates the establishment of robust governance frameworks and risk management practices.
- Regulatory Reporting: XAI can help generate explainable reports and documentation required for regulatory reporting purposes. By providing insights into the decision-making process of AI systems, XAI techniques enable the creation of audit trails, explanations of model behavior, and evidence of compliance with regulatory requirements. This simplifies the reporting process and enhances the ability to demonstrate compliance to regulatory authorities.
- Customer Transparency and Consent: XAI techniques can contribute to enhancing customer transparency and obtaining informed consent in the use of AI systems. By providing understandable explanations of how customer data is processed and used, XAI helps financial institutions communicate the benefits, risks, and implications of AI-driven services to their customers. This supports compliance with data protection regulations, such as the General Data Protection Regulation (GDPR).
- Regulatory Validation and Approval: XAI techniques can assist in the validation and approval of AI models by regulatory bodies. By providing transparent explanations and justifications for model behavior, XAI helps regulators understand and assess the compliance, fairness, and reliability of AI systems. This promotes regulatory oversight and ensures that AI applications in finance meet regulatory requirements.

XAI techniques, such as rule extraction, feature importance analysis, and model-agnostic interpretability methods, provide insights into AI models, enhance transparency, and support regulatory compliance in the finance industry. By enabling understandable explanations of AI systems' behavior, XAI helps bridge the gap between the complexity of AI algorithms and the regulatory and ethical expectations of financial institutions.

13.3.3 Human-Machine Collaboration in Financial Decision Making for Finance

Human-machine collaboration in financial decision making refers to the integration of human expertise and AI technologies to enhance the decision-making process in the finance industry (Goldstein et al. 2017). Here are some key aspects of human-machine collaboration in financial decision making:

- Augmented Decision Making: AI technologies can augment human decision-making capabilities by providing data analysis, pattern recognition, and predictive insights. By leveraging AI algorithms, financial professionals can make more informed and data-driven decisions, complementing their domain expertise with the computational power and analytical capabilities of AI systems.
- Risk Assessment and Management: AI can assist in risk assessment and management by analyzing large volumes of data, identifying patterns, and predicting potential risks or market trends. Human experts can then interpret and contextualize the AI-generated insights to make well-informed decisions. The collaboration between humans and AI helps to enhance risk management strategies, improve accuracy, and mitigate potential losses.
- Trading and Investment Strategies: Human-machine collaboration can be applied to develop trading and investment strategies. AI algorithms can analyze market data, identify trends, and generate potential trading signals, while human experts can add their domain knowledge, intuition, and risk preferences to refine and validate the strategies. This collaboration allows for more robust and adaptive trading approaches that combine the strengths of both humans and machines.
- Customer Service and Personalization: These AI systems can handle routine inquiries, provide personalized recommendations, and assist with basic financial tasks. Human experts can step in for more complex or sensitive interactions, ensuring a seamless customer experience and resolving complex issues that require human judgment and empathy.

The effective collaboration between humans and machines in financial decision making leverages the strengths of both parties. Humans contribute domain knowledge, intuition, ethical judgment, and critical thinking, while AI technologies provide computational power, data analysis capabilities, and pattern recognition. The synergy between human expertise and AI technologies leads to more informed, accurate, and efficient financial decision making in various areas of the finance industry.

13.4 Emerging Applications in Banking and Finance (Fig. 13.2)

13.4.1 Predictive Analytics for Market Volatility in Banking and Finance

Predictive analytics plays a crucial role in forecasting and managing market volatility in banking and finance. By analyzing historical data and identifying patterns and trends, predictive analytics models can provide insights and predictions on future market volatility (Idrees et al. 2019). Here are some key aspects of using predictive analytics for market volatility in banking and finance:

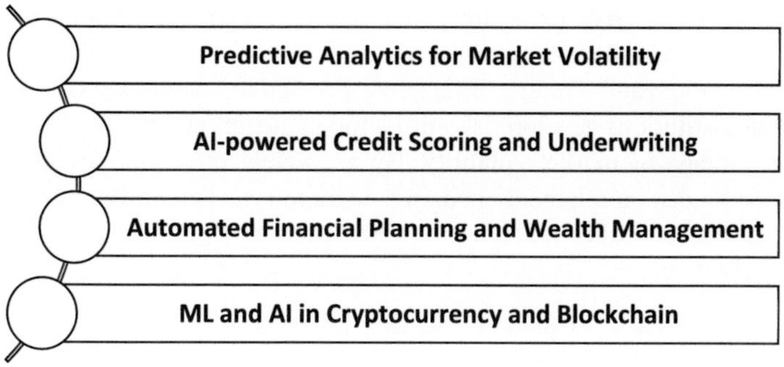

Fig. 13.2 Various Applications in Banking and Finance

- Historical Data Analysis: Predictive analytics models analyze historical market data, including prices, volumes, and other relevant variables, to identify patterns and correlations. By examining historical volatility patterns, the models can establish relationships between market indicators and volatility levels.
- Volatility Modeling: Predictive analytics models use various statistical techniques, such as autoregressive integrated moving average (ARIMA), generalized autoregressive conditional heteroskedasticity (GARCH), or stochastic volatility models, to model and forecast market volatility. These models capture the dynamics and characteristics of volatility, allowing for predictions of future volatility levels.
- Input Variables: Predictive analytics models consider a range of input variables that can impact market volatility. These variables may include macroeconomic indicators, industry-specific factors, news sentiment, market sentiment indicators, and financial market data. By incorporating a comprehensive set of variables, the models can capture the complex dynamics of market volatility.
- Machine Learning Techniques: Machine learning algorithms, such as random forests, support vector machines (SVM), or neural networks, can be employed for market volatility prediction. These algorithms can discover intricate relationships between input variables and volatility patterns, allowing for more accurate and adaptive predictions.
- Real-Time Data Analysis: Predictive analytics models can leverage real-time data feeds and news sentiment analysis to capture up-to-date market information. By integrating real-time data, the models can adapt to changing market conditions and provide more timely volatility predictions.
- Stress Testing and Scenario Analysis: Predictive analytics models can be used for stress testing and scenario analysis to assess the impact of potential market shocks or events on volatility levels. By simulating various scenarios, financial institutions can evaluate their risk exposure, adjust their risk management strategies, and make informed decisions to mitigate volatility-related risks.
- Portfolio Optimization and Risk Management: Predictive analytics models' volatility predictions are valuable inputs for portfolio optimization and risk management strategies. Financial institutions can use these predictions to rebalance

portfolios, adjust risk exposures, and optimize asset allocations to manage volatility-related risks effectively.
- Risk Mitigation and Trading Strategies: Predictive analytics models can help financial institutions develop risk mitigation strategies and trading strategies that take advantage of market volatility. By accurately predicting volatility levels, institutions can implement hedging techniques, derivatives strategies, or volatility-based trading strategies to protect their portfolios or generate profits in volatile markets.

Predictive analytics for market volatility in banking and finance enhances risk management practices, facilitates informed decision making, and enables institutions to navigate volatile market conditions effectively. However, it is important to note that market volatility prediction is challenging due to the inherent uncertainty and complexity of financial markets. Models should be regularly validated, monitored, and refined to ensure their accuracy and reliability in capturing market dynamics and volatility patterns.

13.4.2 AI-Powered Credit Scoring and Underwriting in Banking and Finance

AI-powered credit scoring and underwriting are transformative approaches in banking and finance that leverage artificial intelligence (AI) technologies to enhance the efficiency and accuracy of credit assessment and lending processes (Malali and Gopalakrishnan 2020). Here are some key aspects of AI-powered credit scoring and underwriting:

- Data-driven Decision Making: AI algorithms can analyze vast amounts of data, including traditional credit data (e.g., credit history, income, and employment information) as well as alternative data sources (e.g., social media activity, transactional data) to assess creditworthiness. This data-driven approach enables more comprehensive and accurate credit assessments compared to traditional credit scoring methods.
- Machine Learning Algorithms: Machine learning algorithms, such as logistic regression, decision trees, random forests, and neural networks, are used in credit scoring and underwriting. These algorithms learn patterns and relationships from historical data to predict credit risk and make informed lending decisions. They can handle non-linear relationships and adapt to changing credit dynamics.
- Enhanced Accuracy and Risk Assessment: AI-powered credit scoring models can provide more accurate risk assessments by leveraging advanced data analytics and predictive modeling techniques. They can identify hidden patterns and correlations in data that human analysts may overlook, leading to improved credit risk prediction and more reliable underwriting decisions.

- Real-time and Dynamic Decision Making: AI enables real-time and dynamic credit decision making by automating the credit scoring process. AI models can assess creditworthiness in near real-time, allowing financial institutions to make faster lending decisions. This speed and agility are particularly valuable in scenarios such as online lending or instant credit approvals.
- Automated Underwriting: AI-powered underwriting automates various aspects of the underwriting process, streamlining operations and reducing manual efforts. By applying pre-defined rules and algorithms, AI systems can assess loan applications, verify documentation, and make recommendations on loan approval or rejection. This automation improves operational efficiency and reduces the time and cost involved in underwriting.
- Improved Efficiency and Scalability: AI-powered credit scoring and underwriting eliminate manual processes, paperwork, and repetitive tasks. This leads to significant efficiency gains, allowing financial institutions to handle a higher volume of loan applications with fewer resources. The scalability of AI systems enables institutions to handle increased customer demands and improve customer experience.
- Fairness and Bias Mitigation: AI algorithms can be designed to address fairness and mitigate biases in credit scoring and underwriting. By analyzing the factors contributing to decisions, AI models can identify and mitigate biases related to gender, race, or other protected characteristics. This promotes fair lending practices and helps ensure equal opportunities for all applicants.
- Regulatory Compliance: AI-powered credit scoring and underwriting systems can assist financial institutions in complying with regulatory requirements. The models can incorporate regulatory guidelines and standards into their algorithms, ensuring compliance with laws such as the Equal Credit Opportunity Act (ECOA) and the Fair Credit Reporting Act (FCRA). The transparency and explainability of AI models support regulatory audits and validations.

AI-powered credit scoring and underwriting bring automation, efficiency, and enhanced risk assessment to the lending process. However, it is important to balance the advantages of AI with the need for human oversight, ethical considerations, and ongoing model monitoring and validation. The collaboration between AI systems and human experts ensures responsible lending practices, reduces risks, and maintains the integrity and fairness of credit decisions in banking and finance.

13.4.3 Automated Financial Planning and Wealth Management in Banking and Finance

Automated financial planning and wealth management in banking and finance refer to the use of technology and artificial intelligence (AI) to streamline and enhance the process of managing personal finances and investments (Waliszewski and

Warchlewska 2020). Here are some key aspects of automated financial planning and wealth management:

- Goal-based Financial Planning: Automated platforms offer goal-based financial planning tools that help individuals set and track their financial goals. These platforms use algorithms to analyze personal financial data, such as income, expenses, assets, and liabilities, and provide personalized recommendations and strategies to achieve specific goals, such as retirement planning, education funding, or saving for a major purchase.
- Robo-Advisors: Robo-advisors are digital platforms that use AI algorithms to provide automated investment advice and portfolio management. They assess an individual's risk tolerance, investment objectives, and time horizon and recommend a diversified investment portfolio tailored to their needs. Robo-advisors often utilize advanced asset allocation techniques and rebalancing algorithms to optimize investment returns while managing risk.
- Data-driven Investment Decisions: Automated financial planning and wealth management leverage data analysis and machine learning algorithms to make data-driven investment decisions. These systems analyze a wide range of financial data, including market trends, historical performance, and economic indicators, to identify investment opportunities, assess risk, and make informed investment recommendations.
- Portfolio Rebalancing: Automated systems monitor and rebalance investment portfolios to maintain the desired asset allocation. By analyzing the performance of different asset classes and assessing market conditions, these systems can identify when portfolio adjustments are needed and automatically execute the rebalancing process, ensuring that the portfolio remains aligned with the individual's investment goals and risk profile.
- Tax Optimization: Automated platforms can incorporate tax optimization strategies into financial planning and wealth management. These systems analyze tax rules and regulations, as well as an individual's financial situation, to identify tax-efficient investment strategies, maximize tax deductions, and minimize tax liabilities. This helps individuals optimize their after-tax returns and manage their tax obligations effectively.

Automated financial planning and wealth management offer convenience, cost efficiency, and personalized solutions to individuals seeking professional financial guidance. While these systems provide valuable assistance, it's important for individuals to understand the underlying assumptions, limitations, and risks associated with automated recommendations. Regular reviews, ongoing communication with financial professionals, and the ability to adjust strategies based on changing circumstances remain essential for successful financial planning and wealth management (Fig. 13.3).

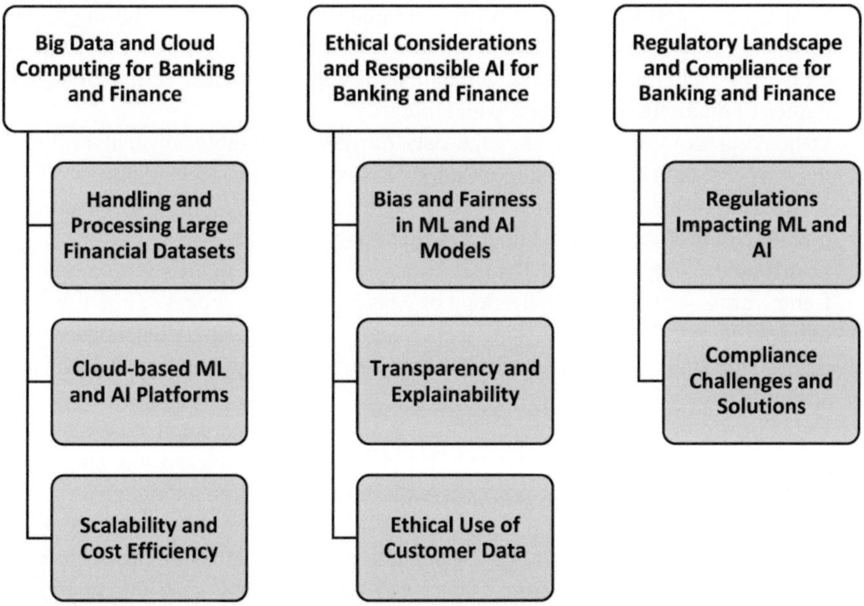

Fig. 13.3 Perspectives in Banking and Finance

13.5 Big Data and Cloud Computing for Banking and Finance

13.5.1 Handling and Processing Large Financial Datasets for Banking and Finance

Handling and processing large financial datasets is a critical task in banking and finance, given the vast amount of data generated by transactions, market feeds, customer information, and other financial activities (Choithani et al. 2022). Here are some key considerations for effectively managing and processing large financial datasets:

- Data Storage Infrastructure: Establish a robust data storage infrastructure capable of handling large volumes of data. This may involve leveraging scalable and distributed storage systems, such as Hadoop Distributed File System (HDFS) or cloud-based storage solutions, to accommodate the growing dataset sizes.
- Data Integration: Integrate data from various sources to create a unified view of financial data. This includes data from internal systems, external market data providers, regulatory filings, and third-party data sources. Implement data integration techniques, such as Extract, Transform, Load (ETL) processes, to standardize and consolidate data into a structured format for analysis.

- Data Quality and Cleansing: Ensure data quality by conducting thorough data cleansing processes. This involves identifying and rectifying data inconsistencies, errors, and missing values. Implement data validation and verification techniques to maintain data accuracy and integrity.
- Data Preprocessing: Preprocess the data to make it suitable for analysis. This may involve data normalization, scaling, feature selection, and handling outliers or missing values. Data preprocessing techniques are essential to ensure the quality and usability of the data for further analysis.
- Distributed Computing and Parallel Processing: Utilize distributed computing frameworks, such as Apache Hadoop or Apache Spark, to process large financial datasets in parallel. These frameworks allow for distributed data processing across multiple machines, significantly improving processing speed and scalability.
- Data Partitioning and Indexing: Partition large datasets into smaller, manageable chunks to enable efficient querying and processing. Implement indexing techniques, such as columnar or compressed indexes, to optimize data retrieval and analysis operations.
- Data Compression and Storage Optimization: Apply data compression techniques to reduce storage requirements and optimize data access. Compressed formats, such as Parquet or ORC, can significantly reduce the storage footprint while allowing for efficient data processing.
- Streaming Data Processing: Incorporate streaming data processing frameworks, such as Apache Kafka or Apache Flink, to handle real-time financial data streams. Streaming data processing enables the analysis of continuous data feeds, such as market tick data or transactional data, in real-time or near real-time.
- Parallelized Algorithms and Distributed Analytics: Utilize parallel processing algorithms and distributed analytics frameworks to analyze large financial datasets efficiently. Techniques like MapReduce, Spark, or distributed machine learning algorithms enable scalable and efficient analysis of financial data across distributed computing resources.
- Data Security and Privacy: Implement robust data security measures to protect sensitive financial data. This includes encryption, access controls, anonymization techniques, and compliance with data protection regulations, such as the General Data Protection Regulation (GDPR) or data privacy laws specific to the banking and finance industry.

Effective handling and processing of large financial datasets require a combination of scalable infrastructure, data integration, preprocessing techniques, distributed computing frameworks, and robust data security practices. By adopting these strategies, banks and financial institutions can unlock valuable insights from their data, improve decision-making processes, and drive innovation in the banking and finance industry.

13.5.2 Cloud-Based ML and AI Platforms for Banking and Finance

Cloud-based machine learning (ML) and artificial intelligence (AI) platforms offer numerous benefits for banking and finance institutions. These platforms provide scalable infrastructure, advanced ML/AI tools, and data management capabilities (Harmon and Psaltis 2021). Here are some popular cloud-based ML and AI platforms used in banking and finance:

- Amazon Web Services (AWS) Machine Learning: AWS offers a comprehensive suite of ML services, including Amazon SageMaker for building, training, and deploying ML models. SageMaker provides a fully managed environment with pre-configured ML algorithms and enables integration with other AWS services for data storage, processing, and analytics.
- Google Cloud Platform (GCP) AI Platform: GCP provides AI Platform, which offers a range of tools for ML and AI development, including scalable infrastructure, pre-built ML models, and AutoML capabilities. It supports TensorFlow, PyTorch, and other popular ML frameworks and provides seamless integration with GCP's data storage and processing services.
- Microsoft Azure Machine Learning: Azure Machine Learning is a cloud-based platform that enables end-to-end ML and AI workflows. It offers features such as automated ML, model training and deployment, and integration with Azure's data services. Azure also provides cognitive services for natural language processing, computer vision, and speech recognition.
- IBM Watson: IBM Watson is an AI platform that provides a suite of services for ML, natural language processing, and computer vision. It offers tools for developing, training, and deploying ML models, as well as APIs for adding cognitive capabilities to applications. Watson also provides industry-specific solutions for banking and finance, such as fraud detection and risk assessment.
- Oracle Cloud Infrastructure (OCI) AI Services: OCI AI Services offers a range of ML and AI capabilities on Oracle's cloud infrastructure. It includes tools for building and deploying ML models, such as AutoML, as well as APIs for computer vision, natural language understanding, and speech recognition. OCI also provides data management and analytics services for comprehensive data processing.
- Salesforce Einstein: Salesforce Einstein is an AI-powered platform specifically designed for customer relationship management (CRM). It offers ML and AI capabilities to analyze customer data, predict customer behavior, and provide personalized recommendations. Einstein can be leveraged by banking and finance institutions to improve customer engagement and sales effectiveness.

These cloud-based ML and AI platforms provide a range of services and tools that enable banking and finance institutions to leverage the power of ML and AI without the need for extensive infrastructure investments. They offer scalability, flexibility, and accessibility to advanced ML algorithms and models, allowing banks and financial institutions to accelerate their AI initiatives, improve customer experiences, automate processes, and gain valuable insights from their data.

13.5.3 Scalability and Cost Efficiency for in Banking and Finance

Scalability and cost efficiency are crucial considerations in the banking and finance industry, and leveraging technology, including machine learning (ML) and artificial intelligence (AI), can play a significant role in achieving these objectives (Moro-Visconti et al. 2020). Here's how scalability and cost efficiency can be addressed:

- Scalability through Cloud Computing: Cloud computing offers scalable infrastructure resources on-demand. Banks and financial institutions can leverage cloud platforms to dynamically allocate computing resources, storage, and networking capabilities as needed. This scalability ensures that the infrastructure can handle increased workloads during peak times, such as high-volume trading-periods or increased customer activity, while avoiding over-provisioning during quieter periods.
- Automated Processes and Workflow Optimization: ML and AI technologies can automate and optimize various banking and finance processes, leading to increased scalability and cost efficiency. For example, automating manual tasks like data entry, document processing, and compliance checks can free up human resources and reduce operational costs. Additionally, optimizing workflows using AI algorithms can streamline processes, reduce processing times, and enhance productivity.
- Data Management and Analytics: Efficient data management and analytics can contribute to scalability and cost efficiency. ML and AI techniques enable banks and financial institutions to analyze large volumes of data to derive valuable insights, make informed decisions, and improve operational efficiency. Advanced analytics can identify patterns, detect anomalies, and predict trends, helping optimize business processes and resource allocation.
- Robotic Process Automation (RPA): RPA involves using software robots or "bots" to automate repetitive and rule-based tasks. This technology can handle data entry, data reconciliation, report generation, and other routine tasks in banking and finance operations. By implementing RPA, organizations can scale their operations without a proportional increase in staffing, leading to significant cost savings.
- Predictive Analytics for Risk Management: ML and AI techniques can improve risk management processes in banking and finance. Predictive analytics models can assess credit risk, market risk, and fraud risk, enabling proactive risk mitigation and cost-effective resource allocation. By accurately predicting risks, financial institutions can allocate capital and resources more efficiently, avoiding unnecessary expenses associated with excessive risk exposure.

By implementing scalable technologies, optimizing workflows, leveraging automation and advanced analytics, and adopting cloud-based ML and AI platforms, banking and finance institutions can achieve greater scalability and cost efficiency. These approaches can lead to streamlined operations, improved customer experiences, enhanced risk management, and reduced operational expenses, ultimately driving business growth and competitiveness.

13.6 Research and Investment Opportunities in Banking and Finance

13.6.1 Collaborative Research Initiatives in Banking and Finance

Collaborative research initiatives in banking and finance bring together industry stakeholders, academia, and research organizations to address complex challenges, foster innovation, and drive advancements in the field. These initiatives aim to promote knowledge sharing, interdisciplinary collaboration, and the development of practical solutions (Jeucken 2001). Here are a few notable collaborative research initiatives in banking and finance:

- Partnership for International Research and Education (PIRE): PIRE is a collaborative effort among multiple universities and research institutions to advance research and education in financial technology (FinTech) and risk management. It focuses on topics such as blockchain, digital currencies, cybersecurity, and financial risk assessment.
- Institute for Quantitative Finance and Insurance (Q-FIN): Q-FIN is an interdisciplinary research institute that brings together experts from academia, industry, and regulatory bodies to conduct research in quantitative finance and insurance. It aims to bridge the gap between theory and practice by addressing practical challenges in risk management, pricing models, and portfolio optimization.
- The Consortium for Data Analytics in Risk (CDAR): CDAR is a collaborative research consortium focused on the development and application of data analytics techniques for risk management in banking and finance. It brings together academic researchers, industry practitioners, and regulatory bodies to explore advanced analytics methods and their implications for risk modeling and decision-making.
- Global Risk Institute (GRI): GRI is an independent research organization that facilitates collaborative research and knowledge exchange among financial institutions, regulators, and academia. It focuses on risk management, financial stability, and emerging risks. GRI conducts research projects, hosts conferences, and provides a platform for sharing insights and best practices.
- Partnership for Research in Markets and the Economy (PRIME): PRIME is a collaborative research initiative that brings together researchers, financial

institutions, and regulators to investigate market structure, trading mechanisms, and regulatory policies. It aims to foster evidence-based research and promote a deeper understanding of market dynamics and their implications for financial stability.
- MIT Golub Center for Finance and Policy (GCFP): GCFP is a research center that promotes collaboration between academia, industry, and policymakers. It focuses on research areas such as financial regulation, systemic risk, and the impact of technology on financial markets. GCFP hosts conferences, sponsors research projects, and facilitates knowledge sharing through publications and events.

These collaborative research initiatives play a crucial role in advancing knowledge, driving innovation, and informing policy decisions in the banking and finance industry. By bringing together diverse expertise and resources, they contribute to the development of practical solutions, enhance industry practices, and address emerging challenges in the evolving financial landscape.

13.6.2 Venture Capital and Investment Trends in Banking and Finance

Venture capital and investment trends in banking and finance reflect the evolving landscape of the industry and the growing interest in innovative solutions and technologies (Gompers 1998). Here are some notable trends in venture capital and investment within banking and finance:

- FinTech Investments: FinTech startups continue to attract significant venture capital investments. These startups leverage technology to disrupt traditional financial services and offer innovative solutions in areas such as payments, lending, wealth management, and insurance. Investors are particularly interested in companies that use ML, AI, blockchain, and data analytics to improve efficiency, enhance customer experiences, and address industry pain points.
- Open Banking and API Platforms: The shift towards open banking, driven by regulatory changes and customer demand for interconnected financial services, has led to increased investment in API platforms. These platforms enable financial institutions to securely share customer data and collaborate with FinTech startups and developers, fostering innovation and the creation of new services.
- Digital Asset and Cryptocurrency Investments: Investments in digital assets, cryptocurrencies, and blockchain-based platforms have been on the rise. Institutional investors and venture capital firms are recognizing the potential of blockchain technology to transform various aspects of finance, including payments, identity verification, and decentralized finance (DeFi) applications.
- InsurTech Investments: InsurTech startups that leverage technology to enhance insurance processes, improve underwriting, and personalize customer experiences

are attracting significant investments. The use of ML, AI, telematics, and IoT devices in the insurance industry is driving innovation and shaping new business models.

- RegTech and Compliance Solutions: Regulatory technology (RegTech) solutions that help financial institutions comply with complex regulatory requirements have gained investor interest. Startups focusing on AML/KYC compliance, fraud detection, data privacy, and cybersecurity are receiving investments as regulatory compliance remains a critical challenge for financial institutions.
- Sustainable and Impact Investing: Environmental, social, and governance (ESG) investing and impact investing are growing trends in the financial industry. Investors are actively seeking opportunities to support companies that promote sustainable practices, social responsibility, and positive societal impact. FinTech startups that address ESG challenges and provide sustainable finance solutions are attracting attention.
- Collaborative Investments: Financial institutions are increasingly engaging in strategic collaborations and investments with FinTech startups through corporate venture capital (CVC) arms and innovation labs. This approach allows financial institutions to access cutting-edge technologies, tap into new markets, and foster innovation within their organizations.

These trends indicate a growing appetite for innovation, digital transformation, and disruption within the banking and finance industry. Venture capital and investments in these areas play a crucial role in driving technological advancements, shaping industry dynamics, and shaping the future of financial services.

13.7 Conclusion and Future Work

The future of banking and finance with machine learning (ML) and artificial intelligence (AI) holds significant potential to revolutionize the industry (Kumar et al. 2023). Here are some possible future developments:

- Enhanced Customer Experience: ML and AI will enable personalized and seamless customer experiences. Chatbots and virtual assistants will become more sophisticated, offering personalized financial advice, assisting with transactions, and addressing customer queries in real-time. ML algorithms will analyze customer data to deliver personalized product recommendations and targeted marketing.
- Advanced Fraud Detection and Security: ML and AI will play a crucial role in detecting and preventing fraud in real-time. Advanced algorithms will continuously analyze large volumes of data to identify patterns and anomalies that indicate fraudulent activities. AI-powered cybersecurity systems will protect against data breaches and cyber threats, ensuring robust security for financial institutions and their customers.

- Intelligent Risk Management: ML and AI will improve risk management in banking and finance. Predictive analytics and ML algorithms will analyze vast amounts of data to assess credit risk, market risk, and operational risk. This will enable more accurate risk assessments, early warning systems, and proactive risk mitigation strategies.
- Automated Financial Decision-making: ML and AI will automate financial decision-making processes, such as loan underwriting, investment portfolio management, and insurance underwriting. Advanced algorithms will analyze data, market trends, and risk factors to make data-driven decisions quickly and efficiently. This will lead to faster, more accurate decisions and increased efficiency (Rusmita et al. 2023).
- Data-driven Insights and Predictive Analytics: ML and AI will provide data-driven insights to drive business decisions. Advanced analytics models will uncover hidden patterns and correlations in large datasets, enabling financial institutions to make more informed strategic decisions, optimize operations, and identify new business opportunities (Panigrahi et al. 2023).
- Regulatory Compliance and Risk Mitigation: ML and AI will help financial institutions comply with complex regulations by automating compliance processes, detecting and mitigating risks, and ensuring adherence to regulatory requirements. AI-powered systems will monitor and interpret regulatory changes, helping institutions stay updated and compliant.
- Blockchain and Distributed Ledger Technology (DLT): ML and AI will work in conjunction with blockchain and DLT to enhance security, transparency, and efficiency in financial transactions. Smart contracts and decentralized applications (DApps) will automate processes such as settlements, trade execution, and identity verification, reducing the need for intermediaries and streamlining operations.
- Ethical Considerations: As ML and AI become more pervasive in the financial industry, ethical considerations will gain prominence. The industry will focus on ensuring fairness, transparency, and accountability in algorithmic decision-making. Regulations and guidelines will be developed to address potential biases, discrimination, and the ethical use of customer data.

While these possibilities paint a promising future, it's important to note that the adoption of ML and AI in banking and finance will require ongoing regulatory oversight, ethical frameworks, and the development of robust risk management strategies. Collaboration between industry stakeholders, regulators, and technology experts will be essential to realize the full potential of ML and AI while ensuring responsible and beneficial outcomes for customers and the financial industry as a whole.

References

Assefa SA, Dervovic D, Mahfouz M, Tillman RE, Reddy P, Veloso M (2020, October) Generating synthetic data in finance: opportunities, challenges and pitfalls. In: Proceedings of the first ACM international conference on AI in finance, pp 1–8

Benbya H, Davenport TH, Pachidi S (2020) Artificial intelligence in organizations: current state and future opportunities. MIS Q Exec 19(4)

Cao L (2020) AI in finance: a review. Available at SSRN 3647625

Choithani T, Chowdhury A, Patel S, Patel P, Patel D, Shah M (2022) A comprehensive study of artificial intelligence and cybersecurity on bitcoin, crypto currency and banking system. Ann Data Sci:1–33

Egger DJ, Gambella C, Marecek J, McFaddin S, Mevissen M, Raymond R et al (2020) Quantum computing for finance: state-of-the-art and future prospects. IEEE Trans Quantum Eng 1:1–24

Erem Ceylan I (2022) The effects of artificial intelligence on the insurance sector: emergence, applications, challenges, and opportunities. Impact Artif Intell Gov Econ Financ 2:225–241

Fares OH, Butt I, Lee SHM (2022) Utilization of artificial intelligence in the banking sector: a systematic literature review. J Financ Serv Mark:1–18

Faris H, Abukhurma R, Almanaseer W, Saadeh M, Mora AM, Castillo PA, Aljarah I (2020) Improving financial bankruptcy prediction in a highly imbalanced class distribution using oversampling and ensemble learning: a case from the Spanish market. Prog Artif Intell 9:31–53

Goldstein IM, Lawrence J, Miner AS (2017) Human-machine collaboration in cancer and beyond: the centaur care model. JAMA Oncol 3(10):1303–1304

Gompers PA (1998) Venture capital growing pains: should the market diet? J Bank Financ 22(6-8):1089–1104

Harmon RL, Psaltis A (2021) The future of cloud computing in financial services: a machine learning and artificial intelligence perspective. In: The essentials of machine learning in finance and accounting, pp 123–138.

Hentzen JK, Hoffmann A, Dolan R, Pala E (2022) Artificial intelligence in customer-facing financial services: a systematic literature review and agenda for future research. Int J Bank Mark 40(6):1299–1336

Huang J, Chai J, Cho S (2020) Deep learning in finance and banking: a literature review and classification. Front Bus Res China 14(1):1–24

Idrees SM, Alam MA, Agarwal P (2019) A prediction approach for stock market volatility based on time series data. IEEE Access 7:17287–17298

Jeucken M (2001) Sustainable finance and banking: the financial sector and the future of the planet. Earthscan

Khan HU, Malik MZ, Alomari MKB, Khan S, Al-Maadid AAS, Hassan MK, Khan K (2022) Transforming the capabilities of artificial intelligence in GCC financial sector: a systematic literature review. Wireless communications and mobile computing, 2022

Kuiper O, van den Berg M, van der Burgt J, Leijnen S (2022) Exploring explainable AI in the financial sector: perspectives of banks and supervisory authorities. In: Artificial intelligence and machine learning: 33rd benelux conference on artificial intelligence, BNAIC/Benelearn 2021, Esch-sur-Alzette, Luxembourg, November 10–12, 2021, Revised Selected Papers 33. Springer International Publishing, pp 105–119

Kumar R, Grover N, Singh R, Kathuria S, Kumar A, Bansal A (2023, March) Imperative role of artificial intelligence and big data in finance and banking sector. In: 2023 international conference on sustainable computing and data communication systems (ICSCDS). IEEE, pp 523–527

Malali AB, Gopalakrishnan S (2020) Application of artificial intelligence and its powered technologies in the Indian banking and financial industry: an overview. IOSR J Human Soc Sci 25(4):55–60

Mogaji E, Nguyen NP (2022) Managers' understanding of artificial intelligence in relation to marketing financial services: insights from a cross-country study. Int J Bank Mark 40(6):1272–1298

Moro-Visconti R, Cruz Rambaud S, López Pascual J (2020) Sustainability in FinTechs: an explanation through business model scalability and market valuation. Sustainability 12(24):10316

Panigrahi RR, Dash M, Shaikh ZH, Irfan M (2023) Review of machine learning techniques in the supply chain management of Indian industry: a future research agenda. In: Advanced machine learning algorithms for complex financial applications. IGI Global, pp 199–219

Rabbani MR, Khan S, Atif M (2023) Machine learning-based P2P lending Islamic Fintech model for small and medium enterprises in Bahrain. Int J Bus Innov Res 30(4):565–579

Ranjan S, Gupta DR, Gupta DA (2020) Artificial intelligence in financial acumen: Challenges and opportunities. Cosmos J Eng Technol 10(1):1–5

Ruiz-Real JL, Uribe-Toril J, Torres JA, De Pablo J (2021) Artificial intelligence in business and economics research: trends and future. J Bus Econ Manag 22(1):98–117

Rusmita SA, An-Nafis MSA, Ramadhani I, Irfan M (2023) The effect of good corporate governance on financial distress in companies listed in sharia stock index Indonesia: machine learning approach. In: Advanced machine learning algorithms for complex financial applications. IGI Global, pp 220–251

Sen J, Sen R, Dutta A (2021) Introductory chapter: machine learning in finance-emerging trends and challenges. Algorithms, Models and Applications, 1

Shah F, Liu Y, Anwar A, Shah Y, Alroobaea R, Hussain S, Ullah SS (2022) Machine learning: the backbone of intelligent trade credit-based systems. Secur Commun Netw 2022:1–10

Waliszewski K, Warchlewska A (2020) Financial technologies in personal financial planning: robo-advice vs. human-advice. Ruch Prawniczy, Ekonomiczny i Socjologiczny 82(4):303–317

Zhang Y, Xiong F, Xie Y, Fan X, Gu H (2020) The impact of artificial intelligence and blockchain on the accounting profession. IEEE Access 8:110461–110477

Chapter 14
Introduction to Machine Learning and Artificial Intelligence in Banking and Finance

Anand Kumar Mishra, Amit Kumar Tyagi, Richa, and Subhra Rani Patra

14.1 Introduction

Machine Learning (ML) and Artificial Intelligence (AI) are rapidly advancing fields that have revolutionized various industries, including finance, healthcare, transportation, and more. These technologies enable computers to learn from data, identify patterns, and make intelligent decisions without explicit programming. At its core, ML focuses on the development of algorithms and models that can learn from data and make predictions or decisions. It leverages statistical techniques and mathematical models to train these algorithms using labeled or unlabeled data. Through a process called "learning," ML algorithms can automatically improve their performance over time. AI, on the other hand, encompasses a broader set of technologies and approaches that aim to create intelligent systems capable of perceiving, reasoning, and acting in a manner that simulates human intelligence. AI seeks to build systems that can exhibit cognitive abilities such as problem-solving, pattern recognition, natural language understanding, and decision-making. Note that ML is a subset of AI, as it focuses specifically on algorithms and models that learn from

A. K. Mishra (✉)
Computer Science and Engineering, NIIT University, Neemrana, Rajasthan, India

A. K. Tyagi
Department of Fashion Technology, National Institute of Fashion Technology, New Delhi, India

Richa
Department of Computer Science and Engineering, Birla Institute of Technology, Mesra, Ranchi, Jharkhand, India

S. R. Patra
University of Texas, Arlington, TX, USA
e-mail: subhra.patra@uta.edu

M. Irfan et al. (eds.), *Applications of Block Chain technology and Artificial Intelligence*, Financial Mathematics and Fintech,
https://doi.org/10.1007/978-3-031-47324-1_14

data. It provides the foundation for many AI applications by enabling machines to automatically analyze and interpret complex data sets, identify patterns and trends, and make predictions or decisions.

The key components of ML and AI include:

- Data: ML and AI systems require high-quality data to train their models. This data can be structured (e.g., numerical data) or unstructured (e.g., text, images, audio), and it serves as the basis for learning and making intelligent decisions.
- Algorithms: ML algorithms are mathematical models that learn from data and make predictions or decisions. They can be classified into various types, such as supervised learning (where labeled data is used for training), unsupervised learning (where patterns and relationships are discovered in unlabeled data), reinforcement learning (where an agent learns through interactions with an environment), and deep learning (which utilizes neural networks with multiple layers).
- Model Training: ML models are trained using algorithms and data. During the training process, the model adjusts its parameters to minimize errors and optimize its performance on specific tasks.
- Evaluation and Testing: After training, ML models need to be evaluated and tested on new, unseen data to assess their performance and generalization capabilities. This step helps determine the effectiveness and reliability of the models.

ML and AI have numerous applications across various domains. In finance, they are used for tasks such as financial forecasting, risk management, fraud detection, algorithmic trading, personalized recommendations, and credit scoring. These technologies have the potential to automate processes, improve accuracy, enhance decision-making, and unlock new insights from vast amounts of financial data. As ML and AI continue to advance, ongoing research and development efforts are focused on improving algorithms, handling big data, ensuring interpretability and transparency, addressing ethical considerations, and integrating these technologies with other emerging fields such as natural language processing and computer vision. In summary, the rapid growth of ML and AI has transformed the way we approach complex problems, and their applications in various industries, including finance, continue to expand, opening up new opportunities for innovation and improvement.

14.1.1 Machine Learning and Artificial Intelligence in Finance

Machine Learning (ML) and Artificial Intelligence (AI) have had a profound impact on the finance industry, revolutionizing how financial institutions operate, make decisions, and manage risks (Ahmed et al. 2022). These technologies enable finance professionals to analyze vast amounts of data, uncover patterns, and derive valuable insights for more accurate predictions and improved decision-making.

- One of the key applications of ML and AI in finance is financial forecasting. ML algorithms can analyze historical financial data, market trends, and external factors to predict stock prices, exchange rates, and other financial indicators. These predictions assist in investment decisions, portfolio management, and risk assessment.
- Risk management is another critical area where ML and AI shine. By analyzing historical data, market conditions, and macroeconomic factors, ML models can assess creditworthiness, detect fraudulent activities, and identify potential risks in financial transactions. This enhances the ability to make informed decisions and mitigate risks.
- Algorithmic trading is a prominent application of ML and AI in finance. ML algorithms can analyze market data, identify patterns, and execute trades automatically, leading to faster and more efficient trading decisions. These algorithms can adapt to changing market conditions and optimize trading strategies for improved profitability.
- ML and AI also play a crucial role in personalized financial recommendations. By analyzing customer data, spending patterns, and financial goals, ML models can provide tailored investment advice, savings plans, and personalized product recommendations. This enhances customer satisfaction and engagement.
- Moreover, natural language processing (NLP) techniques enable the extraction and analysis of information from vast amounts of textual data, such as news articles, financial reports, and social media sentiments. This information can be used to assess market sentiment, predict market movements, and make more informed trading decisions.
- Despite the significant benefits, ML and AI in finance come with challenges. These include ensuring the privacy and security of sensitive financial data, addressing algorithmic bias and fairness, and maintaining interpretability and transparency of ML models, especially in regulatory contexts. Ethical considerations surrounding the use of AI in finance, such as responsible lending practices and customer privacy, also need careful attention.

In summary, ML and AI have transformed the finance industry, enabling more accurate predictions, efficient operations, and improved risk management. As technology continues to advance, the integration of ML and AI with finance is expected to evolve, leading to further innovation and efficiency gains in the industry.

14.1.2 Applications of ML and AI in Banking and Finance

Machine Learning (ML) and Artificial Intelligence (AI) have a wide range of applications in banking and finance, revolutionizing various processes and decision-making capabilities (Ng and Shah 2020). Some key applications include:

- Fraud Detection: ML algorithms can analyze large volumes of transactional data, identify patterns, and detect fraudulent activities in real-time. This helps financial institutions prevent and mitigate fraud, enhancing security and protecting customers.
- Credit Scoring and Underwriting: ML models can assess creditworthiness by analyzing diverse data sources, including credit history, income, and demographic information. This enables more accurate credit scoring and automated underwriting processes, leading to improved efficiency and risk assessment.
- Risk Assessment and Management: ML and AI techniques help financial institutions assess and manage risks more effectively. By analyzing historical data, market trends, and economic indicators, models can identify potential risks and optimize risk management strategies.
- Customer Relationship Management: ML algorithms can analyze customer data, behavior patterns, and preferences to provide personalized recommendations and targeted marketing campaigns. This enhances customer satisfaction, loyalty, and engagement.
- Algorithmic Trading: ML algorithms can analyze market data, identify patterns, and execute trades automatically. This enables faster and more efficient trading decisions, improving profitability and liquidity in financial markets.
- Robo-advisors: ML and AI-powered robo-advisors provide automated investment advice based on customer preferences, risk tolerance, and financial goals. These platforms offer personalized investment portfolios and guidance at lower costs compared to traditional advisory services.
- Chatbots and Virtual Assistants: ML and NLP techniques enable the development of intelligent chatbots and virtual assistants that can handle customer inquiries, provide support, and offer personalized recommendations. This enhances customer service and reduces response times.
- Anti-Money Laundering (AML): ML models can analyze transactional and customer data to detect suspicious activities and potential money laundering patterns. This helps financial institutions comply with regulatory requirements and combat financial crimes.
- Regulatory Compliance: ML and AI systems assist in regulatory compliance by automating processes, analyzing vast amounts of data, and identifying potential non-compliance issues. This reduces manual efforts and ensures adherence to regulations.
- Market Analysis and Forecasting: ML algorithms can analyze market data, news articles, and social media sentiments to predict market trends, stock prices, and other financial indicators. This aids in investment decisions, risk management, and portfolio optimization.

These applications demonstrate the transformative impact of ML and AI in banking and finance, leading to improved efficiency, accuracy, and customer experiences. As technology advances, the integration of ML and AI is expected to further expand, enabling innovative solutions and shaping the future of the industry.

14.1.3 Importance of ML and AI in Financial Decision Making

Machine Learning (ML) and Artificial Intelligence (AI) have become increasingly important in financial decision-making processes due to their ability to process large volumes of data, uncover complex patterns, and generate valuable insights (Ranjan et al. 2020). Here are some key reasons why ML and AI are crucial in financial decision making:

- Data Analysis and Pattern Recognition: ML and AI algorithms excel at analyzing vast amounts of structured and unstructured financial data, including market data, customer data, and economic indicators. By identifying patterns and relationships in these datasets, ML models can provide valuable insights to support decision-making processes.
- Improved Accuracy and Efficiency: ML and AI techniques can process and analyze data at a much faster rate and with higher accuracy than traditional manual methods. This leads to more efficient decision-making processes, reducing human error and enhancing overall accuracy.
- Risk Management and Fraud Detection: ML and AI models play a critical role in assessing and managing risks in financial institutions. They can identify potential risks, anomalies, and fraudulent activities by analyzing historical data, market trends, and transactional patterns. This enables proactive risk mitigation and fraud prevention.
- Personalized Recommendations and Customer Experience: ML and AI enable financial institutions to deliver personalized recommendations, tailored products, and customized services to individual customers. By analyzing customer behavior, preferences, and financial goals, ML models can provide personalized insights, improving the customer experience and satisfaction.
- Portfolio Optimization and Asset Management: ML algorithms can analyze historical market data, economic indicators, and risk profiles to optimize investment portfolios. They can identify potential opportunities, assess risks, and suggest optimal asset allocation strategies to achieve desired financial objectives.
- Automation and Efficiency Gains: ML and AI technologies automate repetitive and time-consuming tasks in financial decision making, such as data processing, risk assessments, and compliance checks. This frees up human resources, reduces operational costs, and allows finance professionals to focus on higher-level strategic tasks.
- Predictive Analytics and Forecasting: ML and AI models can generate accurate predictions and forecasts for financial markets, stock prices, exchange rates, and other relevant indicators. These predictions aid in investment decisions, market timing, and risk management, improving overall financial performance.
- Regulatory Compliance and Reporting: ML and AI systems help financial institutions ensure compliance with regulatory requirements. They can automate compliance checks, monitor transactions for suspicious activities, and generate

accurate and timely reports, facilitating regulatory compliance and reducing compliance-related risks.
- Real-time Decision Making: ML and AI enable real-time data analysis and decision-making processes. This is particularly crucial in dynamic and fast-paced financial markets, where timely decisions can significantly impact outcomes.

In summary, ML and AI provide powerful tools and techniques to analyze data, mitigate risks, automate processes, and optimize financial decisions. They enhance the accuracy, efficiency, and effectiveness of financial decision making, leading to improved outcomes and a competitive edge in the financial industry.

14.1.4 Review of Literature and Organization of the Work

The study presented by Mhlanga (2021) found that artificial intelligence and machine learning have a significant impact on credit risk assessments using alternative data sources, such as public data, to address the issues of information asymmetry, adverse selection, and moral hazard. The study used the literature review approach through documentary and conceptual analysis to investigate the impact of machine learning and artificial intelligence in credit risk assessment. This makes it possible for lenders to do thorough analyses of credit risk, gauge consumer behavior, and afterwards confirm the borrowers' capacity to repay the loans, enabling those who are less fortunate to obtain credit. To ensure that those who are financially excluded, this study advises financial organizations like banks and credit lending institutions to invest more in artificial intelligence and machine learning. The econometric features and their implications cannot be disregarded in a literature review (Buchanan 2019) of AI and financial services. Algorithms, not asymptotic statistical processes, constitute the core of machine learning (ML) methodologies. Maximum likelihood estimation has a more cohesive framework than machine learning. To that purpose, I will talk about supervised and unsupervised learning methodologies in machine learning. A summary of financial AI and ML research is presented in Goodell et al. (2021). We infer the subject organization of AI and ML research in finance for the period of 1986 to April 2021 using co-citation and bibliometric-coupling studies. We further identify three overarching groups of finance scholarship that are roughly equivalent for both types of analysis by identifying nine (co-citation) and eight (bibliometric coupling) specific clusters of finance that apply AI and ML: portfolio construction, valuation, and investor behavior; financial fraud and distress; and sentiment inference, forecasting, and planning. We also highlight trends and future research topics for AI and ML in finance research using co-occurrence and confluence studies. Our findings offer an evaluation of AI and ML in financial research.

All areas of the economy have recently seen an increase in the usage of artificial intelligence tools, in part due to the rising volume of digital data and improved computer power. Applying these techniques to the delivery of financial services could have significant advantages for financial firms as well as society at large. The article (Fernández 2019) outlines some of these advantages as well as some of the key applications that financial firms and central banks are making of these technologies. Additionally, it highlights the technology's primary drawbacks and their potential effects on the proper operation of the financial system. The sustainability of technology, which tends to replace people and the associated personal touch that is frequently the essence of the financial services industry thriving on the art of customization and customer satisfaction, is a major question that needs to be investigated. This is in addition to the ethics-neutral nature of technology and its attendant threats like cybercrimes and macro-financial risks. The current paper (Mehrotra 2019) looks at the understudied possibility of AI taking the place of people in the banking and financial services sector, unwittingly ushering in the demise of the personal touch and service customization that form the basis of client satisfaction and delight in sectors like banking and financial services, which are known for their fiduciary and responsible nature. The whole financial sector has seen a significant transformation thanks to artificial intelligence technology, which has led to the development of several cutting-edge financial services like intelligent consulting, intelligent lending, monitoring, and warning systems, and intelligent customer assistance when necessary. The paper by Xie (2019) seeks to provide a concise overview of the development, implementation, and effects of artificial intelligence and machine learning in the financial system. In the meanwhile, it has become clear that artificial intelligence has caused several issues and concerns throughout its use. Finally, based on the issues highlighted by financial risk management raised by artificial intelligence, some recommendations and techniques are offered for the responsible use of artificial intelligence in financial risk management.

Chatbots are available around-the-clock to address client concerns and are quite effective at answering consumers' questions. 'Chatbot as Islamic Finance Expert' (CaIFE), an interactive chatbot powered by artificial intelligence is proposed by Khan and Rabbani (2020). By allowing users to converse with a robot that has information gathered through machine learning, our interactive chatbot CaIFE receives automatic robot support in relation to Islamic banking and finance. It provides immediate responses to all questions on Islamic banking and finance. The case study of CaIFE is then presented, together with an explanation of its strengths and weaknesses. The type of issues that arise in accounting and auditing, as well as the necessity to use artificial intelligence (AI) technologies in the field (Ukpong et al. 2019). The subject discusses current accounting challenges, particularly auditing, for which new AI research should be beneficial. Both a qualitative and a quantitative research design were used in this study. Utilizing secondary quantitative data, the study was carried out using a descriptive survey research approach. Participants in this study included bank executives and academics with economics and accounting degrees from universities in Nigeria's Akwa Ibom State. 45 stakeholders were chosen to be a part of the sample using a purposeful sampling method. Data were gathered using a development tool created by the researchers called the "Artificial

Intelligence and the Future of Accounting in Africa Questionnaire." The study (Kaur et al. 2020) largely focuses on the idea of AI in the banking industry, how it has revolutionized banking, and how it affects human labor. We are all aware that people make mistakes, but because the world is changing and innovations are following suit, there aren't enough trained workers to handle the automation. Many routine and laborious jobs that were formerly completed by people are increasingly being replaced by automated machines equipped with cutting-edge technology. This report offers a glimpse of the present applications of AI in the banking industry and how it is transforming the face of banking in India given that the industry is undergoing notable transformation at a rapid rate.

To give a methodical evaluation of the model preparation, input data, and model evaluation, the work (Huang et al. 2020) reviews and analyses the literature on the application of deep learning models in the major financial and banking areas. Finally, we talk about three factors that potentially influence how financial deep learning models perform. This study offers guidance and insight on the state-of-the-art deep learning model application in banking and finance to academics and practitioners. The unique promise of AI, which combines cost savings and improved differentiation, makes it alluring overall. These advantages, probably with the exception of fraud detection, are scale-dependent. Nonrepresentative data, bias in representative data, algorithm selection, and human judgements based on AI interpretations are all risk factors (as is the question of whether humans are still involved at all once AI is unleashed). For the foreseeable future, risk mitigation calls for a watchful division of labor between humans and AI (Ashta and Herrmann 2021). Artificial intelligence (AI) and big data help organizations make decisions. They assist businesses in developing fresh goods and procedures or enhancing current ones. AI is expected to offer enormous potential for banks as data volume increases rapidly and data storage and computing power prices decrease. The effects of big data and AI on the banking sector are covered in the chapter (Yu and Song 2021). We start by giving a brief history of big data and AI. Second, we pinpoint the uses of big data and AI that banks may use to their advantage and assess these applications. The effects of big data and AI on oversight and compliance with regulations are covered in our third section. The constraints and difficulties associated with the employment of large data-based AI come last. The presented study (Mogaji and Nguyen 2022) offers empirical understanding of the possibilities, chances, and difficulties associated with applying AI to the marketing of financial services. Additionally, it specifically challenges some assumptions about AI's function in financial services, the use of chatbots in the provision of financial services, and the contribution of marketing managers to the development of AI. The results of the study (Mhlanga 2020) show that AI has a significant impact on digital financial inclusion in areas such as risk identification, measurement, and management, addressing the information asymmetry issue, providing customer support and help-desk through chatbots, fraud detection, and cybersecurity. To ensure that vulnerable groups of people who are not financially active can participate in the formal financial market with the fewest obstacles and greatest rewards, it is advised that financial institutions, non-financial institutions, and governments around the world adopt and scale up the use of AI tools and applications. According to presented research (Königstorfer and Thalmann 2020), commercial banks may use AI to automate compliance-related tasks, decrease loan losses, increase payment security, and enhance consumer targeting. The realization of technological benefits, the integration of AI into corporate processes, assuring user acceptance through openness, privacy, and appropriate documentation are all issues that researchers are concerned about. Finally, we suggest a behavioral finance research agenda. According to Suhel et al. (2020), commercial banks may use AI to automate compliance-related tasks, decrease loan losses, increase payment security, and enhance consumer targeting. The realization of technological benefits, the integration of AI into corporate processes, assuring user acceptance through openness, privacy, and appropriate documentation are all issues that researchers are concerned about. Finally, we suggest a behavioral finance research agenda.

The chapter is organized as the next section discusses the review of literature in the area. Further, fundamentals of finance that discusses about the market and instruments, data and analysis, various approaches in finance. Further it discusses the overview of machine learning approaches in finance that includes supervised learning, unsupervised learning, reinforcement learning and deep learning. The chapter discusses the involvement of artificial intelligence in banking and finance. The next section informs the Data Collection and Preparation for Banking and Finance. The Evaluation and Performance Metrics for Banking and Finance have been discussed in the next section. The next section discusses Ethical Considerations in ML and AI for Banking and Finance has been discussed followed by the Challenges and Future Trends for Banking and Finance. The next section discusses the conclusion.

14.2 Fundamentals of Finance

14.2.1 Financial Markets and Instruments

Financial markets are platforms where buyers and sellers trade financial assets such as stocks, bonds, currencies, commodities, and derivatives (Vasista 2019). These markets facilitate the flow of capital and provide opportunities for investors to buy and sell securities. There are various types of financial markets, including:

- Stock Market: A stock market is where shares of publicly traded companies are bought and sold. Investors can trade stocks through exchanges like the New York Stock Exchange (NYSE) or electronic platforms such as NASDAQ.
- Bond Market: The bond market is where debt securities are bought and sold. These securities represent loans made by investors to governments, municipalities, corporations, and other entities. The bond market includes government bonds, corporate bonds, municipal bonds, and more.

- Foreign Exchange Market: The foreign exchange (forex) market is where currencies are traded. It is the largest and most liquid market globally, enabling participants to exchange one currency for another. The forex market facilitates international trade and investment.
- Commodities Market: In commodities markets, raw materials or primary goods such as oil, gold, wheat, natural gas, and metals are bought and sold. Commodities markets can be physical markets, where actual delivery of goods takes place, or futures markets, where contracts for future delivery are traded.
- Derivatives Market: The derivatives market involves financial contracts whose value is derived from an underlying asset or benchmark. Derivatives include options, futures, forwards, and swaps. These instruments allow investors to hedge against risks or speculate on price movements.
- Money Market: The money market deals with short-term borrowing and lending of funds. It includes instruments such as Treasury bills, certificates of deposit (CDs), commercial paper, and short-term government securities. The money market provides liquidity for financial institutions and corporations.
- Real Estate Market: The real estate market involves buying, selling, and renting properties, including residential, commercial, and industrial real estate. Real estate investments offer opportunities for capital appreciation and rental income.

Financial instruments are tradable assets that represent ownership, debt, or contractual rights. Some common financial instruments include:

- Stocks: Stocks represent ownership in a company and provide shareholders with ownership rights and potential dividends.
- Bonds: Bonds are debt instruments where investors lend money to issuers, typically governments or corporations, for a fixed period at a predetermined interest rate.
- Mutual Funds: Mutual funds pool money from multiple investors to invest in a diversified portfolio of stocks, bonds, or other assets.
- Exchange-Traded Funds (ETFs): ETFs are investment funds that trade on stock exchanges, tracking the performance of an index or a specific sector or asset class.
- Options: Options provide the right, but not the obligation, to buy or sell an underlying asset at a predetermined price within a specified period.
- Futures: Futures contracts obligate parties to buy or sell an asset at a predetermined price on a future date, providing a way to speculate or hedge against price fluctuations.
- Foreign Exchange (Forex): Forex instruments involve trading currencies, such as buying one currency while selling another.
- Commodities: Commodity instruments allow investors to trade physical goods or contracts representing commodities like oil, gold, or agricultural products.

These financial markets and instruments provide opportunities for investors, traders, and businesses to manage risks, raise capital, and invest in various asset classes. They play a crucial role in the global economy, facilitating economic growth and capital allocation.

14.2.2 *Financial Data and Analysis*

Financial data and analysis play a crucial role in understanding and evaluating the financial performance and health of companies, industries, and economies. Here are some key aspects related to financial data and analysis:

- Financial Statements: Financial statements, including the balance sheet, income statement, and cash flow statement, provide a comprehensive overview of a company's financial position, performance, and cash flows. These statements are prepared based on accounting principles and regulations.
- Key Financial Ratios: Financial ratios are used to analyze and compare different aspects of a company's financial performance. Examples include profitability ratios (such as gross profit margin and return on equity), liquidity ratios (such as current ratio and quick ratio), and leverage ratios (such as debt-to-equity ratio and interest coverage ratio).
- Trend Analysis: Trend analysis involves examining financial data over multiple periods to identify patterns and trends. This analysis helps assess the company's financial performance over time and identify areas of improvement or concern.
- Comparative Analysis: Comparative analysis involves comparing a company's financial data with that of its competitors or industry peers. This analysis provides insights into a company's relative performance, market position, and competitive advantages or disadvantages.
- Financial Forecasting: Financial forecasting involves projecting future financial performance based on historical data, market trends, and other relevant factors. These forecasts assist in budgeting, strategic planning, and investment decision-making.
- Valuation Techniques: Valuation techniques are used to determine the intrinsic value of a company or its securities. Common valuation methods include discounted cash flow (DCF) analysis, comparable company analysis, and asset-based valuation.
- Risk Assessment: Financial analysis also includes assessing and managing financial risks. This includes analyzing credit risk, market risk, liquidity risk, and operational risk to evaluate potential impacts on financial performance and stability.
- Data Sources: Financial data is sourced from various channels, including company filings, financial statements, market data providers, government reports, and industry databases. It is essential to ensure the accuracy and reliability of data sources for meaningful analysis.
- Financial Modeling: Financial modeling involves creating mathematical representations of financial situations or scenarios to support decision-making processes. This may include building models for revenue projections, investment analysis, or capital budgeting.
- Data Visualization: Data visualization techniques, such as charts, graphs, and dashboards, are used to present financial data in a visually compelling and

easy-to-understand manner. Effective data visualization enhances the interpretation and communication of financial analysis results.

Financial data and analysis provide insights into the financial performance, trends, and risks associated with companies and financial markets. It assists investors, analysts, and decision-makers in making informed investment decisions, assessing creditworthiness, and understanding the overall financial landscape.

14.2.3 Traditional Approaches in Finance

Traditional approaches in finance refer to established methods and theories that have been widely used in the field for many years. These approaches typically rely on well-established principles, frameworks, and models (Battiston et al. 2021). Here are some key traditional approaches in finance:

- Modern Portfolio Theory (MPT): Developed by Harry Markowitz, MPT focuses on constructing portfolios that maximize returns for a given level of risk. It emphasizes diversification across different asset classes to reduce risk and achieve optimal portfolio allocation.
- Capital Asset Pricing Model (CAPM): CAPM is a model that quantifies the relationship between the expected return of an investment and its systematic risk. It provides a framework for calculating an asset's expected return based on its beta (systematic risk) and the risk-free rate.
- Efficient Market Hypothesis (EMH): The EMH suggests that financial markets efficiently incorporate all available information, making it impossible to consistently outperform the market. It implies that stock prices reflect all relevant information, and it is not possible to consistently beat the market through active trading or stock selection.
- Dividend Discount Model (DDM): The DDM is used to value stocks by discounting future expected dividends. It assumes that the value of a stock is the present value of all expected future cash flows in the form of dividends.
- Financial Ratio Analysis: Financial ratio analysis involves evaluating a company's financial performance and health by examining various ratios derived from its financial statements. Common ratios include profitability ratios, liquidity ratios, and leverage ratios.
- Time Value of Money (TVM): TVM is a concept that recognizes the principle that money today is worth more than the same amount of money in the future due to its earning potential. TVM is used in discounting cash flows, calculating present value, and determining the value of investments.
- Black-Scholes Model: The Black-Scholes model is a mathematical model used to price options. It assumes that stock prices follow a geometric Brownian motion and that options can be hedged using a portfolio of the underlying asset and the risk-free rate.

- Fundamental Analysis: Fundamental analysis involves analyzing a company's financial statements, industry trends, management quality, and competitive position to determine its intrinsic value. It seeks to identify undervalued or overvalued stocks based on their fundamental characteristics.
- Technical Analysis: Technical analysis focuses on analyzing historical price and volume data to identify patterns, trends, and support/resistance levels in financial markets. It aims to forecast future price movements based on historical price behavior.
- Capital Structure Theory: Capital structure theory explores the optimal mix of debt and equity financing for a company. It considers factors such as the cost of capital, tax implications, risk tolerance, and financial flexibility to determine the ideal capital structure.

These traditional approaches have formed the foundation of financial theory and practice for many years. However, it is important to note that finance is a dynamic field, and newer approaches and models, including those based on machine learning and artificial intelligence, have emerged to complement and enhance traditional approaches in recent years.

14.3 Overview of Machine Learning Techniques in Finance

14.3.1 Supervised Learning

Supervised learning is a type of machine learning algorithm in which an AI model learns from labeled training data to make predictions or classify new, unseen data. It is called "supervised" because the model is provided with a supervisor or a ground truth during the training phase (Fig. 14.1). This supervisor consists of input data paired with corresponding correct output labels or target values (Cunningham et al. 2008).

The goal of supervised learning is to train a model that can generalize from the labeled training data and accurately predict the correct output for new, unseen inputs. The model learns patterns, relationships, and correlations between input

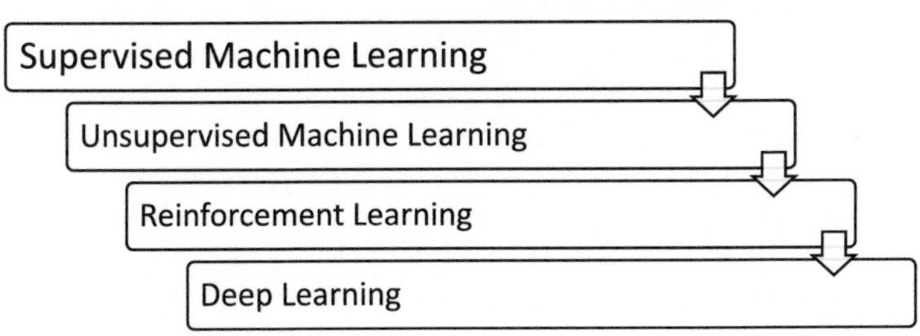

Fig. 14.1 Machine learning techniques in finance

features and output labels by iteratively adjusting its internal parameters during the training process.

The process of supervised learning typically involves the following steps:

- Data Collection: Gathering a dataset that consists of input features and their corresponding labels. The dataset is divided into a training set and a separate validation or test set.
- Model Selection: Choosing an appropriate model or algorithm for the specific task. This selection depends on the nature of the problem, the available data, and the desired output.
- Feature Engineering: Preprocessing and transforming the input features to make them suitable for the chosen model. This may involve scaling, normalization, encoding categorical variables, or extracting relevant features.
- Model Training: The model is trained using the labeled training data. The algorithm iteratively adjusts its internal parameters to minimize the difference between predicted outputs and the true labels.
- Model Evaluation: The trained model is evaluated using the validation or test set to assess its performance and generalization ability. Various metrics such as accuracy, precision, recall, and F1 score can be used to evaluate the model's performance.
- Model Tuning: Fine-tuning the model by adjusting hyperparameters, such as learning rate, regularization parameters, or the number of layers in a neural network, to optimize its performance.
- Prediction and Deployment: Once the model is trained and evaluated, it can be used to make predictions or classify new, unseen data. The model can be deployed in production systems to provide real-time predictions.

Supervised learning algorithms include various approaches such as linear regression, logistic regression, decision trees, random forests, support vector machines (SVM), and neural networks. The choice of algorithm depends on the nature of the problem, the size and complexity of the dataset, and the desired output. Supervised learning is widely used in various applications, including image and speech recognition, sentiment analysis and other useful applications.

14.3.2 Unsupervised Learning

Unsupervised learning is a branch of machine learning where the algorithm learns patterns and structures from unlabeled data. Unlike supervised learning, there are no predefined target labels or outputs provided to the algorithm. Instead, the algorithm discovers hidden patterns, relationships, or structures within the data on its own (Ghahramani 2003).

Here's an overview of how unsupervised learning works:

- Unlabeled Data: In unsupervised learning, the algorithm is given a dataset that consists of only input samples without any corresponding output labels. For example, a dataset containing customer demographic data without specific target labels.

- Pattern Recognition and Clustering: The algorithm aims to identify inherent patterns or clusters within the data. It analyzes the input samples and groups them based on similarities, dissimilarities, or shared characteristics. This process is known as clustering.
- Dimensionality Reduction: Unsupervised learning algorithms can also be used for dimensionality reduction. They aim to reduce the number of input features while preserving the essential information and structure of the data. Dimensionality reduction techniques such as principal component analysis (PCA) and t-distributed stochastic neighbor embedding (t-SNE) are commonly used for this purpose.
- Anomaly Detection: Another application of unsupervised learning is anomaly detection. The algorithm learns the normal patterns or behavior of the data and can identify any instances that deviate significantly from the norm. Anomalies can represent potential outliers, fraud, or unusual events in the data.
- Recommendation Systems: Unsupervised learning algorithms can also be used in recommendation systems. By analyzing patterns and similarities among users or items, the algorithm can make personalized recommendations based on the preferences and behaviors of users.
- Data Exploration and Visualization: Unsupervised learning techniques can help in exploring and understanding the data. By visualizing the clustered groups or reduced dimensions, patterns and insights can be revealed, aiding in data exploration and decision-making processes.

Unsupervised learning is valuable when there is limited or no prior knowledge about the data or when exploring new and unstructured datasets. It helps to uncover hidden patterns, discover unknown relationships, and gain a deeper understanding of the data, which can then be used for various purposes such as clustering, anomaly detection, and recommendation systems.

14.3.3 Reinforcement Learning

Reinforcement learning is a branch of machine learning that deals with how an agent can learn to interact with an environment in order to maximize its cumulative reward. It is inspired by the concept of learning through trial and error, like how humans and animals learn (Wiering and Van Otterlo 2012). Here's an overview of how reinforcement learning works:

- Agent and Environment: In reinforcement learning, there is an agent and an environment. The agent is the learner or decision-maker, and the environment is the external system with which the agent interacts.
- State, Action, and Reward: At each time step, the agent perceives the state of the environment, takes an action, and receives feedback in the form of a reward sig-

nal from the environment. The state represents the current situation or observation, the action represents the decision or behavior of the agent, and the reward signal indicates the desirability or quality of the action taken.

- Markov Decision Process (MDP): Reinforcement learning is often formulated as a Markov Decision Process, which mathematically represents the interaction between the agent and the environment. An MDP consists of a set of states, a set of actions, transition probabilities that define how the environment state changes based on agent actions, and a reward function that assigns rewards based on state-action pairs.
- Policy: The agent follows a policy, which is a mapping from states to actions. The policy guides the agent in selecting actions based on the current state of the environment. The goal of the agent is to find an optimal policy that maximizes the cumulative reward over time.
- Exploration and Exploitation: In reinforcement learning, there is a trade-off between exploration and exploitation. Exploration involves taking actions to gather more information about the environment and discover potentially better actions. Exploitation involves choosing actions that the agent believes will maximize the expected cumulative reward based on its current knowledge.
- Learning and Updating: The agent learns through a trial-and-error process. It uses various algorithms, such as Q-learning or policy gradients, to update its policy based on the observed rewards and the resulting state-action sequences. The learning process aims to improve the agent's decision-making abilities over time.
- Reward Optimization: In reinforcement learning, the agent aims to optimize the long-term cumulative reward. This may involve finding a balance between short-term rewards and long-term goals, as well as dealing with delayed rewards and trade-offs between different actions.

Reinforcement learning has been successfully applied in various domains, including robotics, game playing, autonomous driving, and recommendation systems. It allows agents to learn optimal strategies and make decisions in complex and dynamic environments, even when explicit instructions or labeled training data are not available.

14.3.4 Deep Learning

Deep learning is a subset of machine learning that focuses on the development and application of artificial neural networks with multiple layers (LeCun et al. 2015). Deep learning models, also known as deep neural networks, are designed to automatically learn and extract complex patterns and representations from large amounts of data.

Key characteristics of deep learning include:

- Neural Networks: Deep learning models are based on artificial neural networks, which are composed of interconnected nodes or neurons. The neurons are organized into layers, with each layer responsible for extracting and transforming the data.
- Multiple Layers: Deep neural networks typically consist of multiple hidden layers between the input and output layers. The depth of the network allows it to capture hierarchical representations of the data, enabling the model to learn intricate patterns and features.
- Feature Learning: Deep learning models can learn feature representations directly from raw data, eliminating the need for manual feature engineering. The model automatically learns to extract relevant features as part of the learning process.
- Large-Scale Data: Deep learning models excel in scenarios where there is a large amount of labeled data available for training. The models can effectively leverage the vast data to learn complex patterns and generalize well to unseen examples.
- Computational Power: Deep learning models require substantial computational power for training due to the large number of parameters and computations involved. GPUs (Graphics Processing Units) are commonly used to accelerate the training process.
- Deep Learning Architectures: There are various architectures used in deep learning, such as Convolutional Neural Networks (CNNs) for image and video processing, Recurrent Neural Networks (RNNs) for sequential data analysis, and Generative Adversarial Networks (GANs) for generating synthetic data.
- Deep learning has found numerous applications in finance, including:
- Financial Market Prediction: Deep learning models can be used to forecast stock prices, predict market trends, and estimate volatility.
- Credit Risk Assessment: Deep learning can assist in assessing creditworthiness by analyzing customer financial data and credit histories.
- Fraud Detection: Deep learning models can identify fraudulent transactions by analyzing patterns and anomalies in financial data.
- Natural Language Processing: Deep learning techniques enable sentiment analysis of financial news and analysis of textual data for investment decision-making.
- Algorithmic Trading: Deep learning models can be used to develop trading strategies based on historical data and real-time market information.

In summary, deep learning has revolutionized the field of machine learning and has become a powerful tool for solving complex problems in finance by automatically learning intricate patterns and representations from data.

14.3.5 Machine Learning and Deep Learning Techniques in Finance

Machine learning and deep learning techniques have been extensively applied in the field of finance, revolutionizing various areas of financial analysis and decision-making (Gogas and Papadimitriou 2021; Heaton et al. 2016). Here are some specific applications of machine learning and deep learning techniques in finance:

- Credit Scoring: Machine learning models can be employed to assess credit risk by analyzing historical credit data, customer financial information, and behavioral patterns. These models can accurately predict the likelihood of default and help financial institutions make informed lending decisions.
- Fraud Detection: Machine learning algorithms can identify fraudulent activities by analyzing patterns, anomalies, and deviations from normal behavior in financial transactions. These models can enhance fraud detection systems and minimize financial losses.
- Algorithmic Trading: Machine learning and deep learning techniques are widely used in algorithmic trading to develop trading strategies, predict market trends, and optimize portfolio allocation. These models can process large amounts of financial data and make rapid, data-driven trading decisions.
- Risk Management: Machine learning models can assist in risk management by predicting market volatility, estimating Value-at-Risk (VaR), and optimizing risk allocation. These techniques provide valuable insights for managing and mitigating financial risks.
- Customer Segmentation and Personalization: Machine learning algorithms can segment customers based on their financial behavior, preferences, and demographics. This segmentation enables personalized marketing campaigns, targeted product offerings, and enhanced customer satisfaction.
- Portfolio Management: Machine learning and deep learning techniques can be applied to portfolio optimization, asset allocation, and rebalancing. These models consider various factors such as historical returns, risk profiles, and market conditions to optimize investment portfolios.
- Sentiment Analysis: Natural language processing (NLP) techniques coupled with machine learning can analyze financial news, social media sentiment, and other textual data to gauge market sentiment. This information can be used to make more informed investment decisions.
- High-Frequency Trading: Machine learning models are employed in high-frequency trading to make rapid trading decisions based on real-time market data. These models analyze market microstructure and patterns to exploit short-term trading opportunities.

Market Prediction: Machine learning algorithms, such as regression and time series analysis, can forecast market trends, stock prices, and other financial variables. These models leverage historical data and market indicators to make predictions.

Customer Relationship Management: Machine learning techniques enable customer churn prediction, customer lifetime value estimation, and personalized customer relationship management. These models help financial institutions enhance customer retention and satisfaction.

Machine learning and deep learning techniques continue to evolve and contribute to advancements in finance, enabling more accurate predictions, improved risk management, enhanced customer experiences, and more efficient decision-making processes.

14.4 Machine Learning Applications in Finance

14.4.1 Role of Machine Learning in Financial Forecasting

Machine learning plays a significant role in financial forecasting by leveraging historical data, patterns, and complex algorithms to make predictions about future market conditions, asset prices, risk assessment, and more (Wasserbacher and Spindler 2022). Here are some key ways in which machine learning is applied in financial forecasting:

- Time Series Analysis: Machine learning algorithms can analyze historical financial data, such as stock prices, exchange rates, or economic indicators, to identify patterns, trends, and seasonality. By recognizing these patterns, models can make predictions about future values, aiding in financial forecasting.
- Risk Assessment and Management: Machine learning algorithms can help financial institutions assess and manage risks more effectively. By analyzing large volumes of data, including credit histories, market data, and customer behavior, algorithms can identify potential risks and predict the probability of default, fraud, or other adverse events.
- Portfolio Optimization: Machine learning techniques can assist in optimizing investment portfolios. By analyzing historical market data and considering various factors such as risk tolerance, investment goals, and market conditions, algorithms can recommend an optimal asset allocation strategy to maximize returns and minimize risk.
- Algorithmic Trading: Machine learning algorithms are used extensively in algorithmic trading, where computers execute trades based on predefined rules and models. These algorithms analyze real-time market data, news sentiment, and other relevant information to identify trading opportunities and make fast, automated decisions.
- Credit Scoring and Lending: Machine learning models can analyze a wide range of financial and non-financial data to predict creditworthiness and assign credit scores. This helps financial institutions assess the risk associated with potential borrowers, make lending decisions, and determine interest rates.

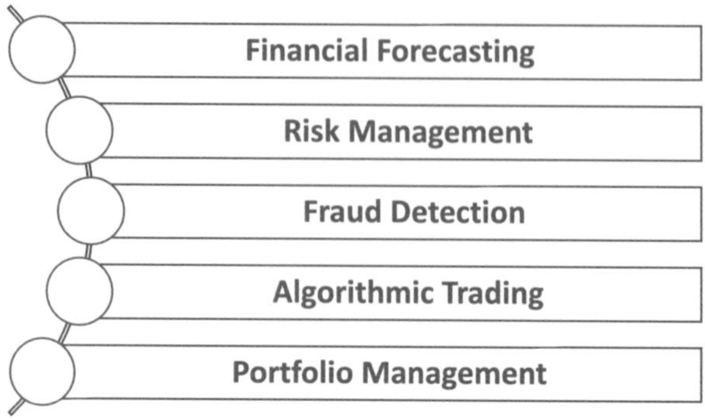

Fig. 14.2 Machine learning applications in finance

- Fraud Detection: Machine learning algorithms can detect fraudulent activities by analyzing vast amounts of transactional and behavioral data. By identifying patterns and anomalies, these algorithms can flag suspicious transactions or activities, helping financial institutions prevent fraud and enhance security.
- Customer Segmentation and Personalization: Machine learning models can segment customers based on their behavior, preferences, and demographic information. This segmentation allows financial institutions to personalize their offerings, provide targeted recommendations, and improve customer satisfaction and retention.
- Economic Forecasting: Machine learning algorithms can analyze macroeconomic indicators, such as GDP, inflation rates, employment data, and consumer sentiment, to predict future economic trends. These forecasts can be valuable for financial institutions, investors, and policymakers in making informed decisions.

Machine learning techniques bring advanced data analysis capabilities to financial forecasting, enabling more accurate predictions, faster processing, and enhanced decision-making in the finance industry. However, note that these models are not infallible, and careful consideration should be given to the quality and relevance of the data used, model validation, and potential biases in the algorithms (Fig. 14.2).

14.4.2 Role of Machine Learning in Risk Management

Machine learning plays a crucial role in risk management by enabling more accurate and efficient assessment, identification, and mitigation of various types of risks (Paltrinieri et al. 2019). Here are some key areas where machine learning is applied in risk management:

- Credit Risk Assessment: Machine learning models can analyze large volumes of data, including credit histories, financial statements, and economic indicators, to predict the creditworthiness of borrowers. By identifying patterns and relationships, these models can assess the probability of default and determine appropriate credit limits and interest rates.

- Fraud Detection: Machine learning algorithms can detect fraudulent activities by analyzing historical transactional data and real-time behavior. These models can identify patterns, anomalies, and suspicious activities, allowing financial institutions to flag potential fraud and take preventive measures.
- Market Risk Analysis: Machine learning algorithms can analyze historical market data, news sentiment, and other relevant information to assess and predict market risks. These models can identify trends, correlations, and volatility patterns, helping financial institutions manage market risks associated with investments, portfolios, and trading strategies.
- Operational Risk Management: Machine learning can help identify and manage operational risks by analyzing various data sources, such as transaction logs, employee behavior, and customer interactions. These models can detect anomalies, identify potential risks, and recommend appropriate actions to mitigate operational disruptions and improve efficiency.
- Cybersecurity: Machine learning techniques are widely used in cybersecurity to detect and prevent cyber threats. These models analyze network traffic, system logs, and user behavior to identify patterns of malicious activities, such as intrusion attempts, malware, or data breaches. By recognizing these patterns, organizations can enhance their security measures and respond proactively to potential threats.
- Compliance and Regulatory Risk: Machine learning algorithms can assist in compliance management by analyzing regulatory requirements, legal documents, and historical compliance data. These models can help identify potential compliance risks and provide recommendations to ensure adherence to regulations, reducing the risk of penalties and reputational damage.
- Risk Prediction and Scenario Analysis: Machine learning models can predict potential risks and assess their impact by analyzing historical data and simulating various scenarios. These models can assist in stress testing, scenario analysis, and evaluating the potential outcomes of different risk events, enabling organizations to develop robust risk mitigation strategies.
- Fraudulent Transaction Monitoring: Machine learning algorithms can monitor transactions in real-time to identify fraudulent activities. These models can analyze patterns, transactional behavior, and contextual information to detect anomalies and flag suspicious transactions promptly, helping organizations prevent financial losses due to fraud.

Machine learning techniques enhance risk management by automating processes, analyzing large volumes of data, and detecting complex patterns that might be challenging for traditional approaches. However, note that these models are regularly monitored, validated, and updated to adapt to evolving risks and changing environments. Human expertise and judgment are also crucial in interpreting the results and making informed risk management decisions.

14.4.3 Role of Machine Learning in Fraud Detection

Machine learning plays a critical role in fraud detection by enabling organizations to identify and prevent fraudulent activities more effectively and efficiently (Lakshmi and Kavilla 2018). Here are some key aspects of how machine learning is applied in fraud detection:

- Anomaly Detection: Machine learning algorithms excel at identifying anomalies or deviations from normal patterns. In fraud detection, these algorithms analyze large volumes of data, such as transactional records, user behavior, or network logs, and learn what constitutes normal behavior. They can then flag transactions or activities that deviate significantly from these learned patterns as potential fraud.
- Pattern Recognition: Machine learning models can recognize complex patterns and correlations in data that may indicate fraudulent behavior. By analyzing historical fraud cases and related data, these models can identify common patterns, trends, and indicators of fraudulent activities. This enables organizations to proactively detect and prevent similar fraudulent patterns from occurring in the future.
- Real-time Monitoring: Machine learning algorithms can continuously monitor transactions, activities, or data streams in real-time. By analyzing incoming data as it arrives, these models can quickly identify suspicious patterns or behaviors, allowing organizations to take immediate action to prevent fraud in real-time.
- Behavioral Analysis: Machine learning techniques can analyze user behavior patterns to identify anomalies that may indicate fraudulent activity. By examining various attributes, such as transaction amounts, locations, or purchasing habits, these models can establish profiles of typical user behavior and flag activities that deviate significantly from these profiles.
- Unsupervised Learning: Unsupervised machine learning algorithms can discover hidden patterns or clusters within data without prior training or labels. In fraud detection, this approach can help identify new and previously unknown types of fraudulent behavior. By analyzing the characteristics of fraudulent instances, these models can uncover emerging fraud patterns and enhance fraud detection capabilities.
- Integration of Multiple Data Sources: Machine learning allows organizations to integrate and analyze data from various sources, such as transactional data, external databases, social media, or geolocation data. By combining and correlating information from multiple sources, these models can gain a more comprehensive view of potential fraud indicators and make more accurate fraud detection decisions.
- Adaptive Learning: Machine learning models can continuously adapt and update themselves based on new data and feedback. As fraudsters evolve their techniques, these models can learn from new fraud patterns and adjust their detection algorithms accordingly. This adaptability helps organizations stay ahead of emerging fraud threats and maintain effective fraud detection systems.

Machine learning techniques bring automation, scalability, and advanced pattern recognition capabilities to fraud detection, enabling organizations to detect fraud more accurately and efficiently. However, note that it is important to regularly validate and update these models to account for changes in fraud patterns and evolving attack techniques. Additionally, human expertise and domain knowledge are essential for interpreting the results, investigating flagged cases, and making informed decisions in fraud detection and prevention efforts.

14.4.4 Role of Machine Learning in Algorithmic Trading

Machine learning plays a crucial role in algorithmic trading by enabling automated decision-making and leveraging vast amounts of data to identify trading opportunities, optimize strategies, and manage risk (Hansen 2020). Here are some key aspects of how machine learning is applied in algorithmic trading:

- Pattern Recognition: Machine learning algorithms excel at recognizing patterns and correlations in large datasets. In algorithmic trading, these algorithms can analyze historical market data, such as price movements, trading volumes, or technical indicators, to identify patterns that may indicate future price movements or market trends. By recognizing these patterns, models can generate trading signals and make informed trading decisions.
- Market Prediction: Machine learning models can be trained to predict future market conditions or asset prices. By analyzing a variety of relevant factors, such as historical market data, news sentiment, economic indicators, or social media sentiment, these models can forecast price movements or market trends. These predictions can guide trading strategies and help identify profitable trading opportunities.
- Automated Trading Decision-making: Machine learning algorithms can automate the decision-making process in trading. Based on predefined rules or learned patterns, these algorithms can generate buy/sell signals, determine the optimal entry and exit points, or allocate assets in a portfolio. This automation allows for faster execution and removes human biases from the decision-making process.
- Risk Management: Machine learning techniques can assist in managing risk in algorithmic trading. By analyzing historical market data and considering factors such as volatility, liquidity, or correlation, models can optimize risk management strategies, including position sizing, stop-loss levels, or hedging techniques. This helps traders mitigate potential losses and manage risk exposure effectively.
- High-Frequency Trading (HFT): Machine learning is widely used in high-frequency trading, where trades are executed within microseconds. Machine learning algorithms can quickly process large volumes of data, identify patterns, and make rapid trading decisions. These models are designed to exploit short-term market inefficiencies and take advantage of fleeting trading opportunities.

- Market Microstructure Analysis: Machine learning techniques can analyze market microstructure data, which includes order book information, bid-ask spreads, or trade volumes. By analyzing this data, models can gain insights into market dynamics, liquidity patterns, or market impact, which can be valuable for optimizing trading strategies, execution algorithms, and order routing decisions.
- News and Sentiment Analysis: Machine learning models can analyze news articles, social media feeds, or other textual data to gauge market sentiment. By understanding the sentiment and impact of news events on asset prices, these models can make trading decisions based on the sentiment analysis of relevant information sources.
- Portfolio Optimization: Machine learning algorithms can optimize portfolio allocation based on various factors, including risk tolerance, return objectives, and market conditions. By analyzing historical data, correlation patterns, and other relevant parameters, models can suggest the optimal allocation of assets to maximize returns while managing risk.

Machine learning techniques in algorithmic trading enable traders to leverage vast amounts of data, identify patterns, make data-driven decisions, and execute trades with speed and precision. However, note that market dynamics are complex and subject to various uncertainties, and careful monitoring, model validation, and risk management practices are essential to ensure the effectiveness and robustness of algorithmic trading strategies.

14.4.5 Role of Machine Learning in Portfolio Management

Machine learning plays a significant role in portfolio management by providing sophisticated data analysis and decision-making capabilities. It enables portfolio managers to optimize asset allocation, assess risk, identify investment opportunities, and enhance overall portfolio performance (Hu and Lin 2019). Here are some key aspects of how machine learning is applied in portfolio management:

- Asset Allocation: Machine learning algorithms can analyze historical market data, economic indicators, and other relevant factors to optimize asset allocation strategies. These algorithms consider risk tolerance, return objectives, and correlation patterns among assets to determine the optimal allocation that maximizes returns while minimizing risk.
- Risk Assessment and Management: Machine learning models can assist in assessing and managing portfolio risk. By analyzing historical data, volatility patterns, and correlations, these models can identify potential risks and estimate the risk exposure of the portfolio. This information helps portfolio managers make informed decisions about risk mitigation strategies, diversification, and hedging techniques.

- Predictive Analytics: Machine learning algorithms can make predictions about future asset prices, market trends, or economic conditions. These predictions assist portfolio managers in identifying investment opportunities and adjusting the portfolio strategy accordingly. By leveraging a wide range of data sources and complex algorithms, machine learning models can provide valuable insights for portfolio decision-making.
- Factor Modeling: Machine learning techniques can help identify and analyze factors that drive asset returns. These models can analyze vast amounts of data, including financial statements, market data, and economic indicators, to uncover hidden patterns and relationships. By understanding the impact of various factors, portfolio managers can make more informed decisions regarding factor-based investing and risk-factor allocation.
- Sentiment Analysis: Machine learning algorithms can analyze sentiment data from news articles, social media, or other textual sources to gauge market sentiment. This sentiment analysis provides insights into the market's perception and sentiment towards specific assets or sectors. Portfolio managers can leverage this information to make informed investment decisions or adjust portfolio positioning based on market sentiment.
- Quantitative Trading Strategies: Machine learning techniques are used in developing quantitative trading strategies. These algorithms analyze historical market data, identify patterns, and generate trading signals. By incorporating machine learning models into trading strategies, portfolio managers can execute trades based on predefined rules, taking advantage of market inefficiencies and generating alpha.
- Risk-adjusted Performance Evaluation: Machine learning algorithms can assist in evaluating the performance of investment portfolios. By considering risk-adjusted metrics such as Sharpe ratio, Sortino ratio, or Value at Risk (VaR), these models provide a more comprehensive assessment of portfolio performance. This evaluation helps portfolio managers compare different strategies, assess the impact of adjustments, and optimize the portfolio composition.
- Robo-Advisory Services: Machine learning is often employed in robo-advisory platforms to provide automated portfolio management services. These platforms use machine learning algorithms to assess clients' risk tolerance, financial goals, and investment preferences. Based on this analysis, the algorithms generate personalized investment recommendations and manage portfolios accordingly, providing cost-effective and efficient portfolio management solutions.

Machine learning brings advanced data analysis capabilities and automation to portfolio management, enabling portfolio managers to make data-driven decisions, optimize asset allocation, and enhance portfolio performance. However, human expertise and judgment are still crucial in interpreting the results, adapting to changing market conditions, and making strategic decisions aligned with investors' goals and preferences.

14.5 Artificial Intelligence in Banking and Finance

Artificial intelligence (AI) is transforming the banking and finance industry by providing advanced data analysis, automation, and personalized customer experiences (Pau et al. 1990). Here are some key areas where AI is applied in banking and finance (Fig. 14.3):

- Customer Service and Support: AI-powered chatbots and virtual assistants are used to provide personalized customer service and support. These AI systems can understand and respond to customer inquiries, provide account information, assist with transactions, and offer financial advice. They enhance customer experiences by providing quick, accurate, and round-the-clock support.
- Fraud Detection and Security: AI algorithms are employed to detect and prevent fraudulent activities in banking and finance. These algorithms analyze large volumes of transactional data, user behavior patterns, and external factors to identify anomalies, patterns, and indicators of fraud. By flagging suspicious activities in real-time, AI helps protect customer accounts and secure financial transactions.
- Risk Assessment and Management: AI models analyze data to assess credit risk, predict default probabilities, and determine appropriate lending decisions. These models consider various factors such as credit history, financial statements, and economic indicators to provide accurate risk assessments. AI also assists in managing market risk, operational risk, and regulatory compliance through advanced data analysis and scenario modeling.
- Personalized Financial Advice: AI-powered platforms provide personalized financial advice based on individual customer data, goals, and preferences. These systems analyze customer profiles, spending patterns, and financial goals to offer tailored recommendations on budgeting, investment strategies, and savings plans. By leveraging AI, banks and financial institutions can offer customized financial guidance to customers at scale.

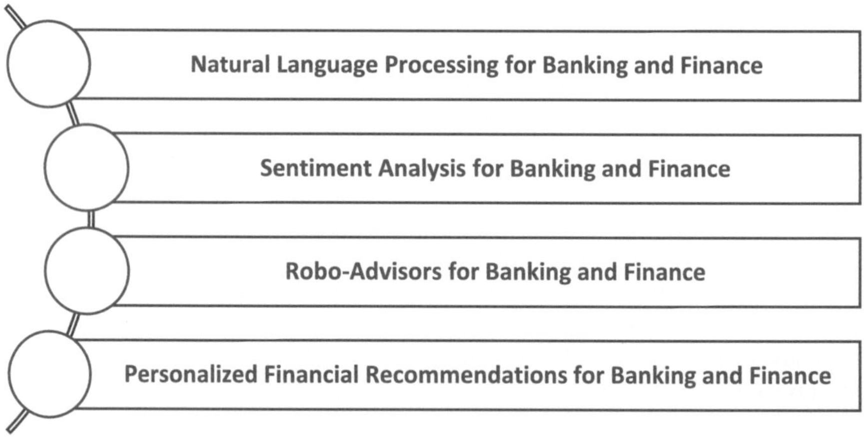

Fig. 14.3 Artificial intelligence in banking and finance

- Credit Scoring and Underwriting: AI algorithms analyze diverse data sources, including credit history, income statements, and alternative data, to assess creditworthiness and streamline the underwriting process. Machine learning models can analyze large volumes of data and identify patterns that traditional credit scoring methods may overlook. This enables faster and more accurate credit decisions.
- Robo-Advisory and Wealth Management: AI-powered robo-advisors assist customers in managing their investments and portfolios. These platforms use algorithms to analyze customer risk profiles, investment goals, and market data to offer automated investment recommendations. Robo-advisors enable cost-effective and accessible wealth management services while providing customized investment strategies to individual investors.
- Trading and Market Analysis: AI algorithms are used for algorithmic trading, high-frequency trading, and market analysis. These algorithms analyze market data, news sentiment, and other relevant factors to identify trading opportunities, optimize trading strategies, and execute trades with speed and efficiency. AI-powered trading systems can process vast amounts of data and recognize complex patterns to make data-driven trading decisions.
- Anti-Money Laundering (AML) and Compliance: AI systems assist in detecting money laundering activities and ensuring regulatory compliance. These systems analyze transactional data, customer profiles, and historical patterns to identify suspicious transactions, monitor compliance with regulations, and generate alerts for further investigation. AI improves the effectiveness and efficiency of AML processes while reducing false positives.

The integration of AI technologies in banking and finance enhances efficiency, accuracy, and customer experiences. However, it is important to consider ethical and privacy concerns, ensure transparency, and have robust data governance practices in place to maintain trust and accountability in AI-driven systems.

14.5.1 Artificial Intelligence in Natural Language Processing for Banking and Finance

Artificial Intelligence (AI) and Natural Language Processing (NLP) have been revolutionizing the banking and finance industry in various ways. Here are some key applications of AI and NLP in this sector:

- Chatbots and Virtual Assistants: Banks and financial institutions have adopted AI-powered chatbots and virtual assistants to enhance customer service. These systems use NLP techniques to understand and respond to customer queries and provide personalized assistance. They can handle routine inquiries, provide balance information, process transactions, and even offer financial advice.
- Sentiment Analysis: NLP algorithms can analyze large volumes of unstructured data, such as social media posts, news articles, and customer feedback, to gauge

public sentiment towards specific financial products, services, or brands. This information helps banks and financial institutions make data-driven decisions and improve their offerings.

- Fraud Detection: AI and NLP are employed to detect fraudulent activities in banking and finance. By analyzing patterns, transactional data, and customer behavior, these systems can identify suspicious transactions, flag potentially fraudulent activities, and minimize risks. NLP techniques are also used to analyze textual data for fraud detection, such as detecting phishing emails or identifying suspicious account activity based on written communication.
- Risk Assessment and Credit Scoring: NLP algorithms can analyze large volumes of financial data, including loan applications, credit reports, and customer profiles, to assess creditworthiness and calculate risk scores. By extracting relevant information from these documents, AI-powered systems can make faster and more accurate credit decisions, streamlining the lending process for both individuals and businesses.
- Compliance and Regulatory Reporting: NLP is used to interpret complex regulatory documents and monitor compliance within the banking and finance industry. By automatically extracting relevant information from regulatory texts, AI systems can assist with regulatory reporting, ensure adherence to legal requirements, and minimize the risk of non-compliance.
- Market Analysis and Trading: AI and NLP techniques are applied in analyzing financial news, social media sentiment, and other textual data to gain insights into market trends, investor sentiment, and company performance. These insights can help traders and investment firms make informed decisions, optimize trading strategies, and predict market movements.
- Personalized Financial Advice: AI-powered systems leverage NLP to understand customer preferences, financial goals, and risk tolerance. Based on this analysis, personalized financial advice and recommendations can be generated, tailored to the individual's needs. This technology enables banks and financial institutions to offer customized wealth management services and investment advice at scale.

Note that while AI and NLP offer significant benefits to the banking and finance industry, there are also considerations around data privacy, security, and ethical use of these technologies. Striking a balance between innovation and responsible deployment is crucial to ensure trust and maximize the potential of AI in this domain.

14.5.2 Artificial Intelligence in Sentiment Analysis for Banking and Finance

Artificial Intelligence (AI) plays a significant role in sentiment analysis for the banking and finance industry. Sentiment analysis involves analyzing and interpreting people's opinions, emotions, and attitudes expressed in text data (Al Ahmed et al. 2022). Here's how AI contributes to sentiment analysis in banking and finance:

- Text Classification: AI techniques, such as machine learning and deep learning, are employed to develop text classification models. These models are trained on labeled data to classify text into positive, negative, or neutral sentiments. In banking and finance, these models can be used to analyze customer feedback, social media posts, online reviews, and news articles to gauge sentiment towards specific financial products, services, or brands.
- Named Entity Recognition: NLP algorithms can identify and extract relevant entities from text data, such as company names, product names, and financial terms. By identifying these entities, AI systems can understand the context and sentiment associated with them. This helps in analyzing sentiments related to specific financial entities, such as banks, insurance companies, or investment products.
- Aspect-Based Sentiment Analysis: AI techniques enable aspect-based sentiment analysis, which focuses on identifying sentiments related to specific aspects or features of a product or service. For example, in the banking industry, aspect-based sentiment analysis can determine customer sentiments about the user experience of a mobile banking app, the quality of customer service, or the interest rates offered by a bank. This provides a more granular understanding of customer sentiment and helps banks address specific areas for improvement.
- Social Media Monitoring: AI-powered sentiment analysis systems can monitor social media platforms to track public sentiment towards financial institutions and their offerings. By analyzing posts, comments, and hashtags, AI systems can provide real-time insights into customer opinions, concerns, and satisfaction levels. This information helps banks and financial institutions understand customer sentiment and take appropriate actions to address issues or capitalize on positive sentiment.
- Risk Management: Sentiment analysis can contribute to risk management in the banking and finance sector. By analyzing sentiments expressed in news articles, market reports, and social media, AI systems can identify early warning signs of potential financial risks. For example, sentiment analysis can help detect negative sentiment towards a particular investment product, signaling a potential decline in its market value. Banks and financial institutions can use this information to adjust their risk exposure and make more informed decisions.
- Brand Reputation Management: AI-powered sentiment analysis can assist in managing brand reputation for banks and financial institutions. By monitoring sentiment across various channels, including news outlets, social media, and review platforms, AI systems can identify and track shifts in public opinion. This enables proactive brand management, allowing banks to address negative sentiment, engage with customers, and take measures to improve brand perception.

AI-driven sentiment analysis in banking and finance enables institutions to gain valuable insights into customer sentiment, market trends, and potential risks. By leveraging these insights, banks and financial institutions can enhance customer experience, tailor their offerings, manage risks effectively, and make data-driven decisions that align with customer needs and market demands.

14.5.3 *Artificial Intelligence in Robo-Advisors for Banking and Finance*

Artificial Intelligence (AI) has played a transformative role in the development and advancement of robo-advisors in the banking and finance industry. Robo-advisors are digital platforms that leverage AI algorithms to provide automated and algorithm-based financial advice and investment management services (Belanche et al. 2019). Here's how AI contributes to robo-advisors:

- Portfolio Construction and Optimization: AI-powered robo-advisors utilize machine learning algorithms to construct and optimize investment portfolios. These algorithms consider various factors such as risk tolerance, investment goals, time horizon, and market conditions to recommend an optimal asset allocation strategy. By analyzing large volumes of historical and real-time financial data, robo-advisors can generate portfolios that are well-diversified and aligned with individual investor preferences.
- Risk Assessment and Suitability: AI-based robo-advisors assess investor risk profiles through sophisticated algorithms. They use techniques like questionnaire-based assessments, analysis of financial information, and behavioral finance models to determine an investor's risk tolerance and investment suitability. Based on this analysis, robo-advisors recommend suitable investment strategies that align with the investor's risk appetite.
- Automated Rebalancing: Robo-advisors use AI algorithms to monitor and automatically rebalance investment portfolios. By continuously analyzing market conditions and portfolio performance, robo-advisors identify deviations from the target asset allocation and execute trades to bring the portfolio back in line with the desired allocation. This automated rebalancing helps maintain a disciplined investment approach and optimizes risk-adjusted returns.
- Tax Optimization: AI-powered robo-advisors consider tax implications when managing investment portfolios. They use algorithms to analyze tax rules, optimize capital gains, and minimize tax liabilities for investors. For example, robo-advisors can implement tax-loss harvesting strategies, which involve selling underperforming assets to offset capital gains and reduce tax obligations.
- Personalized Financial Planning: AI-driven robo-advisors offer personalized financial planning capabilities. They analyze individual financial goals, income, expenses, and other relevant data to provide tailored investment recommendations and financial advice. Robo-advisors can also simulate different scenarios and provide projections on achieving financial goals, retirement planning, and other aspects of personal finance.
- Natural Language Processing (NLP) for Client Interaction: Some robo-advisors incorporate NLP techniques to enhance client interaction. By analyzing and understanding natural language inputs from clients, AI-powered robo-advisors can respond to questions, provide explanations, and offer personalized advice in a conversational manner. This improves the user experience and makes the interaction with the robo-advisor more intuitive and user-friendly.

- Behavioral Finance Insights: AI-based robo-advisors can leverage behavioral finance principles to understand investor biases and behaviors. By analyzing historical investment patterns, risk tolerance changes, and emotional responses to market events, robo-advisors can provide personalized guidance to investors and help mitigate behavioral biases that may impact investment decisions.

Robo-advisors powered by AI offer several advantages, including accessibility, cost-effectiveness, and personalized financial advice. They enable individuals to access investment management services that were traditionally available to high-net-worth individuals or institutional investors. However, note that while robo-advisors provide convenience and automation, some investors may still prefer the guidance of human financial advisors for complex financial situations or unique circumstances.

14.5.4 Artificial Intelligence in Personalized Financial Recommendations for Banking and Finance

Artificial Intelligence (AI) plays a crucial role in delivering personalized financial recommendations in the banking and finance industry. By leveraging AI techniques, banks and financial institutions can analyze vast amounts of data, extract meaningful insights, and provide tailored recommendations to individual customers (Ranjan et al. 2020). Here's how AI enables personalized financial recommendations:

- Data Analysis and Customer Profiling: AI algorithms can analyze diverse data sources, including transaction history, spending patterns, investment preferences, credit scores, and demographic information, to build comprehensive customer profiles. By understanding each customer's financial behavior and needs, AI systems can generate personalized recommendations that align with their unique circumstances.
- Machine Learning for Recommendation Engines: AI-powered recommendation engines utilize machine learning algorithms to analyze customer data and historical trends. These algorithms can identify patterns, correlations, and preferences to make accurate predictions about the financial products or services that are most suitable for each customer. These recommendations can cover various areas, such as investment options, loan products, insurance plans, and credit cards.
- Natural Language Processing (NLP) for Customer Interactions: NLP techniques enable AI systems to understand and interpret natural language inputs from customers. By analyzing customer queries, emails, chat interactions, and voice recordings, AI systems can extract relevant information and provide personalized responses and recommendations. This enhances customer engagement and enables more effective communication between customers and financial institutions.
- Contextual and Real-Time Recommendations: AI algorithms can consider real-time market data, economic indicators, and customer-specific factors to deliver

contextual and timely recommendations. For example, based on changes in market conditions, an AI-powered system may suggest adjusting investment portfolios or refinancing options to optimize returns or save on interest costs. By considering current circumstances, AI systems can provide recommendations that are relevant and up-to-date.
- Risk Assessment and Mitigation: AI techniques can assess individual customer risk profiles and provide recommendations that align with their risk appetite. By analyzing historical data, market trends, and risk indicators, AI systems can offer personalized investment strategies that balance risk and reward. Additionally, AI-powered systems can suggest risk mitigation measures, such as insurance products or diversification strategies, based on each customer's specific needs.
- Financial Goal Planning: AI-driven systems can assist customers in setting and achieving their financial goals. By analyzing customer data and understanding their objectives, AI algorithms can provide recommendations on savings plans, investment strategies, and milestones to help customers reach their targets. AI systems can continuously track progress, adapt recommendations as circumstances change, and provide ongoing guidance and support.
- Predictive Analytics for Life Events: AI can leverage predictive analytics to anticipate significant life events, such as buying a house, getting married, having a child, or planning for retirement. By analyzing customer data and patterns, AI systems can provide personalized financial recommendations tailored to these life events. For example, AI systems can suggest appropriate mortgage options, insurance coverage, or retirement planning strategies based on the expected timing and financial impact of these events.

AI-powered personalized financial recommendations enable banks and financial institutions to offer customized solutions, enhance customer satisfaction, and improve financial outcomes for individuals. However, it is important to balance AI-driven recommendations with human expertise and ethical considerations to ensure transparency, fairness, and accountability in the delivery of financial services.

14.6 Data Collection and Preparation for Banking and Finance

Data collection and preparation are crucial steps in leveraging data analytics and AI techniques in the banking and finance industry (Phan et al. 2008). Here's an overview of the process:

- Data Identification: Identify the types of data required for analysis. In banking and finance, this may include transactional data, customer profiles, market data, economic indicators, credit scores, and more.
- Data Sources: Determine the sources of data, which may include internal systems (core banking systems, CRM platforms), external data providers, regula-

tory databases, social media, and other relevant sources. Ensure data privacy and compliance with data protection regulations when accessing and integrating external data sources.

- Data Extraction: Extract data from various sources and consolidate it into a centralized data repository. This may involve extracting data from databases, APIs, file formats (CSV, Excel), or streaming sources. Consider automation and data integration tools to streamline this process.
- Data Cleaning: Perform data cleaning and preprocessing to ensure data quality. This involves removing duplicate records, handling missing values, standardizing formats, correcting errors, and ensuring consistency across datasets. Apply data cleansing techniques such as outlier detection and data validation to improve data accuracy.
- Data Transformation: Transform the data into a suitable format for analysis. This may involve aggregating transactional data, normalizing data values, converting data types, and creating derived features or variables as needed. Feature engineering techniques can be applied to extract meaningful insights from the data.
- Data Integration: Integrate data from multiple sources to create a unified view. This process involves aligning data structures, resolving data inconsistencies, and establishing relationships between different datasets. Utilize data integration tools or database management systems to efficiently handle large volumes of data.
- Data Security and Privacy: Ensure data security and privacy measures are in place. Anonymize or pseudonymize personally identifiable information (PII) to protect customer privacy. Implement encryption, access controls, and secure data transfer protocols to safeguard sensitive data from unauthorized access.
- Data Governance and Compliance: Establish data governance frameworks to ensure data quality, integrity, and compliance with regulatory requirements. Implement data governance policies, data retention guidelines, and audit trails to maintain data accuracy, traceability, and accountability.
- Data Validation and Quality Assurance: Validate and verify the prepared data to ensure its accuracy and integrity. Perform data quality checks, validate against predefined rules or business logic, and compare data against expected results. Conduct data quality audits to identify and rectify any data issues.
- Data Documentation: Document the data collection and preparation process, including data sources, transformations, cleaning procedures, and metadata. This documentation provides transparency, facilitates reproducibility, and aids in understanding the data for future analysis.

Effective data collection and preparation lay the foundation for accurate and reliable analysis, insights, and decision-making in the banking and finance sector. It ensures that the data used for AI algorithms and analytics is of high quality, trustworthy, and aligned with the specific business needs and objectives.

14.6.1 Data Sources in Banking and Finance

In the banking and finance industry, there are various data sources that organizations rely on to support their operations, decision-making processes, risk management, and regulatory compliance. Here are some commonly used data sources:

- Transactional Data: Banks and financial institutions capture and analyze vast amounts of transactional data, including customer deposits, withdrawals, transfers, payments, and investment transactions. This data helps track financial flows, detect patterns, and generate insights.
- Customer Data: Customer data includes personal information, demographics, account details, credit history, and transactional behavior. It helps banks in customer segmentation, profiling, risk assessment, and developing personalized financial products and services.
- Market Data: Market data encompasses information about securities, commodities, currencies, interest rates, and other financial instruments. It includes real-time and historical pricing data, trading volumes, market indices, and news. This data is crucial for pricing, trading, portfolio management, and risk analysis.
- Economic Data: Economic indicators, such as GDP growth rates, inflation rates, employment data, interest rates, and consumer sentiment, provide insights into the overall health and trends in the economy. Financial institutions leverage this data to make informed investment decisions, assess risks, and adjust their strategies.
- Credit Bureaus: Credit bureaus maintain credit reports and scores for individuals and businesses. Banks and lenders access this data to evaluate creditworthiness, assess the risk of default, and make lending decisions. Common credit bureaus include Equifax, Experian, and TransUnion.
- Regulatory Data: Regulatory bodies require banks and financial institutions to report data for compliance purposes. This may include data related to anti-money laundering (AML), know your customer (KYC), fraud detection, risk management, and financial reporting. Compliance data ensures adherence to legal and regulatory obligations.
- External Data Providers: Financial organizations often leverage external data providers that specialize in collecting and analyzing specific types of data. These providers offer data on alternative assets, sentiment analysis, social media data, satellite imagery, weather data, and other non-traditional sources. Examples of external data providers include Bloomberg, FactSet, and S&P Global Market Intelligence.
- Internal Data Sources: Banks and financial institutions also generate valuable data internally through their own systems and operations. This includes data from core banking systems, customer relationship management (CRM) platforms, trading platforms, risk management systems, and other proprietary databases. Internal data helps organizations understand their own performance, monitor operational risks, and improve processes.

Note that the use of data in banking and finance is subject to regulatory and privacy considerations. Financial institutions must adhere to data protection laws and regulations to ensure the privacy and security of customer information.

14.6.2 Data Preprocessing Techniques for Banking and Finance

Data preprocessing plays a crucial role in banking and finance to ensure the quality, accuracy, and usability of data for various tasks such as risk assessment, fraud detection, investment analysis, and decision making (Sadatrasoul et al. 2013). Here are some common data preprocessing techniques used in the banking and finance industry:

- Data Cleaning: Data cleaning involves identifying and handling missing values, outliers, inconsistencies, and errors in the dataset. This step is important to ensure the accuracy and integrity of the data. Missing values can be imputed using techniques such as mean, median, or regression-based imputation. Outliers can be detected and treated by using statistical methods or domain knowledge.
- Data Integration: In banking and finance, data often comes from multiple sources and systems. Data integration involves combining data from different sources and resolving any inconsistencies or conflicts. It ensures that all relevant data is available for analysis and decision making.
- Data Transformation: Data transformation involves converting data into a suitable format for analysis. Common techniques include normalization, scaling, and logarithmic transformations. Normalization can be used to scale numerical data to a common range, while scaling can be used to standardize data to have zero mean and unit variance.
- Feature Selection: Feature selection aims to identify the most relevant and informative features for a given task. In banking and finance, this step helps reduce dimensionality, improve model performance, and avoid overfitting. Techniques such as correlation analysis, information gain, and regularization methods like L1 regularization (Lasso) can be used for feature selection.
- Feature Engineering: Feature engineering involves creating new features or transforming existing ones to improve the predictive power of models. In banking and finance, this can include deriving new variables such as ratios, moving averages, or financial indicators from raw data. Domain knowledge and expertise are often critical for effective feature engineering.
- Data Discretization: Data discretization is the process of converting continuous variables into categorical or ordinal variables. This technique can help simplify data representation, handle nonlinear relationships, and improve model interpretability. Discretization methods include binning, equal frequency or equal width partitioning, and decision tree-based discretization.

- Handling Imbalanced Data: In banking and finance, fraud detection or default prediction tasks often involve imbalanced datasets where the number of positive cases (frauds, defaults) is significantly lower than the negatives. Techniques such as oversampling, undersampling, and synthetic sample generation (e.g., SMOTE - Synthetic Minority Over-sampling Technique) can be employed to address the class imbalance issue.
- Data Encryption and Anonymization: In scenarios where sensitive customer information is involved, data encryption and anonymization techniques are applied to protect privacy and comply with data protection regulations. Encryption techniques like AES (Advanced Encryption Standard) can be used to secure data at rest and in transit. Anonymization techniques, such as removing personally identifiable information (PII) or applying data masking, can be employed to de-identify sensitive data.

Note that the specific preprocessing techniques employed can vary based on the task at hand, the characteristics of the dataset, and the requirements of the problem domain. Moreover, domain expertise and collaboration between data scientists and domain experts are often crucial to effectively preprocess data in banking and finance.

14.6.3 Feature Engineering for Banking and Finance

Feature engineering is a critical step in the data preprocessing phase for banking and finance applications. It involves creating new features or transforming existing ones to enhance the predictive power of models and improve the accuracy of predictions (Khadivizand et al. 2020). Here are some common feature engineering techniques used in banking and finance:

- Time-based Features: In banking and finance, time is often a crucial factor. You can create various time-based features from timestamps or date variables. Examples include day of the week, month, quarter, or year, time since the last transaction, average transaction frequency, or the number of transactions within a specific time window.
- Aggregation and Statistical Features: Aggregating and summarizing data can provide valuable insights. You can compute statistical features such as mean, median, standard deviation, minimum, maximum, or skewness for numerical variables. For categorical variables, you can generate features like the count of occurrences, unique categories, or mode.
- Ratio and Proportion Features: Ratios and proportions can capture meaningful relationships between variables. For instance, you can calculate the debt-to-

income ratio, loan-to-value ratio, or the proportion of credit card usage to the credit limit. These features can help assess risk, financial health, and affordability.

- Moving Averages: Moving averages are useful for capturing trends and patterns over time. By calculating the average of a variable over a specific window, you can create features that reflect short-term or long-term trends. Moving averages can be applied to variables such as account balances, stock prices, or interest rates.
- Difference Features: Difference features involve calculating the difference between two related variables. For example, you can calculate the difference between the current account balance and the average balance over a specific period. These features can help identify changes or anomalies in the data.
- Interaction Features: Interaction features capture the combined effect of two or more variables. You can create interaction features by multiplying, dividing, or adding different variables. For instance, multiplying the credit utilization ratio by the number of missed payments can reveal the impact of both factors on credit risk.
- Binning and Discretization: Binning or discretization involves dividing continuous variables into distinct bins or categories. This can be useful for capturing non-linear relationships or reducing the impact of outliers. For example, you can create bins for income ranges or age groups to better understand customer segments.
- Domain-Specific Features: Leveraging domain knowledge can lead to valuable features. For example, in credit scoring, you can create features such as credit utilization ratio, payment-to-income ratio, or the length of credit history. Understanding the specific characteristics and drivers of the banking and finance domain can help generate meaningful features.
- Text-Based Features: If you have textual data, extracting relevant information from text can be valuable. Techniques like natural language processing (NLP) can be used to create features such as sentiment scores, keyword presence, or document similarity. Textual features can be used for sentiment analysis, customer feedback analysis, or fraud detection.
- External Data Integration: Integrating external data sources can enrich the feature set. For example, incorporating economic indicators, market trends, or news sentiment can help capture the impact of external factors on financial outcomes. External data can provide contextual information and improve the predictive power of models.

Note that the choice of feature engineering techniques should be driven by a combination of domain knowledge, data exploration, and experimentation. It's also crucial to consider the interpretability, computational complexity, and robustness of the engineered features while balancing the trade-off between feature richness and overfitting.

14.7 Evaluation and Performance Metrics for Banking and Finance

14.7.1 Accuracy, Precision, and Recall for Banking and Finance

Accuracy, precision, and recall are important evaluation metrics used in banking and finance, particularly for tasks like fraud detection, credit risk assessment, and default prediction (Shoumo et al. 2019). Here's how these metrics are relevant in the banking and finance context:

- Accuracy: Accuracy measures the overall correctness of a model's predictions by comparing the number of correct predictions to the total number of predictions. In banking and finance, accuracy can be used to assess the overall performance of a model in correctly identifying fraud cases, predicting defaults, or classifying customers into risk categories. However, accuracy alone may not provide a complete picture, especially when dealing with imbalanced datasets where the number of positive cases (frauds, defaults) is significantly lower than the negatives. In such cases, accuracy can be misleading, and other metrics like precision and recall should be considered.
- Precision: Precision is the proportion of true positive predictions out of the total predicted positive instances. It measures the accuracy of positive predictions made by a model. In banking and finance, precision is particularly relevant in fraud detection, where correctly identifying fraudulent transactions is crucial to prevent financial losses. A high precision indicates a low false positive rate, meaning that when the model flags a transaction as fraudulent, it is likely to be correct. High precision is desirable to minimize false alarms and ensure efficient fraud detection.
- Recall (Sensitivity): Recall, also known as sensitivity or true positive rate, measures the proportion of true positive predictions out of the total actual positive instances. In banking and finance, recall is valuable in identifying all positive cases accurately, especially when the cost of missing a positive instance is high. For instance, in credit risk assessment, recall is important in correctly identifying customers who are likely to default or have a high risk of delinquency. High recall indicates a low false negative rate, meaning that the model can successfully identify the majority of positive instances.

Note that precision and recall often have an inverse relationship. Increasing precision typically results in lower recall and vice versa. This trade-off needs to be carefully considered based on the specific requirements and priorities of the banking and finance application. Additionally, in banking and finance, other evaluation metrics such as F1 score, which combines precision and recall, and area under the receiver operating characteristic (ROC) curve, which measures the trade-off between true positive rate and false positive rate, are commonly used to assess model performance. These metrics provide a more comprehensive evaluation of the model's effectiveness in handling imbalanced datasets and capturing both true positives and true negatives.

14.7.2 F1 Score for Banking and Finance

The F1 score is a commonly used evaluation metric in banking and finance, particularly for tasks such as fraud detection, credit risk assessment, and default prediction. The F1 score combines precision and recall into a single measure, providing a balanced assessment of a model's performance. In the context of banking and finance, the F1 score is especially useful when dealing with imbalanced datasets, where the number of positive cases (frauds, defaults) is significantly lower than the negatives (Krauss et al. 2015). This is because accuracy alone can be misleading in such scenarios, as a high accuracy may be achieved by simply predicting the majority class (non-fraudulent transactions, non-default cases) most of the time. The F1 score considers both precision and recall, providing a more comprehensive evaluation of the model's ability to correctly identify positive instances while minimizing false positives and false negatives.

The F1 score is calculated as the harmonic mean of precision and recall:

$$\text{F1 score} = 2^{*}\left(\text{precision}^{*}\ \text{recall}\right)/\left(\text{precision} + \text{recall}\right)$$

A high F1 score indicates a good balance between precision and recall, meaning the model performs well in correctly identifying positive instances while keeping false positives and false negatives low. The F1 score ranges from 0 to 1, with 1 representing the best performance.

When using the F1 score, it's important to note that the choice between precision, recall, and the F1 score depends on the specific requirements and priorities of the banking and finance task. For example, if the cost of false positives (e.g., flagging a non-fraudulent transaction as fraud) is higher than false negatives (missing a fraudulent transaction), a higher precision may be prioritized. On the other hand, if the cost of false negatives is more significant, a higher recall may be desired. The F1 score provides a balanced assessment, but it's crucial to consider the specific context and consequences of both types of errors in the banking and finance domain.

14.7.3 Receiver Operating Characteristic (ROC) Curve for Banking and Finance

The Receiver Operating Characteristic (ROC) curve is a commonly used evaluation tool in banking and finance, particularly for tasks such as fraud detection, credit risk assessment, and default prediction (Yıldırım and Sanyal 2022). The ROC curve provides a graphical representation of the trade-off between the true positive rate (TPR) and the false positive rate (FPR) at various classification thresholds.

- In the context of banking and finance, the ROC curve helps assess the performance of a binary classifier model by examining its ability to distinguish between positive and negative instances. The ROC curve is particularly useful when dealing with imbalanced datasets, where the number of positive cases (frauds, defaults) is significantly lower than the negatives.

- To construct an ROC curve, the model's predictions are sorted by their predicted probabilities or scores, and different classification thresholds are applied. At each threshold, the TPR (also known as sensitivity or recall) and FPR are calculated. The TPR represents the proportion of actual positive instances correctly classified as positive, while the FPR represents the proportion of actual negative instances incorrectly classified as positive.
- The ROC curve is created by plotting the TPR against the FPR at various classification thresholds. The resulting curve typically starts from the bottom-left corner (TPR = 0, FPR = 0) and moves towards the top-right corner (TPR = 1, FPR = 1). A random classifier would produce a diagonal line from the bottom-left to the top-right, indicating no predictive power.
- The performance of a model can be assessed based on the shape and position of the ROC curve. The closer the curve is to the top-left corner, the better the model's performance. The area under the ROC curve (AUC-ROC) is a commonly used metric to quantify the overall performance of the model. AUC-ROC ranges from 0.5 (random classifier) to 1.0 (perfect classifier). A higher AUC-ROC indicates better discrimination power, with values above 0.7 generally considered good and values above 0.8 considered excellent.

In banking and finance, the ROC curve and AUC-ROC provide insights into the model's ability to balance true positives and true negatives while minimizing false positives and false negatives. They help in evaluating and comparing different models, selecting an optimal classification threshold, and understanding the trade-off between the detection of positive instances and the occurrence of false alarms. Note that the choice of the optimal threshold depends on the specific requirements and priorities of the banking and finance task. It should consider the consequences of false positives and false negatives based on the domain and business context.

14.7.4 Mean Absolute Error (MAE) and Root Mean Squared Error (RMSE) for Banking and Finance

Mean Absolute Error (MAE) and Root Mean Squared Error (RMSE) are commonly used evaluation metrics in banking and finance, particularly in tasks related to forecasting, risk modeling, and investment analysis. These metrics are used to measure the accuracy and precision of predictive models by quantifying the differences between predicted and actual values.

- Mean Absolute Error (MAE): MAE represents the average absolute difference between predicted and actual values. It measures the average magnitude of errors, irrespective of their direction. In banking and finance, MAE can be used to assess the accuracy of models in predicting financial metrics such as asset prices, loan defaults, or customer churn rates. A lower MAE indicates better accuracy, with zero representing a perfect prediction.

$$MAE = (1/n)^* \Sigma \mid y - \hat{y} \mid$$

Where:
n is the number of observations.
y represents the actual values.
ŷ represents the predicted values.

- Root Mean Squared Error (RMSE): RMSE is similar to MAE, but it penalizes larger errors more heavily due to the squaring of differences. RMSE represents the square root of the average of squared differences between predicted and actual values. In banking and finance, RMSE is widely used in risk modeling, portfolio optimization, and financial forecasting tasks. Like MAE, a lower RMSE indicates better accuracy, with zero representing a perfect prediction.

$$RMSE = sqrt\left((1/n)^* \Sigma\left(y - \hat{y}\right)^\wedge 2\right)$$

Where:

n is the number of observations.
y represents the actual values.
ŷ represents the predicted values.

Both MAE and RMSE provide a measure of the average prediction error, with RMSE being more sensitive to outliers and larger errors due to the squaring operation. These metrics are often used together to assess the performance of predictive models in banking and finance. They provide insights into the magnitude of errors, help compare different models, and assist in model selection and improvement.

Note that the choice between MAE and RMSE depends on the specific context and requirements of the banking and finance task. While MAE is more straightforward to interpret as it represents the average absolute error, RMSE can be more useful when larger errors are of greater concern or when the distribution of errors is not symmetric.

14.8 Ethical Considerations in ML and AI for Banking and Finance

14.8.1 Bias and Fairness for Banking and Finance

Bias and fairness are important considerations in banking and finance to ensure equitable treatment, avoid discrimination, and promote ethical practices (Bajracharya et al. 2022). Here's how these concepts are relevant in the context of banking and finance:

- Bias: Bias refers to systematic errors or inaccuracies in decision-making processes that lead to unequal or unfair treatment of individuals or groups. In banking and finance, bias can manifest in various forms:
  - Data Bias: Bias can be introduced during data collection, processing, or analysis, resulting in unequal representation or treatment of certain groups. For example, if historical loan data predominantly includes certain demographics and excludes others, it can lead to biased lending decisions.
  - Algorithmic Bias: Bias can also arise from the use of algorithms in decision-making processes. Machine learning models trained on biased data or with biased features can perpetuate and amplify existing biases. For instance, if a credit scoring model is trained on historical data that reflects discriminatory lending practices, it can lead to unfair loan approval or denial decisions.
  - Behavioral Bias: Bias can be present in human decision-making as well. Biases such as confirmation bias, anchoring bias, or availability bias can influence judgments and decisions made by individuals in the banking and finance industry, potentially leading to unfair outcomes.

    Addressing bias requires careful examination and mitigation at various stages of the decision-making process. This includes data collection, data preprocessing, model development, and ongoing monitoring to identify and rectify potential biases.

- Fairness: Fairness aims to ensure that individuals or groups are treated equitably and without discrimination. In banking and finance, fairness is crucial to prevent discriminatory practices and promote equal access to financial services. Fairness can be approached from different perspectives:
  - Procedural Fairness: Procedural fairness focuses on ensuring fairness in the decision-making process itself. This involves transparent and clear procedures, consistent application of rules and criteria, and opportunities for individuals to provide input or appeal decisions.
  - Outcome Fairness: Outcome fairness concerns the fairness of the outcomes or results of decisions. It aims to ensure that the decisions do not lead to disparate impacts on protected groups or result in systemic discrimination.
  - Group Fairness: Group fairness focuses on avoiding discrimination or unfair treatment of specific demographic or protected groups, such as gender, race, ethnicity, or age. It aims to prevent biased outcomes or disparities that disproportionately affect certain groups.

Promoting fairness in banking and finance requires a combination of careful data collection, unbiased model development, regular monitoring, and an understanding of legal and ethical obligations. It also involves promoting diversity and inclusion within financial institutions and fostering a culture of fairness and ethical decision-making. To address bias and promote fairness, financial institutions can employ techniques such as dataset auditing, algorithmic auditing, fairness-aware machine learning, and regular evaluation of decisions for potential disparities. It's essential to involve multidisciplinary teams, including domain experts, data scientists, ethicists, and legal professionals, to ensure a comprehensive and balanced approach to bias and fairness in banking and finance.

14.8.2 Privacy and Data Protection for Banking and Finance

Privacy and data protection are of utmost importance in the banking and finance industry to safeguard sensitive information, maintain customer trust, and comply with regulatory requirements (Lai et al. 2019). Here's an overview of privacy and data protection considerations in banking and finance:

- Personal Data Protection: Banks and financial institutions collect and process vast amounts of personal data from their customers, including names, addresses, financial transactions, credit histories, and more. It is crucial to handle this data with care and ensure compliance with relevant data protection regulations, such as the General Data Protection Regulation (GDPR) in the European Union or the Gramm-Leach-Bliley Act (GLBA) in the United States. This involves obtaining proper consent for data collection, implementing robust security measures, and limiting data access to authorized personnel.
- Data Security and Encryption: Banks and financial institutions are responsible for safeguarding customer data against unauthorized access, theft, or misuse. This involves implementing strong data security measures, including encryption, firewalls, secure networks, and access controls. Regular security audits and vulnerability assessments should be conducted to identify and address potential weaknesses in data protection systems.
- Confidentiality and Non-Disclosure: Financial institutions have a legal and ethical obligation to maintain the confidentiality of customer information. This includes implementing strict policies and procedures to control access to sensitive data and ensuring that employees are trained on data handling practices and the importance of confidentiality. Non-disclosure agreements (NDAs) may be used when sharing information with third parties or business partners.
- Third-Party Data Sharing: Financial institutions often engage with third-party service providers, such as payment processors, credit bureaus, or cloud providers. When sharing customer data with these entities, proper due diligence should be performed to ensure their compliance with data protection regulations and adherence to industry best practices. Contracts or agreements should clearly outline the responsibilities and obligations of the parties involved.
- Data Retention and Destruction: Banks and financial institutions should establish policies regarding data retention and destruction to ensure that personal data is not kept longer than necessary. This includes defining specific retention periods for different types of data and implementing secure processes for data disposal, such as shredding physical documents or securely deleting electronic data.
- Customer Consent and Transparency: Transparency is essential in maintaining customer trust. Financial institutions should provide clear and easily understand-

able privacy policies that outline the types of data collected, the purposes for which it is used, and any third parties with whom it may be shared. Customers should be given the option to provide informed consent for data processing, and they should have the right to access, correct, or delete their personal information as required by applicable regulations.
- Regulatory Compliance: The banking and finance industry is subject to numerous regulatory requirements, such as the Payment Card Industry Data Security Standard (PCI DSS), Basel III, or the Financial Industry Regulatory Authority (FINRA) rules. Financial institutions must stay up-to-date with the evolving regulatory landscape and ensure compliance with all relevant data protection, privacy, and security regulations specific to their jurisdiction.

By prioritizing privacy and data protection, banks and financial institutions can establish a foundation of trust with their customers, mitigate the risk of data breaches, and demonstrate their commitment to ethical and responsible data handling practices.

14.8.3 Regulatory Compliance for Banking and Finance

Regulatory compliance is a critical aspect of the banking and finance industry to ensure adherence to laws, regulations, and industry standards. Compliance helps maintain the integrity of the financial system, protect consumers, and mitigate risks (Engdahl 2014). Here are some key areas of regulatory compliance in banking and finance:

- Anti-Money Laundering (AML) and Know Your Customer (KYC): AML regulations require financial institutions to implement robust policies, procedures, and controls to prevent money laundering and the financing of terrorism. KYC procedures involve verifying the identity of customers and conducting due diligence to understand their activities and assess potential risks.
- Consumer Protection: Consumer protection regulations aim to safeguard the interests of customers and ensure fair and transparent practices by financial institutions. These regulations govern areas such as disclosure requirements, fair lending practices, handling of customer complaints, and prevention of abusive or unfair practices.
- Data Protection and Privacy: Data protection regulations, such as the GDPR in the European Union, impose obligations on financial institutions regarding the collection, storage, processing, and transfer of personal data. Institutions must ensure the security and confidentiality of customer data and obtain appropriate consent for data processing activities.
- Capital Adequacy and Risk Management: Regulations, such as the Basel III framework, establish minimum capital requirements for banks to ensure their financial stability and ability to absorb losses. Financial institutions are also required to implement robust risk management frameworks to identify, measure,

monitor, and control various types of risks, including credit risk, market risk, operational risk, and liquidity risk.

- Market Integrity and Conduct: Regulations related to market integrity and conduct govern fair and orderly trading, prevention of market abuse, insider trading, market manipulation, and conflicts of interest. Financial institutions must adhere to these regulations to ensure the integrity and transparency of financial markets.
- Financial Reporting and Accounting Standards: Financial institutions are required to comply with financial reporting standards, such as the International Financial Reporting Standards (IFRS) or Generally Accepted Accounting Principles (GAAP), to ensure accurate and transparent reporting of their financial positions and performance.
- Cybersecurity and Information Security: Financial institutions must implement robust cybersecurity measures to protect sensitive customer data and critical infrastructure from cyber threats. Regulatory bodies often provide guidelines and requirements for information security practices, incident response, and cybersecurity risk management.
- Risk-Based Compliance Programs: Financial institutions are expected to develop risk-based compliance programs that assess and manage compliance risks effectively. These programs involve conducting regular risk assessments, establishing internal controls, monitoring compliance activities, and providing training to employees on regulatory requirements.

It's important for financial institutions to have a dedicated compliance function or department responsible for overseeing regulatory compliance. Compliance officers ensure that the institution understands and adheres to applicable regulations, conducts internal audits, and liaises with regulatory authorities. Non-compliance with regulatory requirements can result in significant penalties, reputational damage, and legal consequences. Therefore, staying up-to-date with regulatory changes, engaging in continuous training and education, and maintaining a strong compliance culture are essential for financial institutions to meet their obligations and operate within the boundaries of the law.

14.9 Challenges and Future Trends for Banking and Finance

14.9.1 Explainability and Interpretability for Banking and Finance

Explainability and interpretability are crucial aspects in the banking and finance industry to understand and trust the decisions made by machine learning models, algorithms, and complex systems. These concepts help provide insights into the reasoning behind decisions, facilitate regulatory compliance, support risk management, and enable effective communication with stakeholders (Demajo et al. 2020). Here's an overview of explainability and interpretability in banking and finance:

- Explainability: Explainability refers to the ability to provide understandable and clear explanations for the outcomes or decisions made by a model or system. In banking and finance, explainability is important for various reasons:
  - Regulatory Compliance: Financial institutions must comply with regulations that require explainability, such as the right to explanation under the General Data Protection Regulation (GDPR) or Fair Lending laws. Explainability helps ensure transparency, fairness, and accountability in decision-making processes.
  - Risk Management: In risk assessment and management, explainability is essential to understand the factors influencing risk scores, credit decisions, or fraud detection. It enables the identification of potential biases, model limitations, and vulnerabilities in risk models, helping institutions make informed decisions and manage risks effectively.
  - Customer Trust and Compliance: Providing explanations for decisions, such as loan approvals or investment recommendations, fosters customer trust and confidence in the fairness and integrity of the financial institution. Clear and understandable explanations help customers understand why certain decisions were made, reducing confusion and enhancing transparency.
  - Model Validation and Auditability: Explainability facilitates model validation and audits by regulatory bodies or internal governance teams. It allows the assessment of model performance, fairness, and compliance with regulations, enabling institutions to identify and address potential issues.
- Interpretability: Interpretability focuses on understanding how a model or algorithm arrives at its predictions or decisions. It involves uncovering the underlying factors, variables, or features that contribute to the model's output. Interpretability is relevant in banking and finance for several reasons:
  - Risk Assessment: Interpretable models allow risk analysts to understand and validate the factors contributing to risk scores or credit ratings. This helps in risk assessment, portfolio management, and identifying the key drivers of risk exposures.
  - Compliance and Auditability: Interpretability aids in explaining model decisions to auditors, regulators, and internal stakeholders. It helps institutions demonstrate that models and algorithms comply with regulations, are free from bias, and align with organizational policies.
  - Fraud Detection and Investigation: Interpretable models assist in fraud detection by providing insights into the features or patterns indicative of fraudulent activities. This enables investigators to understand the reasoning behind model alerts and supports more effective fraud mitigation strategies.
  - Investment Analysis: In investment management, interpretability helps portfolio managers and analysts understand the factors contributing to investment recommendations, asset valuations, or risk exposures. It supports investment decision-making and the evaluation of model-generated insights.

Promoting explainability and interpretability often involves adopting transparent and interpretable machine learning techniques, such as rule-based models, decision trees, or linear regression. Additionally, techniques such as feature importance analysis, sensitivity analysis, or surrogate models can provide further insights into model behavior.

Note that the level of explainability or interpretability required may vary depending on the context, stakeholders involved, and specific regulatory requirements. Financial institutions must strike a balance between model complexity, accuracy, and the need for transparency and comprehensibility to meet the objectives of explainability and interpretability in banking and finance.

14.9.2 Integration of ML and AI with Traditional Approaches for Banking and Finance

The integration of machine learning (ML) and artificial intelligence (AI) with traditional approaches in banking and finance has the potential to revolutionize the industry by improving efficiency, accuracy, risk management, and customer experience (Aziz et al. 2022). Here are some key areas where ML and AI can be integrated with traditional approaches:

- Risk Assessment and Management: ML and AI techniques can enhance traditional risk assessment models by analyzing large volumes of structured and unstructured data to identify patterns, detect anomalies, and predict risk. They can be used to build credit scoring models, fraud detection systems, and models for assessing market and operational risks. By combining traditional risk management frameworks with ML and AI, financial institutions can achieve more accurate risk assessments and make better-informed decisions.
- Customer Relationship Management: ML and AI algorithms can analyze customer data, transaction history, and behavior patterns to provide personalized recommendations, targeted marketing campaigns, and customer segmentation. By integrating ML and AI with traditional customer relationship management (CRM) systems, banks and financial institutions can improve customer satisfaction, increase cross-selling and upselling opportunities, and deliver tailored products and services.
- Trading and Investment Strategies: ML and AI techniques can be used to develop algorithmic trading systems that analyze market data, news sentiment, and historical trends to make automated trading decisions. By combining ML and AI with traditional investment strategies, financial institutions can enhance portfolio management, optimize trading execution, and improve investment decision-making.
- Fraud Detection and Prevention: ML and AI algorithms can analyze large datasets to detect fraudulent activities and patterns in real-time, enabling faster and more accurate fraud detection and prevention. By integrating ML and AI with

traditional fraud detection approaches, financial institutions can strengthen their security measures and mitigate potential financial losses.
- Customer Support and Chatbots: ML and AI-powered chatbots and virtual assistants can enhance traditional customer support systems by providing personalized assistance, answering customer queries, and automating routine tasks. These systems can improve response times, reduce operational costs, and enhance the overall customer experience.
- Compliance and Regulatory Reporting: ML and AI techniques can automate compliance processes, monitor transactions for suspicious activities, and generate regulatory reports. By integrating ML and AI with traditional compliance and reporting approaches, financial institutions can enhance efficiency, reduce errors, and ensure compliance with regulatory requirements.
- Data Analysis and Decision-Making: ML and AI algorithms can analyze large datasets to uncover insights, identify trends, and support data-driven decision-making. By combining ML and AI with traditional data analysis approaches, financial institutions can improve their forecasting capabilities, optimize resource allocation, and enhance strategic decision-making.

14.9.3 Continued Innovation in Financial Technology (FinTech) for Banking and Finance

Continued innovation in financial technology (FinTech) is transforming the banking and finance industry, leading to improved services, increased efficiency, enhanced customer experience, and expanded access to financial products (Romanova et al. 2018). Here are some areas where we can expect continued innovation in FinTech for banking and finance:

- Digital Payments and Mobile Banking: The rise of digital payments, mobile wallets, and contactless transactions will continue to shape the future of banking and finance. FinTech companies are developing secure, convenient, and user-friendly payment solutions that enable consumers and businesses to make transactions anytime, anywhere. This includes advancements in mobile banking apps, peer-to-peer (P2P) payments, QR code payments, and tokenization technologies.
- Open Banking and APIs: Open Banking initiatives are driving innovation by enabling secure data sharing and collaboration between banks and third-party providers. Application Programming Interfaces (APIs) play a crucial role in facilitating the integration of financial services, allowing customers to access and manage their financial data across multiple platforms and providers. This fosters competition, encourages the development of new services, and promotes personalized financial solutions.
- Artificial Intelligence (AI) and Machine Learning (ML): AI and ML are being leveraged in various areas of banking and finance. Chatbots and virtual assistants are enhancing customer support and engagement. AI-powered algorithms are

improving fraud detection, credit scoring, and risk management. ML models are being utilized for investment analysis and algorithmic trading. Continued innovation in AI and ML will lead to more advanced predictive analytics, automation of manual tasks, and personalized financial recommendations.
- Blockchain and Distributed Ledger Technology (DLT): Blockchain technology offers secure and transparent transactions, making it valuable for applications such as cross-border payments, smart contracts, supply chain finance, and digital identity verification. FinTech innovations in blockchain and DLT are exploring scalability, interoperability, and regulatory frameworks to drive mainstream adoption and revolutionize traditional financial processes.
- Robo-advisors and Wealth Management: Robo-advisors are automated investment platforms that provide personalized investment advice based on algorithms and customer preferences. They are gaining popularity due to their lower costs, accessibility, and convenience. FinTech companies are continually enhancing robo-advisory services, incorporating AI and ML for portfolio rebalancing, tax optimization, and goal-based financial planning.
- RegTech and Compliance Solutions: Regulatory technology (RegTech) aims to streamline compliance processes and improve regulatory reporting. FinTech companies are developing solutions that leverage automation, data analytics, and AI to help financial institutions comply with complex regulatory requirements. This includes know-your-customer (KYC) solutions, anti-money laundering (AML) software, and risk management tools.
- Financial Inclusion and Access to Capital: FinTech is playing a vital role in promoting financial inclusion by providing access to financial services for underserved populations. Digital lending platforms, microfinance solutions, and crowdfunding platforms are expanding access to capital for small businesses and individuals who may have difficulty obtaining traditional loans. Innovations in alternative credit scoring, such as leveraging non-traditional data sources, are also improving credit access for unbanked or underbanked individuals.

14.10 Conclusions

Today Machine learning (ML) and artificial intelligence (AI) role in the banking and finance industry has brought significant advancements and opportunities for transformation. ML and AI techniques offer the potential to improve accuracy, efficiency, risk management, and customer experience across various aspects of banking and finance. By leveraging ML and AI, financial institutions can enhance risk assessment and management processes, detect and prevent fraud, personalize customer interactions, optimize investment strategies, automate compliance procedures, and improve operational efficiency. These technologies enable financial institutions to analyze vast amounts of data, extract valuable insights, and make data-driven decisions in real-time. However, the adoption of ML and AI in banking and finance also comes with challenges and considerations. Ensuring data quality, addressing bias and fairness, maintaining privacy and data protection, ensuring model explainability and interpretability, and complying with regulatory requirements are crucial factors that need to be carefully addressed. To successfully integrate ML and AI, financial institutions need to foster a culture of innovation, collaborate with domain experts and data scientists, invest in infrastructure and technology, and prioritize ongoing monitoring, validation, and governance of ML and AI models. It is important to strike a balance between technological advancements and regulatory compliance, ensuring that the benefits of ML and AI are harnessed while maintaining trust and accountability. In summary, the introduction of ML and AI in banking and finance has the potential to revolutionize the industry, driving operational efficiency, risk mitigation, and enhanced customer experiences. Continued exploration, innovation, and responsible implementation of ML and AI technologies will shape the future of banking and finance, leading to more personalized, efficient, and inclusive financial services.

References

Ahmed S, Alshater MM, El Ammari A, Hammami H (2022) Artificial intelligence and machine learning in finance: a bibliometric review. Res Int Bus Financ 61:101646

Al Ahmed A, Ayub SA, Kurniawan IMGA, Anantadjaya SPD, Krishnan C (2022) Business boosting through sentiment analysis using Artificial Intelligence approach. Int J Syst Assur Eng Manag 13(Suppl 1):699–709

Ashta A, Herrmann H (2021) Artificial intelligence and fintech: An overview of opportunities and risks for banking, investments, and microfinance. Strateg Chang 30(3):211–222

Aziz S, Dowling M, Hammami H, Piepenbrink A (2022) Machine learning in finance: a topic modeling approach. Eur Financ Manag 28(3):744–770

Bajracharya A, Khakurel U, Harvey B, Rawat DB (2022) Recent advances in algorithmic biases and fairness in financial services: a survey. In Proceedings of the future technologies conference (FTC) 2022, Volume 1. Springer International Publishing, Cham, pp 809–822

Battiston S, Dafermos Y, Monasterolo I (2021) Climate risks and financial stability. J Financ Stab 54:100867

Belanche D, Casaló LV, Flavián C (2019) Artificial Intelligence in FinTech: understanding robo-advisors adoption among customers. Ind Manag Data Syst

Buchanan, Bonnie. "Artificial intelligence in finance." (2019)

Cunningham P, Cord M, Delany SJ (2008) Supervised learning. In Machine learning techniques for multimedia: case studies on organization and retrieval, pp 21–49

Demajo LM, Vella V, Dingli A (2020) Explainable ai for interpretable credit scoring. arXiv preprint arXiv:2012.03749

Engdahl O (2014) Ensuring regulatory compliance in banking and finance through effective controls: the principle of duality in the segregation of duties. Regul Gov 8(3):332–349

Fernández A (2019) Artificial intelligence in financial services. Banco de Espana Article 3:19

Ghahramani Z (2003) Unsupervised learning. In Summer school on machine learning. Springer, Berlin, Heidelberg, pp 72–112

Gogas P, Papadimitriou T (2021) Machine learning in economics and finance. Comput Econ 57:1–4

Goodell JW, Kumar S, Lim WM, Pattnaik D (2021) Artificial intelligence and machine learning in finance: Identifying foundations, themes, and research clusters from bibliometric analysis. J Behav Exp Financ 32:100577

Hansen KB (2020) The virtue of simplicity: On machine learning models in algorithmic trading. Big Data Soc 7(1):2053951720926558

Heaton JB, Polson NG, Witte JH (2016) Deep learning in finance. arXiv preprint arXiv:1602.06561

Hu Y-J, Lin S-J (2019) Deep reinforcement learning for optimizing finance portfolio management. In: 2019 Amity international conference on artificial intelligence (AICAI). IEEE, pp 14–20

Huang J, Chai J, Cho S (2020) Deep learning in finance and banking: A literature review and classification. Front Bus Res China 14(1):1–24

Kaur DN, Sahdev SL, Sharma DM, Siddiqui L (2020) Banking 4.0:'the influence of artificial intelligence on the banking industry & how ai is changing the face of modern day banks'. Int J Manag 11(6)

Khadivizand S, Beheshti A, Sobhanmanesh F, Sheng QZ, Istanbouli E, Wood S, Pezaro D (2020) Towards intelligent feature engineering for risk-based customer segmentation in banking. In Proceedings of the 18th international conference on advances in mobile computing and multimedia, pp 74–83

Khan S, Rabbani MR (2020) Chatbot as Islamic finance expert (CaIFE) When finance meets artificial intelligence. In Proceedings of the 2020 4th international symposium on computer science and intelligent control, pp 1–5

Königstorfer F, Thalmann S (2020) Applications of artificial intelligence in commercial banks–a research agenda for behavioral finance. J Behav Exp Financ 27:100352

Krauss C, Krüger T, Beerstecher D (2015) The Piotroski F-score: a fundamental value strategy revisited from an investor's perspective. No. 13/2015. IWQW Discussion Papers

Lai S-T, F-Y Leu, Lin J-W (2019) A banking chatbot security control procedure for protecting user data security and privacy. In Advances on broadband and wireless computing, communication and applications: proceedings of the 13th international conference on broadband and wireless computing, communication and applications (BWCCA-2018). Springer International Publishing, pp 561–571

Lakshmi SVSS, Kavilla SD (2018) Machine learning for credit card fraud detection system. Int J Appl Eng Res 13(24):16819–16824

LeCun Y, Bengio Y, Hinton G (2015) Deep learning. Nature 521(7553):436–444

Mehrotra A (2019) Artificial intelligence in financial services–need to blend automation with human touch. In: 2019 International conference on automation, computational and technology management (ICACTM). IEEE, pp 342–347

Mhlanga D (2020) Industry 4.0 in finance: the impact of artificial intelligence (ai) on digital financial inclusion. Int J Financ Stud 8(3):45

Mhlanga D (2021) Financial inclusion in emerging economies: the application of machine learning and artificial intelligence in credit risk assessment. Int J Financ Stud 9(3):39

Mogaji E, Nguyen NP (2022) Managers' understanding of artificial intelligence in relation to marketing financial services: insights from a cross-country study. Int J Bank Mark 40(6):1272–1298

Ng J, Shah S (2020) Hands-on artificial intelligence for banking: a practical guide to building intelligent financial applications using machine learning techniques. Packt Publishing Ltd

Paltrinieri N, Comfort L, Reniers G (2019) Learning about risk: machine learning for risk assessment. Saf Sci 118:475–486

Pau LF, Gianotti C, Pau LF, Gianotti C (1990) Applications of artificial intelligence in banking, financial services and economics. Springer, Berlin, Heidelberg

Phan X-H, Nguyen L-M, Horiguchi S (2008) Learning to classify short and sparse text & web with hidden topics from large-scale data collections. In Proceedings of the 17th international conference on World Wide Web, pp 91–100

Ranjan S, Gupta DR, Gupta DA (2020) Artificial intelligence in financial acumen: Challenges and opportunities. Cosmos J Eng Technol 10(1):1–5

Romanova I, Grima S, Spiteri J, Kudinska M (2018) The payment services Directive II and competitiveness: the perspective of European fintech companies. Eur Res Stud 21(2):3–22

Sadatrasoul SM, Gholamian M, Siami M, Hajimohammadi Z (2013) Credit scoring in banks and financial institutions via data mining techniques: A literature review. J AI Data Mining 1(2):119–129

Shoumo SZ, Hasan MI, Dhruba M, Hossain S, Ghani NH, Arif H, Islam S (2019) Application of machine learning in credit risk assessment: a prelude to smart banking. In: TENCON 2019–2019 IEEE region 10 conference (TENCON). IEEE, pp 2023–2028

Suhel SF, Shukla VK, Vyas S, Mishra VP (2020) Conversation to automation in banking through chatbot using artificial machine intelligence language. In 2020 8th International conference on reliability, Infocom technologies and optimization (Trends and Future Directions) (ICRITO), pp. 611–618. IEEE

Ukpong EG, Udoh II, Essien IT (2019) Artificial intelligence: opportunities, issues and applications in banking, accounting, and auditing in Nigeria. Asian J Econ Bus Account 10(1):1–6

Vasista K (2019) A review on the various options available for investment. Int J Creat Res Thoughts 7(2)

Wasserbacher H, Spindler M (2022) Machine learning for financial forecasting, planning and analysis: recent developments and pitfalls. Digit Financ 4(1):63–88

Wiering MA, Van Otterlo M (2012) Reinforcement learning. Adapt Learn Optim 12(3):729

Xie M (2019) Development of artificial intelligence and effects on financial system. J Phys Conf Ser 1187(3):032084. IOP Publishing

Yıldırım Y, Sanyal A (2022) Evaluating the effectiveness of early warning indicators: an application of receiver operating characteristic curve approach to panel data. Sci Ann Econ Bus 69(4):557–597

Yu TR, Song X (2021) Big data and artificial intelligence in the banking industry. In Handbook of financial econometrics, mathematics, statistics, and machine learning, pp 4025–4041

Chapter 15
Applying AI & TOPSIS-MCDM Tool in Evaluating Top Five Private Indian Bank Performances

N. Mohan and Mohammad Irfan

15.1 Introduction

The contribution of the banking industry to both the domestic and international economies is limitless. Banks serve as the primary sources of finance for infrastructure, innovation, and development because they turn consumer deposits and savings into company investments. Additionally, banks and other financial institutions help to create jobs and improve society as a whole. Therefore, a healthy banking industry is crucial to maintaining the stability of the financial markets. A robust and effective banking system promotes economic growth, acts as an early warning system for financial crises, and aids in mitigating their effects.

Banks must efficiently employ their resources in the highly competitive business world of today. Due to these circumstances, it is now necessary for bank management, investors, clients, and policymakers to compare how banks are operating to one another. Accurate evaluation of bank performance is crucial for comparing good and negative practises and observing general improvement in the banking sector. Improvement of management performance, information on comparisons between the best and worst practises for stakeholders, and research into the effects of certain macroeconomic changes on bank performance are just a few of the reasons why measuring a bank's financial performance is crucial (Berger and Humphrey 1997).

N. Mohan (✉)
CMR Institute of Technology, Bangalore, Karnataka, India
e-mail: mohan.n@cmrit.ac.in

M. Irfan
NSB Academy, Business School, Bangalore, Karnataka, India

M. Irfan et al. (eds.), *Applications of Block Chain technology and Artificial Intelligence*, Financial Mathematics and Fintech,
https://doi.org/10.1007/978-3-031-47324-1_15

Researchers are very interested in determining the efficiency and efficacy of banks for the aforementioned reasons. Although it is common knowledge how vital it is to assess bank performance, there are many different approaches and performance metrics.

Financial ratios offer helpful data for assessing the effectiveness and performance of banks. Return on equity (ROE) and return on assets (ROA) are the fundamental financial ratios associated with efficiency (Chuang-Min et al. 2015). Although assets quality, capital strength, profitability, and liquidity are typically among the variables, the number of variables considered in study can vary greatly throughout the literature (Fethi and Pasiouras 2009). The approach and variable selection are intimately related to the goal of the model.

Since India's independence, its economic structure has seen substantial changes in addition to quantitative development.

1. Changing Domestic Product Sectoral Distribution Change in the relative importance (share) of various economic sectors is referred to as a change in the composition of the domestic product or a change in national income by industry of origin. The essential, optional, and tertiary areas are the overall divisions of an economy. The auxiliary area's Gross domestic product commitment moved from 13.6% in 1950–51 to 23.2% in 1990–91 and afterward to 24.7% in 2007–08. The tertiary area's commitment to Gross domestic product moved from 29.9% in 1950–51 to 55.6% in 2007–08, and afterward to more than 7% in 2009–10.
2. Development of Fundamental Capital Merchandise Ventures: When the country acquired autonomy, essential and capital merchandise businesses represented around one-fourth of all modern creation. The capital merchandise businesses were given huge need under the second arrangement since it was accepted that their advancement was fundamental for the economy's general development. Accordingly, various fundamental businesses that make capital merchandise and helpful unrefined components have been laid out, reinforcing the country's modern structure.
3. Increase in Social Overhead Capital: In general, social overhead capital consists of transportation infrastructure, irrigation systems, energy generation, an organised educational system, and medical facilities. Their expansion produces suitable conditions for development as well as for improved human existence.
4. Progress in the Banking and Financial Sector: Since India's independence, the country's banking and financial system has seen a number of notable progressive improvements. The prominence of local bankers and moneylenders has decreased as a result of the really spectacular rise of commercial banks and cooperative credit societies.
5. These banks' credit strategies have seen a critical change since being nationalized. Need enterprises including agribusiness, private companies, transportation, and so on now get more noteworthy financing.
6. Right when we look at the improvement of capital items organizations, the advancement of the structure, the reasonability of the public region, etc., we can see that the Indian economy has fundamentally progressed. These qualities are

remembered to have given the country's economy some dynamism throughout the long term, and one can now probably anticipate that it would support development later on.

Through the use of financial criteria that are frequently found in the related literature, this study seeks to offer a model for comparing the financial performance of the top five private banks. For the analysis, representative financial data from each of the five private banks for the years 2019 to 2023 was used.

The general bank execution is evaluated utilizing a two-stage TOPSIS (Method for Request Execution by Closeness to Ideal Arrangement) investigation and stepwise calculations. The presentation of banks is at first inspected utilizing a bunch of monetary boundaries (measures). Following the underlying examination, information is exposed to stepwise calculations to recognize the standards that best portray execution. The best performing banks and the most exceedingly terrible performing banks are then recognized and differentiated utilizing a changed positioning of the banks in view of more grounded execution examination.

By inspecting the synchronous commitments of every part to a more adaptable and sagacious model, this study develops the collection of earlier examination. To the creators' information, this is the initial time a cross breed TOPSIS and stepwise computation technique has been utilized in the Writing.

The assessment is facilitated as follows: A survey of the composing right now in presence on bank suitability and viability can be found in the underlying fragment. There are introduced examples of various techniques. The investigation approach is analyzed in the ensuing region. In the third section, the tables that summarize the aftereffects of the data assessment are gotten a handle on and examined. The audit's revelations and future investigation are covered in the last part.

15.2 Review of Literature

According to a literature review, researchers typically use ratio analysis and the DEA technique to rank the effectiveness of various options. But in this study, we employ MCDM techniques like AHP, TOPSIS, and Grey Relational analysis to rank the various choices and measure efficiency.

We can highlight the Analytic Hierarchy Process (AHP) (Saaty 1980) among the many MCDM methods that have been proposed to assist decision makers in the process of determining weight, evaluating performance, and dealing with complicated problems. AHP is a decision-making tool used to address pressing issues. Typically, it is used to create multilevel hierarchical structures, different criteria, sub-criteria, and alternatives. We employed pairwise comparisons, which are often used to calculate weight (Triantaphyllou and Mann 1995). By integrating AHP and operational competitiveness evaluation methodologies, Ozbek (2015) aims to evaluate the performance of public banks in Turkey from 2005 to 2014.

According to the survey, Vakifbart had the best performance up until 2012, when Ziraat bank took its spot. Galankashi et al.'s (2016) goal is to build an integrated, balanced approach for choosing suppliers in the automobile industry. Author has taken into account four factors, and each supplier has been assessed according to each factor. Supplier A is the best alternative, according to supplier rating. Islam et al.'s (2016) goal is to apply AHP to prioritise the five service quality aspects for the healthcare industry.

The findings indicate that the two most important service quality indicators in the healthcare industry are reliability and assurance. Gurcan et al.'s (2016) goal is to build an integrated, balanced approach for choosing suppliers in the automobile industry. Author has taken into account four factors, and each supplier has been assessed according to each factor. Supplier A is the best alternative, according to supplier rating. Islam et al.'s (2016) goal is to apply AHP to prioritise the five service quality aspects for the healthcare industry.

The findings indicate that the two most important service quality indicators in the healthcare industry are reliability and assurance. Gurcan et al.'s (2016) goal is to model the selection of a third-party logistics provider (3PL). The outcome reveals that Company B (second options) has the highest score of 0.444 and is regarded as having strong infrastructure technology. The key factors (CFs) for applying reverse logistics (RL) practises in the supply chain were identified and analysed using AHP by Luthra et al. (2017). Four aspects are included in the study for the implementation of RL practises: the competitiveness factor (CP), the policy factor (RP), the supply chain component (SC), and the financial factor (FN). The outcome indicates that the competitiveness component is the most crucial one for implementing RL practises. Wang et al. (2017) makes an effort to confirm the use of data from multiple official sources to urban construction and supervision in Tianjin (China) using AHP and new correlation algorithm.

The findings demonstrate that there is a relationship between construction projects and the goods that are approved for them. According to the survey, the majority of construction projects that have a good chance of meeting the pertinent legal standards also have good supervision. With the aid of the AHP approach, Ratna et al. (2018) make an effort to evaluate and rate the lean performance of four distinct production cells based on three criteria: material flow, visual control, and matrices. Overall results indicate that manufacturing cell 4 is the best and that manufacturing cell 3 has room for improvement. The goal of Petrovic and Kankaras (2018) is to identify and assess the nine key factors for choosing the air traffic for aircraft defence.

To evaluate the data and assess the weights employed in the AHP, the researchers used the DEMATEL approach. The outcome demonstrates that while choosing air traffic for aircraft protection, criteria A1 (aerodynamics and mechanics of the combat) has the largest weight value and is ranked first. Popovic et al. (2018) conducted a survey of the literature on multi-criteria decision making. Pamucar and co.

AI in banks today: In the banking sector, artificial intelligence is being used to advance customer relationship management. Customer relationship management, which includes communication with clients, is a key component of the banking sector. With the advent of ATMs, customer preferences for visiting the sites changed for banking. Banks and other financial institutions are increasingly utilising artificial intelligence (AI) technology for a range of objectives, such as enhancing customer service through the use of virtual assistants or credit scoring to accurately assess a borrower's risk.

Application Areas: Human interactions with machines will become more personalised thanks to artificial intelligence. The goal of the robots replacing the front-office workers in the banking industry is to offer the consumer in front a 24–7 uninterrupted, diligent, and unwavering level of knowledge.

Benefits: With the use of artificial intelligence in finance, banks can manage enormous amounts of data at lightning-fast speeds to gain insightful knowledge and better understand the behaviour of their customers. For the sake of restructuring the description.

The following advantages are in use:

1. Enhancing client base through boosting happiness and confidence.
2. Improving service replies, reducing human error.
3. Developing personalised options.
4. Shortening travel times between locations.

The following is a list of some of the banking industry's artificial intelligence application areas:

1. Improving Consumer Participation; 2. Monitoring Wealth; 3. Analysing Data to Strengthen Defence; 4. Strengthening Security; 5. Dealing with Emotions; 6. Making use of a knowledge database; 7. Rigorousness Control; 8. Growing via Front-Office.

Banking is on the territories, just like every other different industry that is concentrating on utilising the revolution to boost revenues. The examples and applications paint a clear picture of what to expect in terms of the advantages of using artificial intelligence in banking. In this study, banks are ranked according to their performance using TOPSIS. It was created by Yoon and Hwang. The decision-making problem is solved using this approach. The project's ideal and non-ideal solutions are located using the normalised matrix, and the project's relative proximity to the ideal solution is then determined.

15.3 TOPSIS-Algorithm

Typically, there are 7 phases that make up the entire TOPSIS process:

1. Make a matrix with M choices and N requirements. An "evaluation matrix" is the common name for this matrix. M will represent the total number of our companies, and N will represent the total number of metrics (ROA, ROE, DR, and CG).
2. Normalise the evaluation matrix: The evaluation matrix is normalised so that each measure, j, for each firm, i, is between 0 and 1, with a greater value indicating a better evaluation matrix.

3. Create a decision matrix that is weighted and normalised. Each criterion needs to be given its own weight in order for them to all add up to (a). The weights can be determined arbitrarily (not advised) or using professional expertise (industry standard). We must normalise each financial statistic after giving it a weight so that they add up to (b). Each normalised metric from step 2 must then be multiplied by the matching normalised weight.
4. Select the best and worst options for each requirement: The highest and lowest values for each financial parameter across all companies are what we're looking for.
5. Determine the best/worst alternative's Euclidean distance from the target alternative: This is a calculation of the geometric separation between each financial metric's value for a particular business i and its best/worst value across all companies.
6. Determine how closely each option resembles the worst option for each. Our TOPSIS scores are the outcomes. We calculate a score for each organisation based on the distances attained in the previous phase.
7. Sort options in descending order by TOPSIS score. A company will receive the highest score and, as a result, be at the top of our list if its measures are the closest to the best. And that is it. Based on predetermined criteria, we came up with a ranked set of possibilities.

MCDM-Tools: Multiple Criteria Decision Making (MCDM), also known as Multiple Criteria Decision Analysis (MCDA), is a branch of the problem of decision making under several criteria and a component of the operation research paradigm. Multi Objective Decision Making (MODM) and Multiple Attribute Decision Making (MADM) are two possible classifications for MCDM.

The capacity of MCDM to compare several solutions based on predetermined criteria allows it to be used as a key tool for assessing complicated situations at the moment. A critical problem can be solved using a variety of MCDM strategies, but different approaches will yield different results for problems with comparable characteristics. Multi-criteria decision-making technologies are employed in several fields of study to assess the best alternative solutions.

Multi-Standards Direction (MCDM) strategies have a set of experiences, including the Logical Order Cycle (AHP), Promethee, Electre, Dematel, Vikor, TOPSIS, and Uta. As indicated by Bilbao-Terol, TOPSIS has been applied to back to assess government security reserves. A famous MCDM system is the TOPSIS strategy, which depends with the understanding that the positive ideal option has the ideal level for all elements considered while the negative ideal is the one with the most obviously terrible levels for all variables.

The most ideal choice is the one that augments the advantage while additionally bringing down the general uses. Conversely, the negative-ideal arrangement is one that expands the cost while limiting the advantage.

The assurance of the loads comparative with the significance of every rule (specifically advantages and expenses or basically results and data sources, separately) is a key stage in TOPSIS while in DEA these loads are characterized inside the extent of the model. This is valid in spite of the overall similitudes between TOPSIS goals and DEA targets, where results might be expanded and additionally inputs limited in non-spiral (outspread) models. DEA and TOPSIS, nonetheless, force no utilitarian structure on the information or make any distributional suspicions for the determined scores, as opposed to SFA.

The essential principle of TOPSIS holds that the chose elective should be both the nearest to the positive-ideal arrangement and the farthest from the negative-ideal arrangement. One more qualification among TOPSIS and DEA models is that, while DEA utilizes simply insightful methods to register Euclidean distance capabilities on standardized vectors of positive (results) and negative (inputs) rules, considering that the chief has previously characterized the loads, DEA improves the separation from each firm to the curved effective creation outskirts. All in all, the two primary benefits of TOPSIS approaches over DEA models are (I) loads are at the tact of the leader, and (ii) there is no supposition requiring convexity of information. The major TOPSIS logical advances are portrayed straightaway.

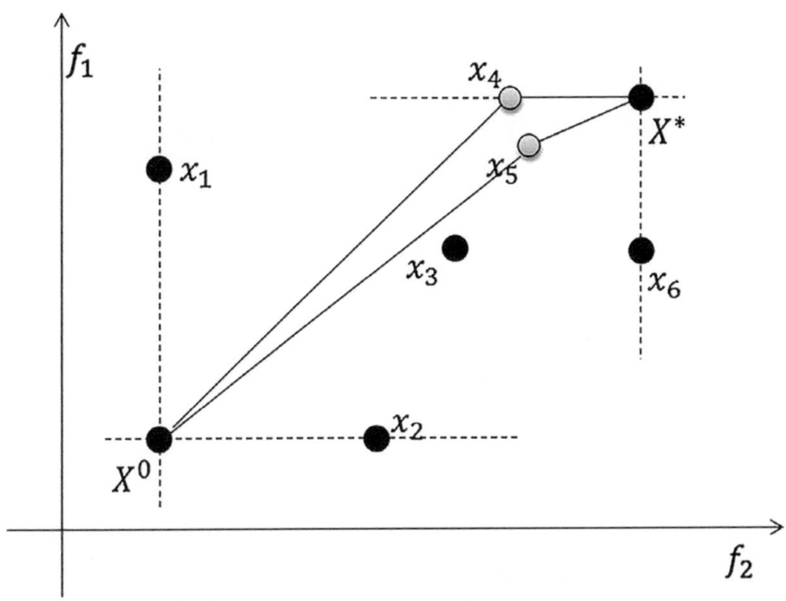

Analytical framework for the TOPSIS method. Score: https://www.researchgate.net/figure/The-ideal-solution-of-TOPSIS-method_fig 1_330291834

In this example, the benefit qualities are represented by f1 and f2 while X* and X0, respectively, are the positive ideal and negative ideal solutions. By comparing the choices of x1, x2, x3, and x6 to X*, it is simple to evaluate each one. Although the distances between x4 and x5 are similar (X*), another factor—the distance between the alternative and the undesirable solution—X0—is chosen to determine the outcome. As a result of this relative greater distance in comparison to X0, x4 has a significantly more efficient score than x5. The issue of the units' inconsistency caused by other factors can also be assessed using this approach. This method has been used extensively in efficiency analysis and risk management due to its benefits in ranking and choosing a variety of externally determined alternatives through a distance measure.

In this study, we focused on MCDM techniques, specifically TOPSIS algorithms at various phases to compare the effectiveness of Indian banks. These two goals are established in order to evaluate performance:

1. To evaluate and rate the top five private sector banks in India.
2. To demonstrate the value of AI in banks by employing TOPSIS-MCDM calculations that assist investors in making better equities company selections.

Solving MCDM problem using TOPSIS Method (Calculate Normalised Matrix)

	Bank	Bank	Bank	Bank	Bank
Year on Year	IDFC	HDFC	ICICI	AXIS	Kotak Bnk
Mar-19	12,887	116,598	77,913	68,116	42,900
Mar-20	17,589	138,073	91,247	78,172	49,018
Mar-21	18,222	146,063	98,087	78,483	63,729
Mar-22	20,395	157,263	104,892	82,597	72,488
Mar-23	27,195	192,800	129,063	101,665	83,520
	$\bar{X}_{ij} = \frac{X_{ij}}{\sqrt{\sum_{i=1}^{n} X_{ij}^2}}$				

From the above Table; Calculating Normalized Matrix and weighted Normalize matrix. We normalize each value by making it: where m is the number of rows in the dataset and n is the number of columns. I vary along rows and j varies along the column. Where we can see that HDFC, ICICI, AXIS banks 5 years shareholding are giving profitable to investors.

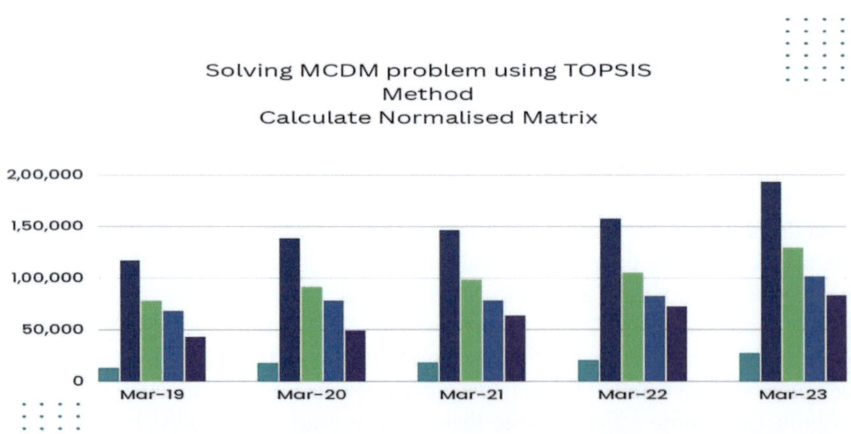

Score: Author Extracted using Power bi

Step-2	Calculate weighted normalised matrix
	$V_{ij} = \overline{X}_{ij} \times \mathrm{W}_{\mathrm{j}}$

From the above Formula: Normalized matrix for the above-given values will be multiply each value in a column with the corresponding weight.

Step-3	Calculate the ideal best and ideal worst value				
Year on Year	IDFC	HDFC	ICICI	AXIS	Kotak Bank
Mar-19	0.2909	0.3425	0.3427	0.3690	0.2994
Mar-20	0.3970	0.4056	0.4014	0.4235	0.3421
Mar-21	0.4113	0.4290	0.4315	0.4252	0.4447
Mar-22	0.4603	0.4619	0.4614	0.4475	0.5058
Mar-23	0.6138	0.5663	0.5677	0.5508	0.5828

From the above Table: Calculating Ideal Best and Ideal worst and Euclidean distance for each row from ideal worst and ideal best value. First, we will find out the ideal best and ideal worst value: Now here we need to see the impact, i.e. is it '+' or '–' impact. If '+' impact Ideal best for a column is the maximum value in that column and the ideal worst is the minimum value in that column, and vice versa for the '–' impact.

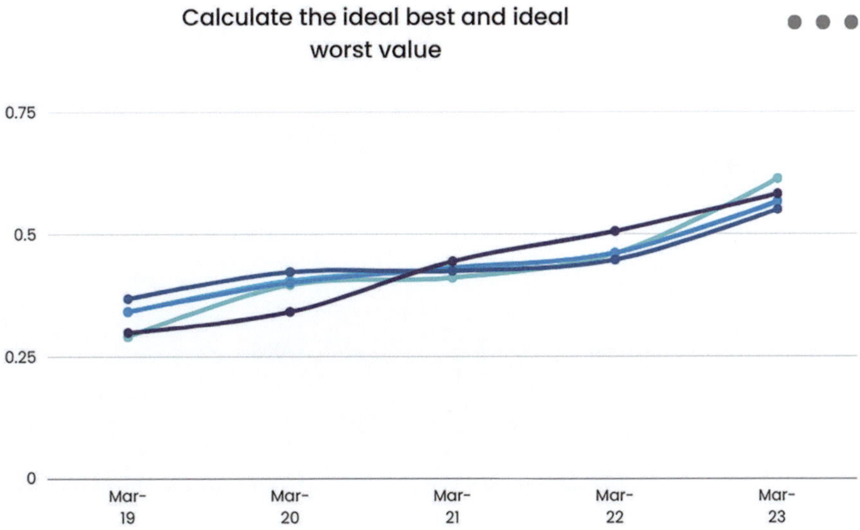

Score: Author Extracted using Power bi

Step-4	Calculate the Euclidean distance from the ideal best
	$S_i^+ = \left[\sum_{j=1}^{m} \left(V_{ij} - V_j^+ \right)^2 \right]^{0.5}$

We have calculate Euclidean distance for elements in all rows from the ideal best and ideal worst, Here d_{iw} is the worst distance calculated of an *i*th row, where $t_{i,j}$ is element value and *tw,j* is the ideal worst for that column. Similarly, we can find d_{ib}, i.e. best distance calculated on an *i*th row. From this table we see ideal best value of HDFC, ICICI, AxiS for five years performance and IDFC & Kotak banks are still needs time to pick up.

	IDFC	HDFC	ICICI	AXIS	Kotak B	Si+	Si-	Pi	Rank
Mar-19	0.10180	0.08562	0.08568	0.05536	0.04491	0.08390	0.11303	0.57396	1
Mar-20	0.13895	0.10139	0.10035	0.06353	0.05131	0.07134	0.07930	0.52641	3
Mar-21	0.14395	0.10726	0.10787	0.06378	0.06671	0.06685	0.07782	0.53790	2
Mar-22	0.16111	0.11549	0.11535	0.06712	0.07588	0.07173	0.06925	0.49121	4
Mar-23	0.21483	0.14158	0.14194	0.08262	0.08743	0.11303	0.08390	0.42604	5

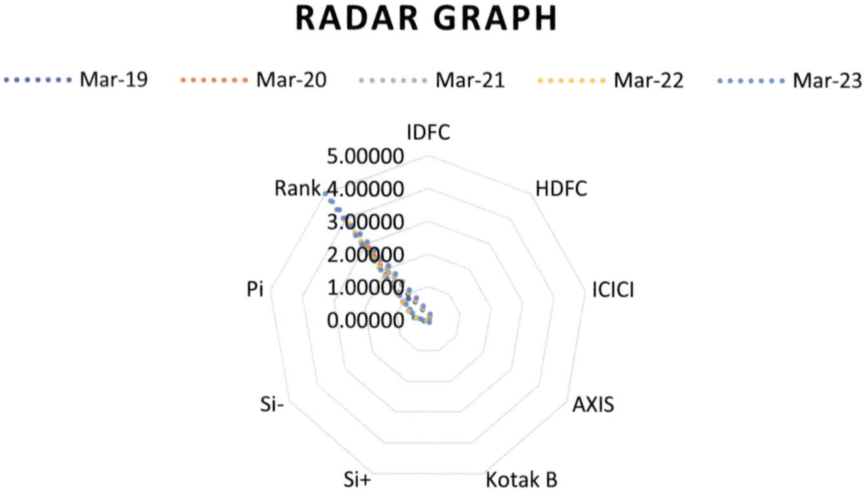

Score: https://www.researchgate.net/figure/Criteria-weights-according-to-all-decision-makers_fig 5_330291834

From the above Radar Graph applied for Euclidean distance ideal test we notice great progress from 2021 till date on words for their investments.

Step-5	Calculate the Euclidean distance from the ideal worst
	$S_i^+ = \left[\sum_{j=1}^{m} \left(V_{ij} - V_j^+ \right)^2 \right]^{0.5}$ This distance is larger than that for a random point near the origin of a 2-D circle.

Step-6	Calculate performance score				
V+	0.21483	0.08562	0.08568	0.05536	0.04491
V−	0.10180	0.14158	0.14194	0.08262	0.08743

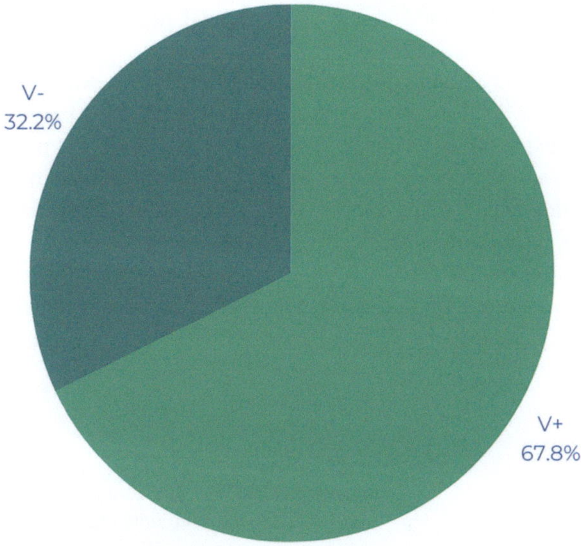

Score: Author Extracted using Power bi

From above table calculating Topsis Score and Ranking. Now we have Distance positive and distance negative with us, let's calculate the Topsis score for each row on basis of them. TOPSIS Score = d_{iw}/ (d_{ib} + d_{iw}) for each row Now rank according to the TOPSIS score, i.e. higher the score, better the rank.

15.4 Conclusion

Any nation's economic development is seen to depend heavily on banks. In a market where there is competition, banks must make the best use of their resources. All five Indian private banks are evaluated in the current study based on a variety of input and output criteria. The report suggests ranking the top five private sector banks based on how well they execute. Additionally, AI has assisted banks in understanding consumer preferences and providing real-time support for their concerns. Along with reducing operational expenses and enhancing customer service, banks have begun implementing AI in risk management. It likewise makes a near report involving TOPSIS calculations for registering weight rules for information and result factors, AHP device is utilized and later this weight is applied in TOPSIS examination are additionally used to rank banks as per their exhibition.

Finally, TOPSIS analysis was used in a comparison study. The best performing banks are identified by the computation of the results, which serves as a standard for other banks to raise their performance. The outcome also shows that HDFC, ICICI, and AXIS banks are placed first regardless of the analysis tools (TOPSIS) applied. It turns into a standard for other banks. The report mainly offers useful information on benchmarking, performance evaluation, weight analysis, and rating of Indian private sector banks.

The performance of the ineffective decision-making units (DMUs) is evaluated in this study using many criteria decision making technologies such the Analytic Hierarchy Process (AHP), Topsis, and algorithms. By carefully weighing the possibilities, this strategy will assist the decision-maker in finding the optimal answer to a difficult situation. Additionally, it helps the top body create better plans and policies. When the variables utilised in a study are typically complicated, these multi-criteria decision-making tools are used. However, in subsequent investigations, alternative approaches like Promethee, Electre, Vikor, etc. may also be used.

References

Berger AN, Humphrey DB (1997) Efficiency of financial institutions: international survey and directions for future research. Eur J Oper Res 98:175–212. http://doi.org/10.1016/S0377-2217(96)00342-6

Chuang-Min C, Ming-Miin Y, Hsiao-Ning W (2015) An application of the dynamic network DEA model: the case of banks in Taiwan. Emerg Mark Financ Trade 51(sup1):S133–S151. https://doi.org/10.1080/1540496X.2014.998887

Fethi MD, Pasiouras F (2009) Assessing Bank efficiency and performance with operational research and artificial intelligence techniques: a survey. Eur J Oper Res 204(2):189–198. https://doi.org/10.1016/j.ejor.2009.08.003

Galankashi MR, Helmi SA, Hashemzahi P (2016) Supplier selection in automobile industry: a mixed balanced scorecard-fuzzy AHP approach. Alex Eng J 55(1):93–100

Gurcan OF, Yazici I, Beyca DF, Arslan CY, Eldemir F (2016) Third party logistics (3PL) provider selection with AHP application. Proc Soc Behav Sci 235:226–234

Islam R, Selim A, Tarique KM (2016) Prioritisation of service quality dimensions for healthcare sector. Int J Med Eng Inform 8(2):108–123

Luthra S, Mangla SK, Kumar S, Garg D (2017) Identify and prioritise the critical factors in implementing the reverse logistics practices: a case of Indian auto component manufacturer. Int J Bus Syst Res 11(1–2):42–61

Ozbek A (2015) Performance analysis of public banks in Turkey. Int J Bus Manag Econ Res 6(13):21–30

Petrovic IB, Kankaras M (2018) DEMATEL-AHP multi-criteria decision making model for the determination and evaluation of criteria for selecting an air trafc protection aircraft. Decis Mak Appl Manag Eng 1(2):93–110

Popovic M, Kuzmanović M, Savić G (2018) A comparative empirical study of analytic hierarchy process and conjoint analysis: literature review. https://doi.org/10.31181/dmame1802160p

Ratna S, Subham A, Saiuddin M (2018) Assessment of lean performance of manufacturing cells in an SME using AHP. Int J Mech Prod Eng Res Dev (IJMPERD) 8(3):435–440

Saaty TL (1980) The analytic hierarchy process. McGraw-Hill, New York

Triantaphyllou E, Mann SH (1995) Using the analytic hierarchy process for decision making in engineering application: some challenges. Int J Ind Eng Appl Pract 2(1):35–44

Wang S, Sheng Z, Xi Y, Ma X, Zhang H, Kang M, Ren F, Du Q, Ke H, Han Z (2017) The application of the analytic hierarchy process and a new correlation algorithm to urban construction and supervision using multi-source government data in Tianjin. Int J Geo-Inf 7(50):3–14

Websites References

https://en.wikipedia.org/wiki/TOPSIS

https://mathewmanoj.files.wordpress.com/2018/08/example-topsis.xlsx

https://mintgenie.livemint.com/news/markets/idfc-first-bank-vs-canara-bank-which-is-a-better-investment-option-for-long-term-we-ask-analysts-151662105490368

https://robertsoczewica.medium.com/what-is-topsis-b05c50b3cd05

https://www.equitymaster.com/stock-research/compare/CNRA-IDBL/Compare-CANARA-BANK-IDFC-FIRST-BANK

https://www.indmoney.com/stocks/dividend/idfc-first-bank-ltd-dividend

https://www.upgrad.com/blog/artificial-intelligence-in-banking/

https://www.yourarticlelibrary.com/economics/4-significant-structural-changes-in-indias-economic-structure-since-independence/2754

MIX
Papier aus verantwortungsvollen Quellen
Paper from responsible sources
FSC® C105338
FSC
www.fsc.org

If you have any concerns about our products,
you can contact us on
ProductSafety@springernature.com

In case Publisher is established outside the EU,
the EU authorized representative is:
Springer Nature Customer Service Center GmbH
Europaplatz 3, 69115 Heidelberg, Germany

Printed by Libri Plureos GmbH
in Hamburg, Germany